RADIATION EFFECTS AND SOFT ERRORS IN INTEGRATED CIRCUITS AND ELECTRONIC DEVICES

SELECTED TOPICS IN ELECTRONICS AND SYSTEMS

Editor-in-Chief: **M. S. Shur**

---

*Published**

Vol. 23: Oxide Reliability: A Summary of Silicon Oxide Wearout, Breakdown, and Reliability
ed. *D. J. Dumin*

Vol. 24: CMOS RF Modeling, Characterization and Applications
eds. *M. Jamal Deen and Tor A. Fjeldly*

Vol. 25: Quantum Dots
eds. *Elena Borovitskaya and Michael S. Shur*

Vol. 26: Frontiers in Electronics: Future Chips
Proceedings of the 2002 Workshop on Frontiers in Electronics (WOFE-02)
eds. *Yoon Soo Park, Michael S. Shur and William Tang*

Vol. 27: Intersubband Infrared Photodetectors
ed. *V. Ryzhii*

Vol. 28: Advanced Semiconductor Heterostructures:
Novel Devices, Potential Device Applications and Basic Properties
eds. *Mitra Dutta and Michael A. Stroscio*

Vol. 29: Compound Semiconductor Integrated Circuits
ed. *Tho T. Vu*

Vol. 30: Terahertz Sensing Technology:
Volume 1: Electronic Devices and Advanced Systems Technology
eds. *Dwight L. Woolard, William R. Loerop and Michael S. Shur*

Vol. 31: Advanced Device Modeling and Simulation
by *Tibor Grasser*

Vol. 32: Terahertz Sensing Technology:
Volume 2: Emerging Scientific Applications and Novel Device Concepts
by *Dwight L. Woolard, William R. Loerop and Michael S. Shur*

Vol. 33: GaN-Based Materials and Devices:
Growth, Fabrication, Characterization and Performance
eds. *Michael S. Shur and R. F. Davis*

Vol. 34: Radiation Effects and Soft Errors in Integrated Circuits and Electronic Devices
eds. *R. D. Schrimpf and D. M. Fleetwood*

*The complete list of the published volumes in the series can be found at
http://www.worldscientific.com/series/stes

Selected Topics in Electronics and Systems – Vol. 34

RADIATION EFFECTS AND SOFT ERRORS IN INTEGRATED CIRCUITS AND ELECTRONIC DEVICES

Editors

R. D. Schrimpf
D. M. Fleetwood
Vanderbilt University, USA

World Scientific

NEW JERSEY • LONDON • SINGAPORE • BEIJING • SHANGHAI • HONG KONG • TAIPEI • CHENNAI

Published by

World Scientific Publishing Co. Pte. Ltd.
5 Toh Tuck Link, Singapore 596224
USA office: Suite 202, 1060 Main Street, River Edge, NJ 07661
UK office: 57 Shelton Street, Covent Garden, London WC2H 9HE

British Library Cataloguing-in-Publication Data
A catalogue record for this book is available from the British Library.

Selected Topics in Electronics and Systems — Vol. 34
RADIATION EFFECTS AND SOFT ERRORS IN INTEGRATED CIRCUITS AND ELECTRONIC DEVICES

ISBN-13 978-981-238-940-4
ISBN-10 981-238-940-7

Editor: Tjan Kwang Wei

Preface

Electronic devices in space, defense, medical, and power systems may be exposed to various types of radiation, including high-energy photons and energetic particles (electrons, protons, neutrons, and ions). The radiation may produce effects in the electronics ranging from temporary loss of data to catastrophic failure. The specific effects produced depend strongly on the specific technology and the radiation environment. Most systems designed for use in radiation environments are designed conservatively using electronic parts that are at least several generations behind the current state of the art. However, the demand for higher performance and reduced time from design to deployment has increased the pressure to use advanced technologies. The effects of radiation in some advanced technologies are poorly understood, or in some cases, completely unknown. In addition, highly scaled devices may be sensitive to the naturally occurring radiation at the earth's surface, even though the atmosphere provides significant protection.

This volume discusses radiation effects at the material, device, and circuit levels. The emphasis is on effects produced by the cumulative energy deposited by the radiation (total ionizing dose effects) and transient effects caused by single ionizing particles (single-event effects). Bipolar (Si and SiGe), metal-oxide-semiconductor (MOS), and compound semiconductor technologies are considered. In addition to considering the specific issues associated with high-speed devices and technologies, the background material necessary to understand radiation effects at a more general level also is included.

The volume opens with a description of single-event effects, with an emphasis on high-speed devices and test procedures. These include papers by Normand on single event effects in avionics and on the ground, by Baumann on soft errors in commercial integrated circuits, and by McMorrow, Melinger, and Knudson on III-V semiconductor electronics. In addition, the investigation of single-event transients in high-speed circuits with a pulsed laser is described by Fouillat, Pouguet, Lewis,

Buchner, and McMorrow. Design techniques for improving radiation tolerance are discussed, including layout approaches, circuit design issues, and error correction. Heidergott describes system level single event upset mitigation strategies; Holman discusses radiation-tolerant design for high performance mixed-signal integrated circuits; and Nowlin, Bailey, Turfler, and Alexander describe a total-dose hardening-by-design approach for high-speed mixed-signal CMOS integrated circuits.

Radiation effects in several specialized areas are considered, including high-energy physics experiments (Faccio), optocouplers (Reed, Marshall, and LaBel), charge-coupled device imagers and CMOS active pixel sensors (Hopkinson and Mohammadzadeh), power devices (Shenai, Galloway, and Schrimpf), silicon-on-insulator electronics (Cristoloveanu and Ferlet-Cavrois), and silicon-germanium heterojunction bipolar transistors (Cressler). Descriptions of total-dose and displacement-damage effects in bipolar devices and circuits are provided in papers by Schrimpf and by Barnaby. The volume closes with papers discussing hardness assurance issues (Pease), basic mechanisms of total ionizing dose effects in MOS devices (Cester and Paccagnella; Rashkeev, Fleetwood, Schrimpf, and Pantelides; and Oldham), and dosimetry (Dusseau and Gasiot).

Radiation effects is a changing and evolving discipline; new issues and effects are identified as devices become smaller and new technologies are introduced. We have endeavored to provide descriptions of the basic phenomena required to understand radiation effects, as well as define important issues relevant to advanced technologies.

Ron Schrimpf
Dan Fleetwood
Nashville, TN

CONTENTS

International Journal of High Speed Electronics and Systems
Vol. 14, No. 2 (2004) 285–298

World Scientific
www.worldscientific.com

SINGLE EVENT EFFECTS IN AVIONICS AND ON THE GROUND

EUGENE NORMAND
The Boeing Company
Seattle, WA 98124, USA

Single event effects in electronics caused by the atmospheric neutrons have been an issue for systems using large blocks of random access memory (RAM) in avionics applications as well as those on the ground. At ground level there are two main sources of single event effects, alpha particles from the packaging materials as well as the neutrons, but at aircraft altitudes, where the neutron flux is about 300 times higher than the ground, the alpha particles make a negligible contribution. We review the trends over the last 5-10 years in the response of COTS computer systems to single event effects, taking into the response of devices as well as fault tolerant measures incorporated into the systems.

Keywords: Single event effect, single event upset, avionics, single event burnout, cosmic ray neutrons, atmospheric neutrons, aircraft altitudes

1. Introduction

We can distinguish four generally distinct environments in which man has placed his technologically developed hardware: under the sea, on land, in the atmosphere via airplanes and balloons and in space. For three of these environments, land, atmosphere and space, when the man-made objects have included microelectronics, what has been observed is that the naturally occurring ionizing radiation field within these environments has interacted with the electronics, causing malfunctions. These effects in microelectronics are called single event effects (SEEs), and they have been discussed in numerous review papers[1, 2]. A single event effect requires that the energy deposited by a single ionizing particle interacting with a microelectronics device be sufficient to cause it to malfunction. Single event upset (SEU), in which the deposited energy causes a single bit to flip its logic state, is the most common form of SEE, and the one for which most data has been accumulated.

In space, the potential for the electronics to undergo SEE from the naturally occurring ionizing radiation environment is widely known and accepted. Thus government and industry standards have been written to assure that the possibility of SEE in the electronics of spacecraft systems be formally addressed following accepted testing and analysis procedures. Elements of these standards or their equivalent have been incorporated into the requirements of almost all satellite programs. In addition, within the radiation effects community several thousand technical papers have been written covering: 1) occurrences of SEE effects in satellites, 2) the results of laboratory SEE tests to simulate the responses of electronics to the space environment and 3) methodologies to calculate the SEE rate of occurrence.

2. Similarities between SEE in Avionics and on the Ground

However, when it comes to SEE on the ground and in aircraft, the situation is drastically different. Until the mid-1990s there had been very limited recognition of the possibility and impact of single event effects in ground-based or airplane-based electronics. Only within the last five years have standards been, or are in the process of being, written that address SEE in electronics on the ground and in avionics.

2.1 *SEU at ground level*

For ground level applications, JESD89, "Measurement and Reporting of Alpha Particles and Terrestrial Cosmic Ray-Induced Soft Errors in Semiconductor Devices,"[3] was issued in 2001, after work was begun on it in the middle of 1999. In this case, NASA, which had the experience of SEE in space for almost two decades, was instrumental in working with the microelectronics vendors through SEMATECH and organizing the working group that produced JESD89. The Symposium On Space Radiation Effects On Terrestrial Electronics, spearheaded by NASA and sponsored by NASA along with the electronics industry (SEMATECH and SRC) in October, 1997, was a key activity that started the process of an organized and industry-wide approach. Previous to this, microelectronics companies had conducted their own proprietary research on SEU, also called soft errors, on an individual basis, often spending millions of dollars[4].

However, based on this old approach, the possibility of SEU in microelectronics was known to only a few hundred people. By the late 1990s more technical papers on this topic had begun to appear, such as an entire issue of IBM Journal of Research and Development[4] and in a few papers each year at the International Reliability Physics Symposium[5]. In all of these, the actual upset rates recorded on the ground were never given in absolute units, until a paper written by someone from outside the microelectronics industry provided such measured SEU rates[6]. Nevertheless, this topic received a great deal more attention when it was written up in the popular press. Thus, in an article in Forbes magazine of November, 2000[7], it was reported that Sun servers were having problems with dozens of machines crashing due to bit flips in the SRAM used for the L2 cache memory which were caused by cosmic rays or alpha particles. Sun Microsystems received a great deal of adverse publicity and hundreds of thousands of people became aware of the fact the cosmic rays can cause errors in memory chips. In this case, the problem was amplified because Sun initially blamed the vendor of the SRAMs. However, knowledgeable people in the industry knew that the real mistake was Sun's solution to SEU in the L2 cache through parity protection, the cheapest way, but with the consequence of system shutdowns whenever a parity miscompare occurred[8].

2.2 *SEU in avionics*

With regard to avionics the situation is generally similar. A standard on the topic was begun at the end of 2001 by Technical Committee 107 within the IEC (International Electrotechnical Commission), with the title of TC107-AR "AVIONICS INDUSTRY- Standard for the Accommodation of Atmospheric Radiation Effects (SEE) Within Avionics Electronic Equipment".

Beginning in the late 1980s, SEE in avionics had evolved from a series of interesting anecdotal incidents (relying exclusively on overall pilot observations) to accepted fact[9]. Following a series of proprietary flight experiments by IBM on aircraft in which upsets in a large array of SRAMs were measured, a joint IBM-Boeing study sponsored by DNA and NRL, collected actual in-flight upset data recorded in the CC-2E flight computer on military aircraft. This study, completed in 1992 [10], demonstrated that SEUs are real, that the measured in-flight rates correlate with the atmospheric neutron flux, and that the rates can be calculated using laboratory SEU data.

Since the IBM-Boeing study more technical papers have been written on SEE in avionics, including a review article[9], but the number of papers on this topic has been just a few a year. Most of these papers have focused on methodologies for calculating the SEU rate and laboratory measurements in devices, so that the number of papers since Ref. 9 in which actual in-flight upsets in avionics have been measured are very rare[11, 12]. Similar to the situation with SEU on the ground, the possibility of SEUs in avionics has received more attention in trade magazines, such as EE Times[13]. Although EE Times has about 15% of the circulation of Forbes, the EE Times article, with the overly dramatic title "Engineer calls off-the-shelf components 'unsafe' for aircraft," did inform thousands of engineers and computer scientists of SEE in avionics as a potential threat. This isn't the first time that articles have appeared in trade magazines that put unrealistic emphasis on the potential problems with SEE in avionics and ignored the built-in fault tolerant features that protect against them.

Nevertheless, the fundamental issue that is in common regarding SEE in electronics used in aircraft and on the ground (sea level) is that the same kind of radiation field is largely responsible for the effects in both environments, and the same types of devices and circuit boards are involved. The first factor is that the radiation environment causing the SEE, primarily the high-energy neutrons created by the interaction of the cosmic rays with the atmosphere, is the same. These are often called atmospheric neutrons or cosmic ray neutrons. The intensity or flux level is about 300 times higher at aircraft altitudes than it is on the ground (sea level), but the energy spectrum is essentially identical. This is discussed further in section 4. For the same reason, accelerated laboratory testing of devices to estimate their SEE rate in both environments have used the same facilities. The neutron beam at the WNR (Weapons Neutron Research) facility within LANSCE (Los Alamos Neutron Science Center), which has an energy spectrum very similar to that of the atmospheric neutrons, has been utilized over the last 10 years for both ground level and avionics applications [14]. One hour in the WNR beam is equivalent to 1E5-1E6 hours at aircraft altitudes, depending on the beam intensity at the time of testing (3E7-3E8 hours at sea level). The second factor is the types of devices that are used are essentially the same, and this is largely dictated by the reality of using commercial-off-the-shelf (COTS) electronics and circuit boards for nearly all applications on the ground and in aircraft.

3. Differences Between SEE in Avionics and on the Ground

Although there are strong similarities between SEE in avionics and on the ground, there are also differences. The differences derive mainly from the fact that the atmospheric neutron flux is so much higher at aircraft altitudes than it is on the ground. In addition, there is a second SEE effect, different from SEU, that is relevant on the ground in high voltage devices, but which to date, has not been an issue in aircraft.

3.1 *Comparison of SEU mechanisms in electronics, neutrons and alpha particles*

At ground level there are really three sources of radiation that cause the SEUs, the high energy cosmic ray neutrons previously discussed, the thermal neutron component of the cosmic ray neutrons and alpha particles from the microelectronics package. The high-energy cosmic ray neutrons of concern are those with E> 10 MeV, for which the neutron cross section in Si for reactions giving rise to energetic recoils is significant, since it is the energy deposited by recoils that cause the SEUs.

Thermal neutrons are those with energies centered at 0.025 eV, (comparable to the energy at room temperature) which is about 8 orders of magnitude lower than the high-energy neutrons. Thermal neutrons can be significant because they have a very high probability of interacting with certain isotopes, like B^{10}. When a B^{10} nucleus reacts with a thermal neutron, two reaction products are created, an alpha particle and a Li^7, which have a combined 2.3 MeV of energy that can be deposited. If this energy is deposited within the sensitive volume of a microelectronic device, it can lead to a single event upset (SEU). Based on the testing of a small number of memories with a neutron source that included a thermal neutron component, some devices were found to be more susceptible to SEU from the thermal neutrons than from fission neutrons ($E_{avg} \sim 1$ MeV, not nearly as high as the atmospheric neutrons), and a few were less susceptible[15]. It depended on the concentration and location of the B^{10} within a particular device. Thus, thermal neutrons are a second source of ionizing radiation that can cause SEU in devices. Based on the best available calculations of the complex spectrum of cosmic-ray neutrons at various altitudes [16], the ratio of thermal/>10 MeV neutrons is much higher on the ground than it is at aircraft altitudes. However, this may change when accurate account is taken of the contribution of the aircraft structure in producing thermal neutrons. Thus, on the ground, the thermal neutrons may be a larger contributor to the SEU rate, compared to that from the > 10 MeV neutrons, than they are at aircraft altitudes, but this conclusion may change once an accurate assessment is made of the thermal neutrons produced inside an aircraft.

It has been known that alpha particles contribute to the SEU rate on the ground for more than 20 years. When SEUs were first discovered in microelectronics on the ground[17] during the late 1970s, the cause was alpha particles emitted from the IC package. The actual source was uranium and thorium impurities found in trace amounts in the various production and packaging materials that emit alpha particles. Once this was recognized, several steps were taken to greatly reduce the problem, with the main one being that alpha particle detectors scanned all raw materials used in the IC manufacturing process. All materials having an emitted alpha particle flux above a threshold level were rejected. During the 1980s and early 1990s this worked very well. However, as packaging technologies changed and solder bumps were incorporated into newer style packages, the alpha problem re-emerged as solder became a packaging component. However, specially purified solder to minimize alpha emitters has been available, and the movement to utilize lead-free solder by IC vendors may reduce the problem. Thus for current devices, the SEU rate contribution on the ground due to alphas may be appreciable, comparable to that due to neutrons, in those devices with packages having standard solder bumps.[18]

The SEU rate from alphas remains the same on the ground and at aircraft altitudes. On the other hand, the high-energy (>10 MeV) neutrons are 300 times higher in the atmosphere compared to the ground. Thus, even if the SEU contribution from the alpha particles relative to the contribution from high-energy neutrons is significant on the ground, in the atmosphere, the SEU contribution from alpha particles is small enough to be neglected, and the entire focus placed on the high-energy neutrons.

3.2 Other SEE Effects in Low Voltage Devices

It should also be noted that there are other single event effects that can be induced by > 10 MeV neutrons in low voltage ICs, including single event latchup (SEL), single event transient (SET), single event hard error (SHE) and multiple bit upset (MBU). These will be summarized very briefly. Although many of these effects were measured primarily with high energy protons, as indicated in section 5, they are assumed to apply to high energy neutrons as well.

Many CMOS devices are susceptible to SEL from heavy ions. A small number of these have also been tested with beams of energetic protons and neutrons, and a few have been shown to be susceptible SEL from the protons and neutrons. Refs. 19 and 20 list a total of seven devices with proton SEL cross sections. Unfortunately, two very recent proton-induced latchup cross sections have been published for SRAMs[21,22] and these proton-induced SEL cross sections both exceed the largest previous proton SEL value[19]. There is thus a clear need to track the SEL susceptibility of new and emerging devices by energetic neutrons and protons.

An SEU-related event in some devices can lead to the generation of a single event transient (SET) that a device may interpret as a new bit of information. These transients are spurious signals or voltages, induced by the deposition of charge by a single particle that can propagate through the circuit path during one clock cycle. Most SET tests have been performed using beams of heavy ions, but SET has also been induced in linear devices with high-energy proton beams[23,24] hence the atmospheric neutrons can also cause this kind of effect. It has long been known that heavy ions can cause "stuck bits" in RAMs, events often referred to as single hard errors (SHE). These hard errors are due to very localized total dose effects in the gate oxide of sensitive transistors. More recently the first occurrence of SHE by protons in a laboratory test was published, indicating that in newer devices[25], protons, and hence atmospheric neutrons, are capable of inducing stuck bits. Some RAM devices are also susceptible to multiple bit upset, MBU, in which more than one bit is upset by a single particle. In general, MBUs can be expected to involve more than a single logic word, however, with MBUs, energetic particles may also upset two or more bits in a single word, depending on the physical arrangement and size, and the distribution of the memory cells within a device. Generally, the MBU rate by protons or neutrons is at most 1-2% of the SEU rate, but some devices can be much more susceptible to MBU than others, even if they have similar SEU rates[26].

3.3 *SEE effects in high voltage devices*

The high-energy neutrons within the atmospheric neutron spectrum have also been shown to induce a different effect, single event burnout (SEB), in high voltage devices[27]. These devices include power MOSFETs and IGBTs (insulated gate bipolar transistors), but also higher voltage devices (>2000V) such as diodes and thyristors[28].

It has been found most helpful to utilize the Kuboyama model[29] to understand the complex mechanism involved in SEB. Although it was developed to describe the SEB caused by heavy ions, it has been adapted to apply, in qualitative terms, to SEB caused by high-energy neutrons and protons (see Ref. 28 for further details). The main result is that the key to SEB susceptibility is the operating voltage. In all of the testing that has been carried out in which power MOSFETs and higher voltage devices (diodes and IGBTs) have been exposed to beams of high-energy neutrons and burnout measured[27,28], there is always a threshold value for the drain-to-source voltage, Vds, which must be exceeded in order for SEB to occur. Therefore, the best way of ensuring that high voltage devices are not susceptible to neutron-induced SEB is to operate devices at a sufficiently reduced voltage compared to their rated voltage, called derating. Good design practice, having nothing to do with radiation effects, dictates that power MOSFETs be operated at a derated condition. However, it appears that additional derating is likely needed in order to adequately reduce the sensitivity of high voltage devices to SEB.

A number of power MOSFETs have been tested with beams of protons and energetic neutrons for SEB, and the results of these, the passing Vds values, have been tabulated[27,28,30]. Relying on these measured results and commonly used derating factors, we can generalize that, based on those devices that were tested, atmospheric neutrons are likely to induce SEB only in power devices rated at >400V. If the devices are derated and operated at a Vds of < 300V, the probability of an SEB should be low enough for SEB to be a negligible concern.

At the present time, avionics systems do not operate at voltages > ~270V DC. Although internal trade studies for future avionics systems have considered operation at much higher voltages, there are several distinct disadvantages to such high voltage operation. These include both SEB and arcing and corona effects, which pose major obstacles before such high voltage operation can be considered feasible. Nevertheless, because the higher voltage operation may lead to an overall reduction in system weight, there is a possibility of aircraft buses operating at voltages > 270 V, which will require careful consideration of SEB effects in the high voltage electronics.

At the ground level, the situation is different; very high voltage devices have been used for many years. The main application of these devices has been for controlling electric motors on trains, with such designs being common in Europe but not in the US. Thus the problem first surfaced in the field in Europe during the early 1990s when burnout failures of newly developed devices were experienced on trains. European and Japanese manufacturers of the high voltage semiconductors were puzzled by these failures, which had not occurred previously. The failure mode was investigated in great detail by the vendors and a variety of causes were proposed[31]. After a series of laboratory tests, including simultaneous testing in laboratories located on a rooftop and in a salt mine, at the end the cause was found to be the energetic cosmic ray neutrons at ground level[31], even though their flux is very low (~20 n/cm^2hr, E> 10 MeV).

Afterwards we demonstrated that accelerated testing with the WNR neutron beam at the LANSCE could be used to assess the SEB susceptibility of high voltage devices[28]. By measuring the SEB neutron cross section in the beam and normalizing to the neutron flux level on the ground (~20 n/cm^2hr), the required reliability (FIT rate, number of failures in 1E9 device-hours) could be determined as a function of the applied voltage. This agreed with measured values from field failure data[28, 32]. Since then, manufacturers of very high voltage devices have been using the WNR beam to measure the susceptibility of new device designs to SEB, and obtain the required curve of the FIT rate as a function of applied voltage. In the US new designs for control systems are being developed that utilize the same kind of very high voltage devices, such as IGBTs and diodes rated at > 2000V. These systems are being considered for advanced solid-state induction modulators and klystron power supply systems at high-energy accelerator facilities, and also for advanced MRI (magnetic resonance imaging) equipment and related imaging controls. These devices, and the more standard power MOSFETs and IGBTs that American manufacturers are now designing for voltages > 1000V, will all be susceptible to SEB by the atmospheric neutrons. This is a failure mode that such vendors will need to pay attention to as the devices are put into the field.

4. Atmospheric and Ground Level Environments

The atmosphere is composed of a mixture of particles created by the interaction of the continuous stream of primary galactic cosmic rays (GCR) with the nitrogen and oxygen atoms in the atmosphere, and so are called secondary cosmic rays. As the primary GCR particles, mainly very high-energy protons, bombard the atmosphere, they create a cascade of secondary, tertiary, etc. particles. Thus, for each primary cosmic ray entering, many more secondary particles are created. At a very approximate level, the flux of incoming primary GCR at the top of the atmosphere is ~0.1 particle/cm^2-sec, and at aircraft altitudes, the flux of all secondary particles is ~100 particle/cm^2-sec. The density of the lower portion of atmosphere is high enough that most of the particles are absorbed, so that at sea level the flux of secondary particles is reduced to ~1 particle/cm^2-sec[33].

The flux of secondary particles is not uniform around the earth due to the effect of the earth's magnetic field, which is strongest at the equator and weakest at the geomagnetic poles. In addition, the space weather can have a significant effect on the production of secondary cosmic rays due to several different effects. First, the activity on the sun follows an approximately 11 year solar cycle, such that during solar minimum conditions, the GCR particles propagate to earth more easily; thus the flux of secondary particles in the atmosphere is higher. In addition, occasional eruptions on the sun, (solar flares and coronal mass ejections), lasting hour-to-days, send out a stream of charged particles, which, when they reach the earth, generate additional secondary particles in the atmosphere like the GCR. The various factors involved and measurements of the particles have been discussed in greater detail in a variety of sources[4,9,34].

Of the various secondary cosmic rays, three can induce SEU, neutrons, protons and pions (the pi-meson particles, with a rest mass of ~140 MeV, that are involved in holding the nucleus together), and of these, the neutrons have the highest flux both in the atmosphere and on the ground. Three aspects of the neutrons are of main interest: 1) the energy spectrum, 2) the flux variation with altitude and 3) variation with latitude. The energy distribution of the atmospheric neutrons is usually presented by plotting the differential flux (flux per unit energy) as a function of energy. In Figure 1 we plot four neutron spectra at an altitude of approximately 40,000 ft. These include the original measurements made by Wilmot Hess in 1959[35], a calculation by T. Armstrong[16], a fit to measurement by made by NASA Ames[36] and 1997 DOE measurements in an ER-2 aircraft[37]. A fit to the NASA Ames data has been used in the past[9] as one component of an overall simplified model of the atmospheric neutrons. However, this data applies only up to an energy of 300 MeV, so a modification has been made to the model for E>300 MeV. The modified spectrum is given, with E in MeV, as:

$$\frac{dN}{dE} = \begin{cases} 0.346E^{-0.922} \times \exp\left[-0.0152(\ln E)^2\right] & E < 300\text{MeV} \\ 340E^{-2.2} & E > 300\text{MeV} \quad \text{n/cm}^2\,\text{sec MeV} \end{cases} \tag{1}$$

When this differential flux is integrated for E> 10 MeV, the integrated neutron flux is ~5600 n/cm^{2}hr. At ground level the flux is approximately a factor of 300 lower than at 40,000 ft, thus on the ground, the flux for E> 10 MeV is ~20 n/cm^{2}hr.

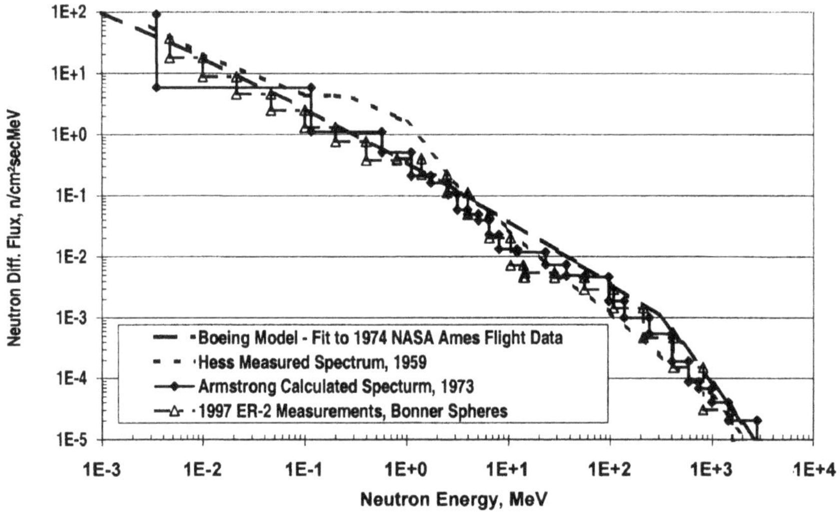

Fig. 1 Energy Spectrum of Atmospheric Neutrons at 40,000 Feet

Models for the altitude and latitude variation of the atmospheric neutron flux have been presented [9] and are still applicable. There are two main models, a simplified Boeing model[38], and a more accurate NASA model[39] that is currently called AIR. To use the NASA model, a distribution of the vertical rigidity cutoff (a measure of the effect of the earth's magnetic field in deflecting the incoming GCR particles) around the globe is needed. Ref. 3 provides a tabulation of the rigidity cutoff data, and how to interpolate it for any arbitrary point (as function of geographical latitude and longitude).

For energies up to about 500 MeV, the secondary protons in the atmosphere comprise about 25% of the flux of the neutrons, thus the effect of the secondary protons in causing SEE in electronics can be included within SEE rates due to the neutrons since these two particles have very similar SEE cross sections in devices. The energy spectrum and altitude and latitude variations of the proton flux have been given previously[9]. In the atmosphere the pion flux is only a small fraction of the neutron and proton flux and so it can be neglected. More specifically, for energies $\leq$ 1 GeV, the pion/proton ratio is estimated to be ~0.1[34] which applies at both aircraft altitudes and at ground level.

5. SEE Data in devices

The key to being able to deal with single event effects in avionics is to quantify the single event effect rate that might be expected in an avionics box. Once an estimate can be made of the event rate, appropriate measures can be considered to accommodate the event. If the event rate is low enough, a decision might be made to ignore it. If it is high enough, a variety of fault tolerant measures can be considered for incorporation, such as error correction and detection (EDAC) or redundancy. SEU, multiple bit upset (MBU, more than one bit being upset at the same time) and single event function interrupt (SEFI, SEU in a complex device such that a control path is corrupted, leading the part to improperly functioning) are the three SEE effects that present the largest potential threat to aircraft avionics. They are listed in order of decreasing likelihood. At a simplified level, the SEU rate depends on the number of bits in a device, and in most cases the bits serve as memories, registers or latches. The kinds of devices that contain the largest number of bits are random access memories, microprocessors and field programmable gate arrays (FPGAs). When the SEU rate at aircraft altitudes or on the ground is considered on a per bit basis, the SEU rate is similar for these various kinds of devices to within a range of 1-3 orders of magnitude.

The SEU response of a device to a beam of particles is characterized as the SEU cross section, which is the number of upsets divided by the fluence of particles (p/cm^2, particle flux integrated over the exposure time) to which the device was exposed. Thus the cross section is given in units of cm^2/device or cm^2/bit. Various scientific and engineering organizations around the world carry out such SEU tests in order to characterize the SEU response of devices to the different particles. The vast majority of these tests are performed for electronics that are being considered for use in space. Therefore, most of the tests are with heavy ion particle beams, and some are with proton beams. A much smaller number are performed using neutron beams for aircraft or ground level applications.

For particle energies > 50 or 100 MeV, SEU cross sections measured with protons and neutrons are very similar, and can be taken to be essentially the same, although at lower energies this isn't true. The main difference is that the proton SEU cross sections are measured using monoenergetic proton beams, whereas it is difficult to achieve a beam of neutrons of a single energy. Proton beams are available at a number of high-energy accelerators around the world. The main source available for exposing devices with a simulated atmospheric neutron environment is the WNR facility at LANL. The relationship between WNR SEU cross sections and monoenergetic proton and neutron SEU cross sections has previously been shown for three older SRAMs[20]. It appears the WNR SEU cross section is nearly equal to the monoenergetic proton and neutron SEU cross section for $E \geq 100$ MeV[20].

Thus even though the best SEU data is from measurements in the WNR beam, the largest source of SEU data that can be used as a substitute for the WNR SEU cross sections are proton SEU cross sections at E > 100 MeV. Recognizing this, tabulations were compiled in 1997 of SEU cross sections, cm^2/bit, applicable to avionics, that were derived from WNR and proton (E> 200 MeV) measurements for both SRAMs and DRAMs[40]. The conclusion reached is that, based on the data that had been compiled, the SEU susceptibility of both SRAMs and DRAMs has been decreasing over time[40]. For SRAMs the SEU sensitivity was decreasing slowly, but for DRAMs the SEU sensitivity was decreasing much more rapidly.

Because of the usefulness of this compilation (SEU cross section/bit as a function of the date of the device) in understanding trends in the SEU susceptibility, we have updated the two curves for SRAMs and DRAMs. The updating consisted of plotting all of the original data, and then adding data from a variety of more recent sources. The SRAM data is shown in Figure 2. Two fits are shown in the figure, the first is from the original paper and the second is a trend curve derived from SER (Soft Error Rates) for ground level applications[41]. Most of the data appears to be centered around a SEU cross section of 1E-13 cm^2/bit, but more broadly within a band of a factor of 3-0.33 of this value, thus within the band of 3E-14-3E-13 cm^2/bit. The trend curve (transformed from the ground-level SER data by Ziegler[41] which isn't shown) also exhibits the same tendency, i.e., for most devices there appears to be an asymptotic value of the SEU cross section within the band of 1E-14-1E-13 cm^2/bit.

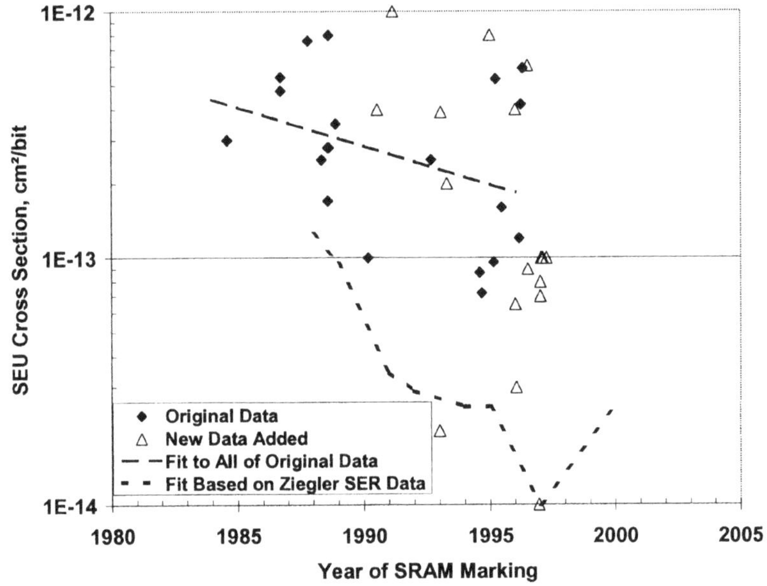

Fig. 2. SEU Cross Section in SRAMs as Function of Manufacture Date

The conclusion regarding the approximate asymptotic behavior of SRAMs over time with respect to their SEU susceptibility from atmospheric neutrons should be treated with caution. It is derived primarily from the data in Figure 2, but there is no way to determine how representative the parts used in the figure are for all SRAMs that may be used in avionics or ground level applications. However, it serves as a useful starting point.

With DRAMs the situation is different. The steep decline of the SEU cross section with year of DRAM production is enhanced by including more recent data. In this case, as microelectronic technologies have been modified to enable individual devices to hold an increasing number of bits, the result for SEU susceptibility is a continuing decline in that susceptibility on a per bit basis. As new capacitor geometries have been implemented, the same capacitance in a smaller chip area, the SEU cross section per bit has decreased. This is seen in Figure 3, which shows the original data (mainly 4Mb and 16Mb DRAMs) along with SEU results from more recent tests[41] (mainly 64-256 Mb DRAMs). It also shows that the stacked capacitance (SC) design used by most DRAM vendors is effective in lowering the SEU susceptibility, but that the more ambitious trench internal capacitance (TIC) design used only by IBM and another vendor, is even more effective.

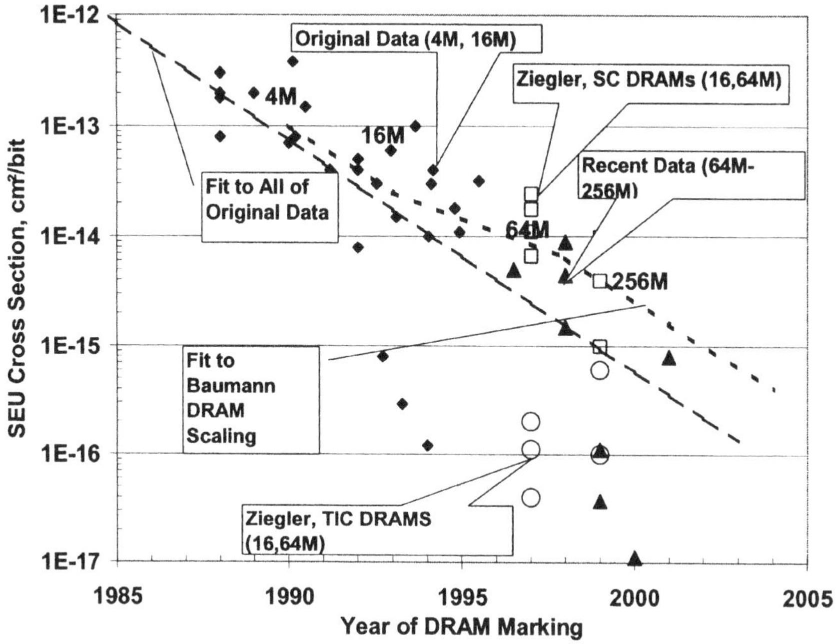

Fig. 3 SEU Cross Section in DRAMs as Function of Manufacture Date

Two trend lines are shown in the Figure 3. One is the original fit to original the data, and the second (Baumann curve[18]) is derived from a trend analysis of other data (SC DRAMs). The slope of the decrease of the SEU cross section with year from the Baumann curve is similar to that from the original data. The main point to be derived from these trend lines is that the SEU per bit cross section has continued to decrease as the size of the DRAM has increased, which appears to be related to the new DRAM technologies that have been developed to achieve the higher density memories.

Independent of the decreasing trend of the SEU cross section per bit in DRAMs as shown in Figure 3, in most applications, the occurrence of SEU in the DRAMs that constitute main memory is protected by EDAC. However, there are also other SEE effects in DRAMs that should be considered, namely their susceptibility to multiple bit upset, stuck bits (single hard errors) and SEFI. The new larger DRAMs appear to be susceptible to these effects, however, the SEE cross sections and therefore the SEE rates are quite low compared to SEU rates.

This review has focused primarily on the SEE response of individual devices, necessarily ignoring system effects which can be very significant. Not all the bits of a device are used in circuit or systems applications, and even for those that are used, if a SEU occurs, it may not affect the operation of the circuit. In addition, fault tolerant measures such as EDAC and various forms of redundancy, can be used to protect overall circuits from SEUs in individual devices. These are issues that need to be dealt with separately.

6. Summary

This review has examined single event effects in electronics used on the ground and in aircraft. We have noted the many similarities between the SEE concerns for electronics at these two locations, with the main one being the major role played by the atmospheric neutrons as the main cause of the SEE effects through nuclear reactions with the ICs. We also noted that there are really two kinds of effects, the most common, SEU, occurring in ICs operated at low voltages (< 5V), along with other effects like MBU and SEL, and single event burnout (SEB), which only affects devices operated at high voltage (> 300V). The key aspects of the atmospheric neutron environment at sea level and within the atmosphere were reviewed. The SEU data was examined for the most common devices, SRAMs and DRAMs, in terms of the per bit SEU cross section as a function of the year of IC manufacture. It was found that the DRAM SEU cross section has been declining much more rapidly over time compared to that for SRAMs, and that for SRAMs the SEU cross section appears to have reached an order-of-magnitude-wide plateau. Highlights of the neutron-induced SEB phenomenon in high voltage devices were reviewed, with prospects for the importance of this effect being noted for future applications.

References

1. D. Fleetwood and R. Gaillard, "Forward to Special Issue," IEEE Trans. Nucl. Sci., 43, 342, 1996 and the 30 invited papers on Single Event Effects in that volume
2. P. E. Dodd, "Basis Mechanism for Single-Event Effects," 1999 IEEE Nuclear and Space Radiation Effects Conference Short Course, Norfolk, VA, July, 1999
3. JEDEC Standard JESD89, Measurement and Reporting of Alpha Particles and Terrestrial Cosmic Ray-Induced Soft Errors in Semiconductor Devices, JEDEC Solid State Technology Association, August, 2001
4. J. F. Ziegler + 26 co-authors, "IBM Experiments in Soft Fails in Computer Electronics (1979-1984)' IBM J. Res. Develop. 40, 3, (1996)
5. W. R. McKee + 13 co-authors, "Cosmic Ray Neutron Induced Upsets as a Major Contributor to the Soft Error Rate of Current and Future Generation DRAMs," 1996 IEEE International Reliability Physics Symposium, April 1996
6. E. Normand, "Single Event Upset at Ground Level," IEEE Trans. Nucl. Sci., 43, 2742, 1996
7. "Sun Screen" Forbes, November 13, 2000
8. P. Baston "Unsafe At Any Speed?," SPARCproductDIRectory.com, January, 2002
9. E. Normand, "Single Event Effects in Avionics," IEEE Trans. Nucl. Sci., 43, 461, (1996)
10. A.Taber and E. Normand, "Investigation and Characterization of SEU Effects and Hardening Strategies in Avionics", IBM Report 92-L75-020-2, August, 1992, republished as DNA-Report DNA-TR-94-123, Defense Nuclear Agency, Feb, 1995
11. K. Johansson et al, "In-flight and Ground Testing of Single Event Upset Sensitivity in Static RAMs," IEEE Trans. Nucl. Sci., 45, 1628, (June, 1998)
12. E. Normand, "Correlation of Inflight Neutron Dosimeter and SEU Measurements with Atmospheric Neutron Model, " IEEE Trans. Nucl. Sci., 48, 1996, (2001)
13. D. Rea, "Engineer calls off-the-shelf components 'unsafe' for aircraft," EE Times, September 23, 2002
14. E. Normand, "Boeing Experience in Using the WNR Neutron Beam for Radiation Testing of Electronics," LANSCE Workshop for High-Energy Neutron Irradiation of Electronics, Manhattan Beach, CA, April 20, 1998
15. P. J. Griffin et al, "The Role of Thermal and Fission Neutrons in Reactor-Induced Upsets in Commercial SRAMs," IEEE Trans. Nucl. Sci, 44, 2079, 1997
16. T. W. Armstrong, K. Chandler and J. Barish, "Calculation of neutron flux spectra in the earth's atmosphere by galactic cosmic rays", J. Geophys. Res., 78, 2715 1973

17. T. C. May and M. H. Woods, "A New Physical Mechanism for Soft Errors in Dynamic Memories, " Proceedings 16 Int'l Reliability Physics Symposium, p. 33, April, 1978
18. R. Baumann, "Silicon Amnesia: Radiation Induced Soft Errors in Commercial Semiconductor Technology," Short Course presentation, RADECS 2001, 6th European Conference on Radiation Effects on Components and Systems, Grenoble, France, Sept. 2001
19. A. Johnston, G. M. Swift and L. D. Edmonds, "Latchup in Integrated Circuits from Energetic Protons," IEEE Trans. Nucl. Sci, 44, 2367, 1997
20 E. Normand, "Extensions of the Burst Generation Rate Method ... to Proton/Neutron-Induced Single Event Effects," IEEE Trans. Nucl. Sci, 45, 2904, 1998
21 R. Koga et al, "Comparison of Heavy Ion and Proton-Induced Single Event Effects (SEE) Sensitivities," IEEE Trans. Nuc. Sci., 49, 3135, (2002)
22 C. Sanderson and R Green, "A comparative experimental study for avionic system application of single event effects induced by neutrons and protons," DERA/KIS/SPACE/TR010322, UK Defence Research Information Centre, Feb. 2001
23 R. Harboe-Sorensen et al, "Single Event Transient Characterization of Analog IC's for ESA's Satellites," RADECS 99, 1999 5th European Conference on Radiation and Its Effects on Components and Systems, Fontevraud, France , p.573
24 D. K. Nichols et al, "Heavy Ion and Proton Induced Single Event Transients in Comparators," IEEE Trans. Nuc. Sci., 43, 2960, (1996).
25 L. Edmonds et al, "Ion-Induced Stuck Bits in 1T/1C SDRAM Cells," IEEE Trans. Nucl. Sci., 48, 1925 (2001)
26 C. S. Dyer et al, "Microdosimetry Code Simulation of Charge-Deposition Spectra, Single-Event Upsets and Multiple Bit Upsets," IEEE Trans. Nucl. Sci., 46, 1486, (1999
27 D. L. Oberg, J. L. Wert, E. Normand, P.P. Majewski and S. A. Wender, "First Observations of Power MOSFET Burnout with High Energy Neutrons," IEEE Trans. Nucl. Sci., 43, 2913, (1996)
28 E. Normand et al, "Neutron-Induced Single Event Burnout in High Voltage Electronics," IEEE Trans. Nucl. Sci., 44, 2358, (1997)
29 S. Kuboyama, S. Matsuda, T. Kanno and T. Ishii, "Mechanism for Single Event Burnout of Power MOSFETs and Its Characterization Technique," IEEE Trans. Nucl. Sci., 39, 1698 (1992)
30 J. R. Coss et al, "Compendium of Single Event Failures in Power MOSFETs," Workshop Record, 1998 IEEE Radiation Effects Data Workshop, p. 15 (1998)
31 H. Kabza, et al, "Cosmic Radiation as a Cause for Power Device Failure and Possible Countermeasures," presented at 6th International Symposium on Power Semiconductor Devices, Davos, Switzerland, May, 1994
32 R. Sheehy, J. Dekter and N. Machin, "Sea Level Failures of Power MOSFETs Displaying Characteristics of Cosmic Radiation Effects," Proceedings 33rd Annual IEEE Power Electronics Specialist Conference, Australia, 2002 pp. 1741-1746
33 J. F Ziegler, "Terrestrial Cosmic Rays," IBM J. Res. Develop. 40, 19, (1996)
34 O. C. Allkofer and P. K. Grieder, Physics Data: Cosmic Rays on Earth, Fachinformationszentrum Energie, Physik, Mathematik GmbH, Karlsruhe, 1984
35 W. N. Hess, et al, "Cosmic-Ray Neutron Energy Spectrum," Phys. Rev., 116, 445 (1959)
36 J. Hewitt et al, "Ames Collaborative Study of Cosmic Ray Neutrons: Mid-Latitude Flights", Health Physics, 34, 375, 1978
37 P. Goldhagen et al, "Measurement of the Energy Spectrum of Cosmic-Ray Induced Neutrons Aboard an ER-2 High Altitude Airplane," Nucl. Inst. Methd., A467, 52 (2002)
38 E. Normand and T. J. Baker, "Altitude and Latitude Variations in Avionics SEU and Atmospheric Neutron Flux", IEEE Trans. Nucl. Sci., 40, 1484 (1993)
39 J. W. Wilson et al, "Radiation Safety in the Earth's Atmosphere" in Transport Methods and Interactions in Space Radiations, NASA Reference Publication 1257, 1991
40 N. Kerness and A. Taber, "Neutron SEU Trends in Avionics," Workshop Record, 1997 IEEE Radiation Effects Data Workshop, p. 67 (1997)

41 J. Ziegler, "Silicon Diagnosis, Review of Accelerated Testing of Modern SRAMs," Short Course presentation, RADECS 2001, 6th European Conference on Radiation Effects on Components and Systems, Grenoble, France, Sept. 2001

International Journal of High Speed Electronics and Systems
Vol. 14, No. 2 (2004) 299–309

World Scientific
www.worldscientific.com

SOFT ERRORS IN COMMERCIAL INTEGRATED CIRCUITS

R. C. BAUMANN

Silicon Technology Development Group, Texas Instruments Inc.,
13560 North Central Expressway, MS 3737, Dallas, Texas, USA

The once-ephemeral soft error has recently caused considerable concern for manufacturers of advanced silicon technology as this phenomenon now has the potential for inducing the highest failure rate of all other reliability mechanisms combined. We briefly review the three radiation mechanisms responsible for causing soft errors in commercial electronics and the basic physical mechanism by which ionizing radiation can produce a soft error. We then focus on the soft error sensitivity trends in commercial DRAM, SRAM, and peripheral logic devices as a function of technology scaling and discuss some of the solutions used for mitigating the impact of soft errors in high reliability systems.

Keywords: soft error rate, radiation effects, single event upset

1. Introduction

Radiation effects have long been one of the most serious issues in spacecraft and aviation electronics[1,2]. The harsh high-energy particle fluxes experienced in orbital and flight altitude environments demand that the electronics be shielded, have redundant components, and that they be radiation hardened, thereby increasing their cost and reducing their performance. In the more benign terrestrial environment where most commercial microelectronics are used, the most common radiation-induced effect is the soft error*. In stark contrast to the many "hard" reliability failure mechanisms such as gate oxide breakdown, metal interconnect stress- and electro-migration, transistor negative bias temperature instability, etc., soft errors corrupt the data state of devices while not permanently damaging the devices themselves. So when new data is rewritten into the device, it operates correctly. As the dimensions and operating voltages of commercial integrated circuits (ICs) are reduced to satisfy the ever-increasing demand for higher density (functionality) and lower power (portability), their sensitivity to radiation has increased dramatically. Soft errors have recently become a huge concern in advanced commercial ICs because they can induce a product failure rate that is higher than all the other reliability mechanisms combined. For example, in a qualified manufacturing silicon process technology the typical failure rate for the hard reliability mechanisms

* Soft errors include a plethora of different failure modes. Single-event upsets (SEUs) are errors in which an individual memory or register component is corrupted by radiation directly. Single-event transients (SETs) are radiation-induced transients in combinatorial logic that when propagated and latched create SEUs. Single-event functional interrupts (SEFIs) occur when an SEU or latched SET corrupts critical data causing a system interrupt or lock-up. Lastly, non-destructive single-event latch-up (SEL) occurs when a latch-up is induced by an ionization event but the current is limited by the circuit parametrics so that it is not permanently damaged. Of course the circuit must be powered-down to release the latch-up condition and this in itself causes system lock-ups. In most microelectronics in terrestrial applications SEUs are the predominant soft error failure mechanism.

collectively is 10-100 FIT while the soft error rate (SER) can easily exceed 50,000 FIT (one FIT or Failure-in-Time is equivalent to one failure in a billion device hours). For single-chip consumer systems even an SER of 50kFIT is usually not problematic, but for high-reliability systems composed of multi-chip assemblies such a failure rate is intolerable.

In the last two decades three radiation mechanisms have been identified to cause soft errors in ICs at terrestrial altitudes[3]. In the late 1970s, alpha particles emitted from the natural radioactive decay of uranium, thorium, and daughter isotopes present as impurities in packaging materials were found to be the dominant cause of SER in DRAMs[4]. During the same period, it was demonstrated that the ionizing reaction products created from the interaction of cosmic high-energy neutrons with device materials could cause soft errors[5], and by the mid-1990s it had been established that high-energy cosmic radiation was the dominant source of soft errors in DRAM devices[6-8]. More recently a third mechanism was identified–soft errors from low-energy cosmic neutron interactions with ^{10}B in device materials[9], specifically in the borophosphosilicate glasses (BPSG)[10] used extensively as insulating layers in integrated circuits. This mechanism has been shown to be the dominant SER mechanism in 0.25 and 0.18 um CMOS SRAMs fabricated with BPSG.[11,12]

Significant improvements in the SER performance of microelectronics can be obtained by eliminating or mitigating the sources of radiation. To reduce alpha emissions in the final packaged IC, high purity materials and processes are employed. In products requiring high reliability, uranium and thorium impurities have been reduced below one hundred parts per trillion (in the case of eutectic lead-tin solder this is not sufficient since radioactive Pb isotopes remain that cause the in-growth of high activity daughter products months after the sample is produced)[13]. Going from conventional IC packaging to ultra-low alpha packaging materials the alpha emission can be reduced from 5-10 alphas/cm^2-hr to less than 0.001 alphas/cm^2-hr. Another method used to reduce the impact of alpha particles is exclusion zones, where materials with high alpha emission (such as solder bumps) are kept physically separated from sensitive circuit components[14]. Thick (5-10 μm) polyimide coatings are often deposited on the IC prior to its encapsulation to reduce package alpha emission[15]. To reduce SER induced by the ^{10}B activation by low-energy neutrons, BPSG is replaced by other insulators which do not contain boron, and any processes using boron precursors are scrutinized for ^{10}B content prior to introducing them into the manufacturing process. When all these mitigation techniques are used, the SER of the IC is reduced significantly, but ultimately limited by cosmic high-energy neutron interactions that cannot be easily shielded.

The mechanism by which a soft error is produced occurs after an energetic ionizing particle has been brought to rest in the silicon near sensitive device nodes (floating or weakly-driven reverse-biased junctions). Along the path it traverses, the charged particle produces a dense radial distribution of electron-hole pairs as illustrated in figure 1a (an n+/p junction is shown, but a similar situation occurs for p+/n junctions except that the collected charge is holes instead of electrons). If the resultant ionization track traverses the depletion region, carriers are rapidly (over several picoseconds) collected by the electric field, thereby compensating the charge stored in the junction (reducing the voltage on the node permanently if the node is floating and temporarily if the node is driven). Outside the depletion region the non-equilibrium charge distribution induces a temporary funnel-shaped potential distortion[16] along the trajectory of the event, further enhancing charge collection by

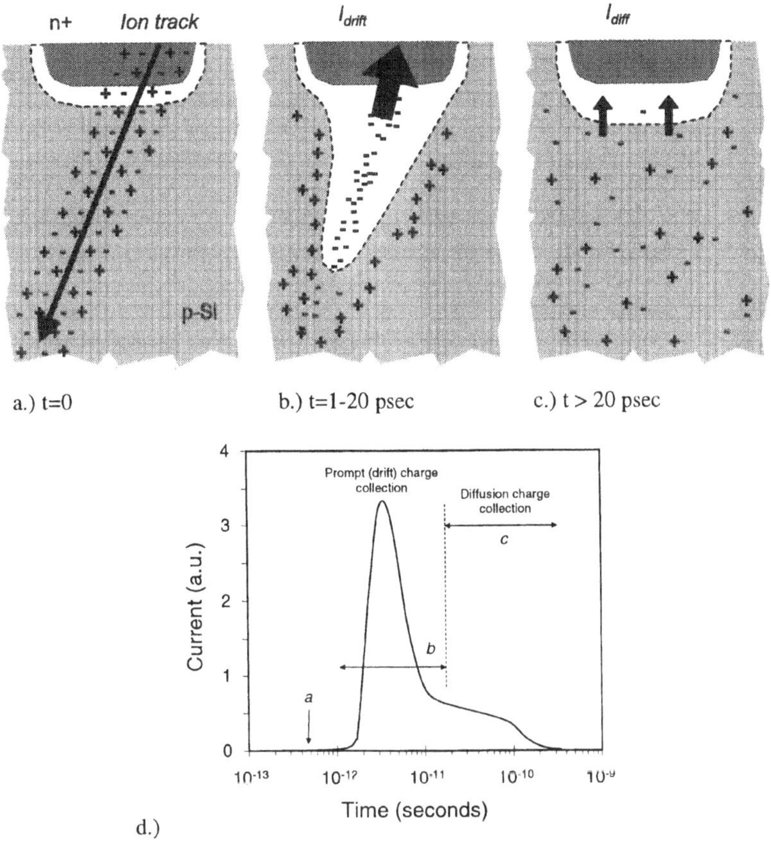

Fig. 1 Schematic representation of charge collection in a silicon junction immediately after an ion strike a.), during prompt (drift) collection (b.), and during diffusion collection (c.). A graph of the junction current induced as a function of time is shown in d.).

drift (figure 1b). This "prompt" collection phase is typically completed within tens of picoseconds, and as the funnel collapses (as the potential distortion relaxes), diffusion then dominates the collection process (figure 1c) until all excess carriers have been collected, recombined, or diffused away from the junction area (~ nanoseconds). The transient charge collected from the radiation event produces a current pulse at the junction as illustrated in figure 1d – note the distinctive shape of the curve defined by the different collection mechanisms. In general, the farther away from the junction that the event occurs, particularly if the track does not cross the junction, the less charge is collected, and the less likely it is that the event will cause a soft error. For modern submicron CMOS technologies, the current transient typically lasts 200 psecs with the bulk of the charge collection occurring within 2-3 microns of the junction region[17]. Of course these time constants depend strongly on the type of ion, its initial energy, and the specifics of the process technology.

If enough of the charge is collected by a node the data state of the circuit may be compromised. The collected charge (Q_{coll}) is a function of the ionizing particle's energy and trajectory, silicon substrate structure and doping, and the local electric field. If Q_{coll} is greater than the critical charge (Q_{crit}) then a soft error occurs. Q_{crit} is defined as[18]:

$$Q_{crit} = C_n \cdot V_n + I_{rstr} \cdot t_{sw} \qquad \text{(eq. 1)}$$

where C_n and V_n are the node capacitance and voltage, and I_{rstr} is the restoring current provided by feedback and t_{sw} is the time required for the circuit to switch to its opposing data state. The second term in equation 1 will obviously be zero for circuits such as dynamic random access memories (DRAM) where there are no external means to maintain the data state. For circuits which have some form of feedback, such as static random access memory (SRAM), flip-flops, or latches, the second term will be non-zero and represents the additional charge provided by the current feedback driving the node – if enough charge is provided before the memory bit's characteristic switching time, then Q_{crit} will be greater than Q_{coll} and a soft error is avoided. Unfortunately in high-speed, low-power SRAM designs, I_{rstr} and t_{sw} are minimized so this restoring term is often negligible.

Several process technologies can be used to improve SER performance either by reducing Q_{coll} and/or increasing Q_{crit}. One common approach is the use of additional well isolation (triple-well and guard-ring structures) which reduce the amount of charge collected by creating potential barriers that limit the efficiency of the funneling effect and reduce the likelihood of parasitic bipolar collection paths[19]. While these isolation methods can improve SER they are expensive because they require additional processing steps (masking and implant) and in the case of guard rings, additional chip area. In deep-submicron technologies where most of the charge collection is reduced to within a micron or two of the junction, triple well-isolation loses its effectiveness. Another popular approach is to replace bulk silicon with silicon-on-insulator (SOI) substrate material. Because the active device volume is greatly reduced (due to the thin silicon device layer on the oxide layer) direct charge collection is significantly reduced in SOI devices[20] and initially drastic reductions in SER were expected. Ultimately though, parasitic bipolar action and/or charge multiplication enhances floating body effects in SOI technologies limiting the SER performance from the same as to about 2x better than bulk[21]. However, recent work with fully depleted SOI substrates is more encouraging and shows a 10x reduction in SER over conventional bulk devices[22]. Unfortunately SOI substrates are more expensive than conventional bulk substrates. For this reason, in high reliability commercial products, design techniques based on adding redundant circuitry and/or software coding methods are often favored to provide much more significant reductions in SER.

2. Scaling trends for memory devices

To create the functionality provided by today's electronic systems and appliances several distinct components must be integrated together. At the core of each system is a microprocessor or digital signal processor with large embedded memories (usually SRAM) interconnected with sequential logic. In larger systems, discrete external main memory (usually DRAM) is also used. Finally, all systems have some analog or digital input/output components to allow the device to respond and interact with the user and the outside world. As we shall see, the SER performance of these various components is affected differently by technology scaling.

It is somewhat ironic that soft errors were first discovered to be a problem in DRAM because, after many generations, it is currently one of the more robust electronic devices. Originally though, as DRAM technologies were scaled (in the late 1980s) DRAM SER was becoming a major problem because the planar capacitor cells that stored the signal charge in large reversed-biased area junctions were becoming more and more sensitive to ionizing radiation as they were scaled down in size (Q_{coll} was not shrinking as fast as Q_{crit} so SER increased). Interestingly, charge coupled devices (CCDs) and CMOS image sensors that are the mainstay of the digital camera industry get their high sensitivity to photon-induced charge by using large two-dimensional arrays of junctions that float during "exposure" similar to old DRAM designs. To address the DRAM soft error sensitivity issue and increase the packing density, manufacturers developed cell designs in which the large storage junction was replaced by a minimal storage contact (source-drain contact of the pass gate transistor used to "write" or "read" charge from the storage cell) connected to a three-dimensional (3-D) capacitor cell. The move to 3-D capacitor cells greatly reduced DRAM SER because the Q_{crit} of the cell was increased significantly while its charge collection efficiency was reduced with the sizable reductions in junction area[23]. In scaling from 1 Mbit (this was the first DRAM node to migrate to 3D cell designs) to 1 Gbit technology, the DRAM bit SER (solid diamonds) has been reduced by nearly five times per generation as illustrated in figure 2. This continuous

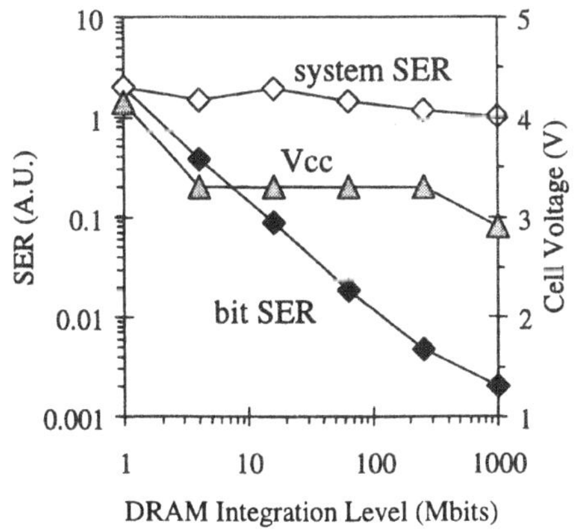

Fig. 2. DRAM bit (solid diamonds) and system SER (open diamonds) and operating voltage (gray triangles) as a function of technology scaling.

reduction in DRAM bit SER is also related to the relatively gradual (compared to logic devices) voltage scaling (gray triangles). Despite the DRAM bit SER having been reduced by nearly 1000x over six generations, the DRAM system SER (open diamonds) has remained essentially unchanged because system memory requirements have increased the memory density (bits/system) almost as fast as the SER reduction provided by technology scaling.

Therefore commercial DRAM system reliability has remained roughly constant over many generations. With DRAM technology scaling, operating frequencies have increased. At frequencies below 50Mhz DRAM SER is dominated by cell hits, but as frequencies are increased, the probability of sense-amplifier and bit-line hits increases since there are more sensing operations in a given time period during which the bit-line is floating (sensitive to soft errors) for a larger percentage of the DRAM cycle time. Therefore, as DRAM is scaled to operate at higher speeds the SER is expected to increase in proportion to the increases in DRAM operating frequencies.

In contrast to early DRAM, SRAM was once far more robust against soft errors because of its higher operating voltage and the fact that data was stored as an active state in the bi-stable circuit made up of two large cross-coupled inverters, each strongly driving the other to keep the SRAM bit in its programmed state. According to equation 1 shown earlier, Q_{crit} for the SRAM cell is defined by the charge on the node capacitance and a second term related to the size of the transistor (its current drive capability) keeping the node voltage at the proper value – the bigger the transistor, the higher Q_{coll} must be for the node to reach the switching threshold. With scaling, the SRAM junction area has been deliberately minimized to reduce capacitance, leakage, and cell area, while simultaneously the SRAM operating voltage has been aggressively scaled down to minimize power. Initially, with each successive SRAM generation, reductions in cell collection efficiency due to shrinking cell depletion volume were swamped out by large reductions in operating voltage and reductions in node capacitance. So SRAM bit SER initially increased with each successive generation as illustrated in figure 3a. However, as feature sizes have been reduced below 0.35 μm the SRAM bit SER has

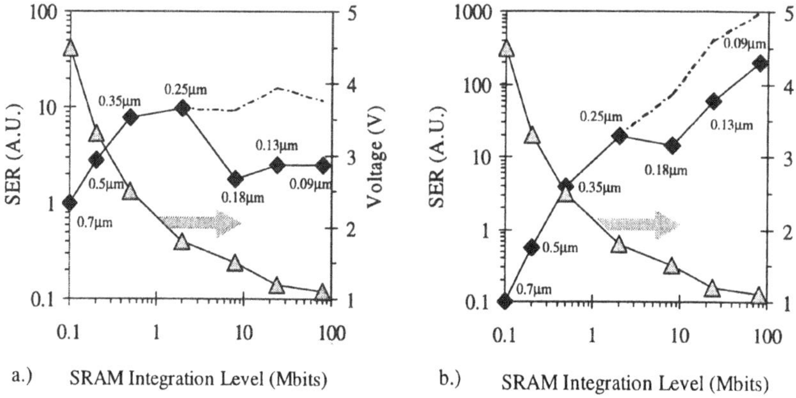

Fig. 3. SRAM voltage (gray triangles), SRAM bit SER (a.) and SRAM system SER (b.) as a function of technology scaling. Dashed line represents SER for SRAM processes still employing BPSG in their production process.

saturated. This saturation is due to the fact that further reduction in operating voltage (gray triangles) is limited by the threshold or turn-on voltage of the transistors, reductions in junction collection efficiency, and increased charge sharing with neighboring nodes. As can be seen by the dashed line, SRAM fabricated with BPSG will have an SER which is about eight times higher than that of SRAMs with no BPSG. Ultimately, as with DRAM, scaling also implies increased density, so saturation in SRAM bit SER does not translate to saturation in the SRAM system SER as is seen in figure 3b. The exponential growth in the amount of embedded SRAM in electronics has led the SRAM system SER to increase with each generation, with no end in sight. This trend is of great concern to chip manufacturers since SRAM constitutes a large part of all advanced integrated circuits today. As SRAM operating frequencies increase, the probability that soft errors will occur during reading/sensing operations increases (since the SRAM cell Q_{crit} is lowered during sensing[24]). In a manner similar to DRAM, above a certain operating frequency, it is expected that the SRAM SER will increase with increasing frequency.

By far the most effective method of dealing with soft errors in memory is error detection and/or correction. In its simplest form, error detection consists of adding a single bit to store the parity (odd or even) of each data word (regardless of word-length). Whenever data is retrieved, a check is run comparing the parity of the stored data to its parity bit. If a single error has occurred the check will reveal that the parity of the data does not match the parity bit. Thus the parity system allows for the detection of a soft error for a minimal cost in terms of circuit complexity and memory width (only a single bit is added to each memory word and the parity checker utilizes minimal additional chip area). The two disadvantages of this system is that the detected error cannot be corrected since the parity cannot localize the error, and if an even number of errors has occurred, the parity check will not register that any error has occurred, since the parity bit will match the parity of the corrupted data in the word (even + first error => odd: odd plus second error => even). The next commonly employed solution is to implement error detection and correction (EDAC) or error correction circuits (ECC)[25]. EDAC/ECC requires that more bits be assigned to each data word. For a 64-bit wide memory, usually eight correction bits are used. These eight correction bits allow up to two errors to be detected and a single error to be corrected (more correction bits can be added to correct for more than a single error). Since most soft error events are single bit errors (it is rare that multiple bits in the same data word are effected since in most memory designs physically adjacent bits do NOT make up the same memory word) EDAC/ECC protection will provide a significant reduction in failure rates. EDAC/ECC incurs a higher cost in terms of design complexity and the additional memory required. Since time is required to check parity and correct for errors, error detection and/or correction schemes trade memory bandwidth for reliability performance. In high-reliability applications where EDAC/ECC has been implemented, the system SER is no longer dominated by memory.

3. Scaling trend for peripheral logic devices

Peripheral logic is usually predominantly made up of sequential logic components. Sequential logic elements include latches and flip-flops that provide storage to hold system event signals and to buffer data before it goes in or out of the chip, and combinatorial logic components whose output is based on a logical relation to the inputs (AND, OR, XOR, etc.) with no capability for retention. The system impact of errors in sequential logic is much harder to quantify since the period of vulnerability (when the devices are sensitive to soft error effects) varies widely depending on the circuit design, frequency of operation, and algorithm.[26] Latches are similar to the SRAM cell in that they use a cross-coupled inverter circuit. However, they tend to be more robust because they are usually designed with larger and/or more transistors (higher Qcrit), that more easily compensate for spurious charge collected during radiation events. Flip-flops are even more robust as they are constructed of two stages, again with larger and more transistors. As illustrated in figure 4, the SER of sequential logic is several orders of magnitude lower than that of SRAM cells. However,

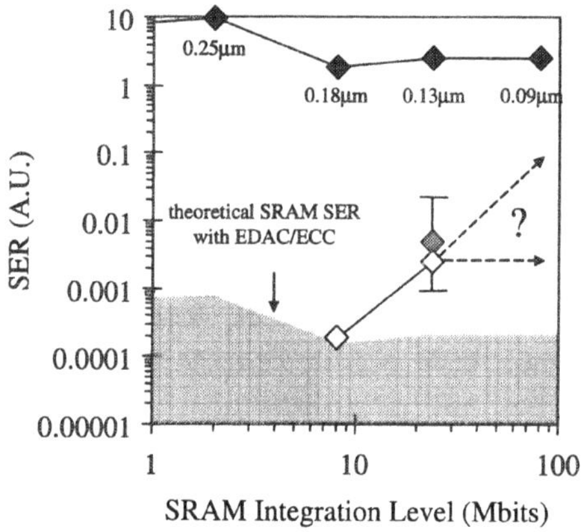

Fig. 4. Simulated (open diamonds) and experimental data (gray diamond with error bars) for flip-flop/latch SER and SRAM SER (solid diamonds) as a function of technology scaling. Note that the shaded region represents the SRAM SER when EDAC/ECC is used.

from the 0.18 to the 0.13 μm node, the logic SER has increased by 20x. Thus sequential logic SER may be increasing with technology scaling. Logic SER is particularly a concern in high reliability systems where the memory has been protected from soft errors with parity or EDAC/ECC. In such systems the peripheral logic failure rate will limit the impact of memory error correction and will ultimately limit the reliability of the product. It should be noted that there is another soft failure mode related to combinatorial logic. If enough radiation-induced charge is collected so that the voltage on the input of a combinatorial gate exceeds the threshold between a "one" and "zero", the event will generate a SET or transient in the output of the device. Since the charge collection occurs on a short timescale, the "glitch" is propagated to the output of the device only as long as the event itself. A soft error is induced by a SET only if the glitch is propagated to the input of a memory element at the same time that the element is being latched or written to. With older technologies this was not a concern since the glitch was shorter than the response of the logic and thus SETs were not readily propagated through the circuit. With the shorter propagation delays of modern technologies, the SET will become a larger component of the observed logic SER.

To mitigate SER in sequential logic involves the use of redundant logic paths[27] and/or time redundancy[28]. In the circuit redundancy approach, three identical logic paths feed into a majority voting circuit in which the two matching inputs (out of three) are assumed to represent the correct data (since the probability of a soft error in two of the circuits is low) and thus this "correct" data is propagated to the next stage. In the temporal approach a single circuit sends the data through the logic pathway multiple times and the results of the multiple passes are then compared to establish the correct result. The most robust method combines the spatial and temporal redundancy scheme so that the same signal is passed down three identical circuits but out of phase – thus the probability that a single event will upset more than one signal with different delays is significantly reduced (assuming the delays are longer than the current transient). These methods unfortunately require three times the chip area and/or a significant reduction in system speed. Since correcting for logic SER/SET is extremely expensive from a performance and area standpoint, simulations to identify the critical logic paths are needed so that only the most critical paths are hardened.

4. Conclusion

We have shown how transient charge generated when an energetic ion is brought to rest in the silicon substrate can lead to soft errors in semiconductor devices. At terrestrial altitudes three mechanisms are responsible for soft errors: alpha particles emitted from trace radioactive impurities in the device materials, the reaction of low-energy cosmic neutrons with high concentrations of ^{10}B in the device, and the reaction of high-energy cosmic neutrons with silicon and other device materials. With proper screening and choice of manufacturing processes, the SER induced by alpha particles and ^{10}B reaction products can be reduced to the point where they are a small fraction of the total SER – dominated by cosmic high-energy neutrons that cannot easily be shielded. The soft error sensitivity of various memory and logic components has been considered as these components are scaled to smaller dimensions, higher integration densities, and lower operating voltages. While shrinking junction sizes have led to reductions in direct charge collection and enhanced charge sharing with neighboring cells (thereby reducing the charge collected by any one device), cell voltage and capacitance reductions have decreased the critical charge as well. The impact of these changes has been dependent on the type of technology. DRAM system SER has been relatively independent of technology scaling, because reductions in cell sensitivity have been matched by increased memory density (more bits/system). In contrast, while SRAM system SER used to be minimal, it has increased rapidly with each new technology node. The SER of both memory types is expected to increase with increasing operating frequencies. Finally, we have shown how sequential logic elements, once insensitive to SER, are now becoming much more sensitive - indeed, in high reliability systems where memory error correction is employed, the failure rate is limited by SER and latched SETs in the sequential logic.

References

1. J. Feynmen and S. Gabriel, "High-Energy Charged Particles in Space at One Astronomical Unit", IEEE Trans. Nucl. Sci., vol. 43, No. 2, pp. 344, April 1996.
2. E. Normand, "Single-Event Effects in Avionics", IEEE Trans. Nucl. Sci., vol. 43, No. 2, pp. 461, April 1996.
3. R. Baumann, "Soft errors in advanced semiconductor devices part I – the three radiation sources", IEEE Trans. Device and Materials Reliability., vol. 1, No. 1, pp. 17, Mar. 2001.
4. T. May and M. Woods, "Alpha-particle-induced soft errors in dynamic memories," IEEE Trans. Elec. Dev., vol. 26, No. 1, pp. 2, Jan. 1979.
5. J. Ziegler and W. Lanford, "The effect of sea level cosmic rays on electronic devices," J.

App. Physics, vol. 52, pp. 4305, 1981.
6. C. Gossett, B. Hughlock, M. Katoozi, G. LaRue, and S. Wender, "Single event phenomena in atmospheric neutron environments," IEEE Trans. Nucl. Sci., vol. 40, No. 6, pp. 1845, Dec. 1993.
7. E. Normand, "Single event upset at ground level," IEEE Trans. Nucl. Sci., vol. 43, No. 6, pp. 2742, Dec. 1996.
8. W. McKee, et al., "Cosmic ray neutron induced upsets as a major contributor to the soft error rate of current and future generation DRAMs," IEEE IRPS, pp. 1, 1996.
9. T. Oldham, S. Murrill, and C. Self, "Single Event Upset of VLSI Memory Circuits Induced by Thermal Neutrons," Rad. Effects, Research, and Engineering., vol. 5, No. 1, pp. 4, 1986.
10. R. Baumann, T. Hossain, S. Murata, and H. Kitagawa, "Boron compounds as a dominant source of alpha particles in semiconductor devices," IEEE IRPS, pp. 297, 1995.
11. R. Baumann and E. Smith, "Neutron-induced ^{10}B fission as a major source of soft errors in high density SRAMs," IEEE IRPS, pp. 211, 2000.
12. R. Baumann and E. Smith, "Neutron-induced ^{10}B fission as a major source of soft errors in high density SRAMs," Elsevier Microelectronics Reliability, vol. 41, No. 2, pp. 211, 2001.
13. "Low-Alpha Lead Symposium", hosted by Lawrence Livermore National Labs. and Johnson Matthey Electronics, Feb. 1997.
14. Y. Tosaka, et al., "Measurement and analysis of neutron-induced soft errors in sub-half-micron CMOS circuits" IEEE Trans. on Electron Devices, vol. 45, No. 7, pp. 1456, July 1998
15. D. Makino, "Application of polyimide resin to semiconductor devices in Japan", Electrical Insulation Magazine, IEEE , vol. 4, No. 2 , pp. 19, March 1988.
16. C. Hsieh, P. Murley, and R. O'Brien, "Dynamics of Charge Collection from Alpha-Particle Tracks in Integrated Circuits", IEEE IRPS, pp. 38, 1981.
17. J. Palau, et al., "Device simulation study of the SEU sensitivity of SRAMs to internal ion tracks generated by nuclear reactions", IEEE Trans. Nucl. Sci., vol. 48, No. 2, pp. 225, April 2001.
18. P. Roche, J.M. Palau, G. Bruguier, C. Tavernier, R. Ecoffet, and J. Gasiot, "Determination of key parameters for SEU occurrence using 3-D full cell SRAM simulations", IEEE Trans. Nucl. Sci., vol. 46, No. 6 , pp. 1355, Dec. 1999
19. D. Burnett, C. Lage, and A. Bormann, "Soft-error-rate improvement in advanced BiCMOS SRAMs", IEEE IRPS, pp. 156, March 1993.
20. O. Musseau, "Single-event effects in SOI technologies and devices" IEEE Trans. Nucl. Sci.,. vol. 43, No. 2, pp. 603-613, 1996.
21. S. Hareland, J. Maiz, M. Alavi, K. Mistry, S. Walsta, and Changhong Dai, "Impact of CMOS process scaling and SOI on the soft error rates of logic processes", IEEE VLSI Tech., pp. 73, 2001.
22. P. Oldiges, et al., "Soft error rate scaling for emerging SOI technology options", IEEE VLSI Tech., pp. 46, 2002.
23. L. Massengill, "Cosmic and terrestrial single-event radiation effects in dynamic random access memories", IEEE Trans. Nucl. Sci., vol. 43, No. 2, pp. 576, April 1996
24. N. Seifert, D. Moyer, N. Leland, and R. Hokinson, "Historical trend in alpha-particle induced soft error rates of the Alpha^{tm} microprocessor," IEEE IRPS, pp. 259, 2001.
25. K. Furutani, K. Arimoto, H. Miyamoto, T. Kobayashi, K. Yasuda, and K. Mashiko, "A built-in Hamming code ECC circuit for DRAMs" IEEE Trans. Solid-State Circ., vol. 24, No. 1, pp. 50, Feb. 1989.

26. L. Massengill, A. Baranski, D. Van Nort, J. Meng, and B. Bhuva, "Analysis of single-event effects in combinatorial logic – simulation of the AM2901 bitslice processor," IEEE Trans. Nucl. Sci., vol. 47, No. 6, pp. 2609, Dec. 2000.
27. M. Baze, S. Buchner, and D. McMorrow, "A digital CMOS design technique for SEU hardening," IEEE Trans. Nucl. Sci., vol. 47, No. 6, pp. 2603, Dec. 2000.
28. D. Mavis and P. Eaton, "Soft error rate mitigation techniques for modern microcircuits", IEEE IRPS, pp. 216, 2002.

International Journal of High Speed Electronics and Systems
Vol. 14, No. 2 (2004) 311–325

World Scientific
www.worldscientific.com

SINGLE-EVENT EFFECTS IN III-V SEMICONDUCTOR ELECTRONICS

DALE McMORROW and JOSEPH S. MELINGER
Naval Research Laboratory, Code 6812, 4555 Overlook Avenue, SW, Washington, DC 20375, USA

ALVIN R. KNUDSON
SFA Inc., Largo, MD, 20774, USA

Single-event effects are a serious concern for high-speed III-V semiconductor devices operating in radiation-intense environments. GaAs integrated circuits (ICs) based on field effect transistor technology exhibit single-event upset sensitivity to protons and very low linear energy transfer (LET) particles. The current understanding of single-event effects in III-V circuits and devices, and approaches for mitigating their impact, are discussed.

Keywords: single-event upset; charge collection; radiation; InP; InGaAs; InAs; AlSb; HEMT; FET

1. Introduction

Devices fabricated from III-V semiconductor compounds offer significant advantages over silicon-based technology in terms of speed, range of power, and versatility of application. A wide range of band gaps and band structures now can be engineered, and applications have been extended to higher and lower power levels. Carriers can be forced into undoped intrinsic regions where, lacking scattering centers, they achieve high mobilities and speeds. III-V semiconductors are used to develop devices such as heterojunction bipolar transistors (HBTs), high electron mobility transistors (HEMTs), tunnel diodes, and multi-quantum-well (MQW) and superlattice devices and, due to their versatility, III-V devices can be found in conventional electronics, optoelectronics, nano-electronics, photovoltaics, high-power, and ultra-low-power applications. The advantages of low power and high speed make many III-V devices ideal for applications in satellite environments where ambient radiation is high. In order to be useful in space, devices must be able to withstand the harmful effects of radiation.

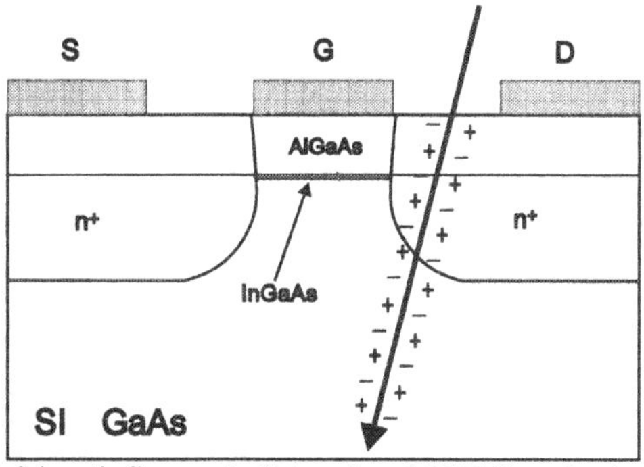

Fig. 1. Schematic diagram of a GaAs n-channel HFET illustrating the source, gate, and drain and the passage of a heavy ion.

Single-event effects (SEEs) refer to phenomena that arise from the interaction of a single energetic particle with a microelectronic device. SEEs can be either transient disruptions or permanent failures, and can have serious consequences for spacecraft, including loss of information, functional failure or total loss of control. Unlike displacement damage effects, SEEs are not related to the cumulative effects of radiation. Rather, they occur via a stochastic process arising from the random interactions of a device with ions of various energies, species, and angles of incidence. With the passage of an energetic ion through a material, carriers are generated as the ion excites electrons, leaving behind a track of electrons and holes (a dense electron-hole plasma). This phenomenon is illustrated schematically in Fig. 1 for the passage of an ion through an n-channel heterojunction field-effect transistor (HFET).

Single-event upset (SEU) is the most common and the most well characterized nondestructive effect. SEUs are caused by ionization tracks through or near junctions of a semiconductor device, in which the logic state of the circuit is changed. Such effects are also called soft errors because they can be corrected and cause no permanent damage. The impact of an SEU is highly system-dependent, and depends on its type, location, timing, and the specific nature of the application. Single-event upset is a significant concern for applications of III-V semiconductors in the space radiation environment, particularly as clock rates increase and critical charges decrease.

In what follows we discuss single-event effects in III-V electronic devices and circuits. A significant amount of data has been accumulated on single-event effects in GaAs microelectronic devices since the initial measurement in 1983.[1] That body of work was reviewed several years ago[2], and updated recently.[3] In contrast, very little data on SEE in other III-V devices is available in the public domain. Because of space limitations, the following discussion is limited primarily to the charge collection processes that occur in III-V devices at the single transistor level.

2. Single-Event Effects in III-V Electronic Devices

Since their discovery,[4,5] single-event effects have been investigated and characterized using terrestrial accelerators in an attempt to reproduce the effects of galactic cosmic rays and protons trapped in the Earth's radiation belts. The ionizing energy loss by the ion to the material is characterized by the mass stopping power for a material, or linear energy transfer (*LET*), which is a measure of energy transfer per unit length of material. *LET* is obtained by dividing the energy loss per unit length, *dE/dx*, by the material density, ρ:

$$LET = \frac{dE}{dx}\frac{1}{\rho}. \tag{1}$$

Units of *LET* are $MeV{\cdot}cm^2/mg$ of material, and this quantity depends on the ion, its energy, and the material that it is passing through. The metric used to measure the sensitivity of an IC to SEU is the error cross section, σ, which is the ratio of number of upsets detected to the total particle fluence, with units of upsets/particles/cm^2. An example of a typical heavy ion cross-section verses LET curve for GaAs MESFET logic is shown in Fig. 2.[6] The two key parameters determined from cross section curves are the threshold LET, LET_{th}, and the saturation cross-section, σ_{sat}. LET_{th} is a measure of the minimum charge required to upset the most sensitive region of the chip; below LET_{th}, no upsets are observed. As the ion LET increases, more regions of the chip become susceptible to upsets, and the cross-section increases until the curve reaches saturation.

The saturation cross-section is often interpreted as approximating the total area of the sensitive regions of the chip. This physically appealing interpretation, however, is often complicated by other factors, such as ionization-induced gain and the dependence of the measured cross section on clock rate, for example. Different definitions of LET_{th} are found in the literature. In addition to the definition noted above (which can be referred to as the *absolute* threshold), a common definition is the LET for which the cross section is given by $0.1\sigma_{sat}$. This latter definition is of practical utility, since the statistics required for determination of the absolute threshold can lead to prohibitive (expensive) data collection times, and are unwarranted for many applications.

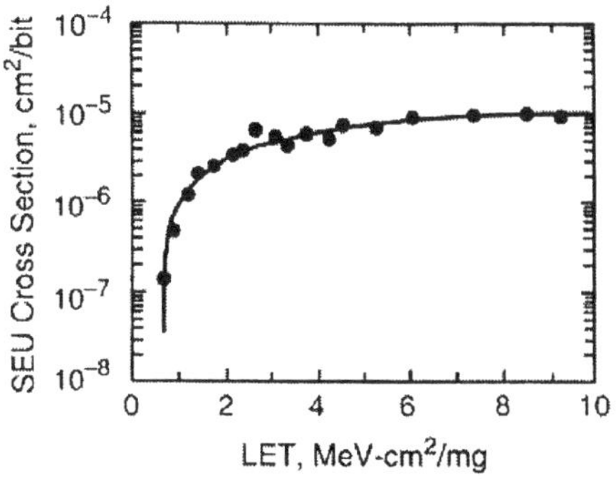

Fig. 2. Typical heavy-ion cross section vs. LET curve for a GaAs DCFL E/D MESFET logic circuit illustrating the low threshold and large saturated cross section that is typical of this technology (after ref. 6).

Single-event upsets in semiconductor integrated circuits are caused by energetic protons, neutrons, and heavy ions. Except for the most sensitive devices (photodiodes and charge-coupled devices, for example),[7] the stopping power of primary protons or neutrons does not generate sufficient charge to cause an upset. The interaction of protons and neutrons with atoms of the semiconductor lattice can produce nuclear reactions, and the spallation products of such reactions can deposit sufficient charge in the sensitive volumes of devices to cause upsets[8-10]. While the probability of such events is quite low, this is compensated for by the high flux of protons trapped in the Van Allen radiation belts, such that proton events typically dominate the upset rates of sensitive devices that pass through the belts.

The saturation cross section and threshold LET determined from terrestrial measurements, such as those of Fig. 2, are used to predict the error rate for a component in a given radiation environment. For such calculations one uses the ion flux (ions/unit area/unit time) as a function of LET for the environment of interest which, for a spacecraft, depends on its orbit. Discussions of the space environment and error rate calculations are beyond the scope of this article and are discussed in detail elsewhere.[11-13]

2.1. *Charge-collection mechanisms of III-V field-effect transistors*

By the early 1980s, the advantages of GaAs technology in terms of its total ionizing dose and latch-up immunity were well known.[14-16] The sensitivity of GaAs technologies to SEU, however, rapidly became apparent.[1,17-20] Heavy-ion SEU testing on GaAs MESFET ICs led to predicted soft error rates on the order of 10^{-3} errors/bit-day for common spacecraft orbits.[6,18-21] This very significant error rate is a function of two factors: a heavy ion LET threshold below 1 MeV·cm^2/mg; and sensitive areas (as determined from the saturation cross sections) much larger than the physical areas of the devices.[6,19] These characteristics are exemplified by the heavy-ion cross-section *vs.* LET curve of Fig. 2. It is now widely recognized that GaAs ICs exhibit SEU sensitivity to protons as well as very low LET particles, making their SEU performance a serious concern for applications in radiation environments.[2] Since those early experiments, significant effort has been expended in an attempt to understand and control the mechanisms of error generation and propagation in GaAs devices and circuits.

The primary events from which all SEU processes originate are charge deposition and its collection at the electrodes of a device (*cf.*, Fig. 1). The most fundamental approach to understanding the origins of the SEU performance of integrated circuits, therefore, involves investigation of the charge transport, recombination, and collection mechanisms that occur at the single transistor level. Charge-collection measurements, when performed as a function of experimentally controlled parameters, provide insight into the physics that occurs when an energetic particle interacts with a semiconductor device. The goal of such investigations is first to understand the charge-collection processes that occur at the transistor level, and then to use that knowledge to improve the SEU performance of complex ICs through changes in the device and/or circuit design. Representative examples of time-resolved and time-integrated charge-collection measurements for an n-channel GaAs HFET are shown in Fig. 3.

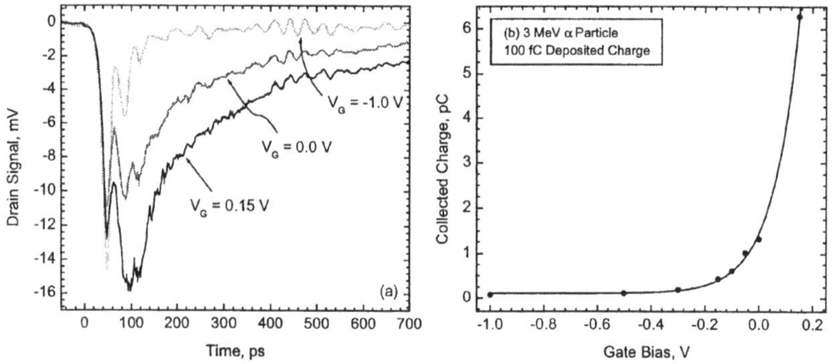

Fig. 3. Gate-bias dependence of drain charge-collection transients of an n-channel HFET: (a) time-resolved charge collection measurements for excitation by above-band-gap laser pulses with 57 fC deposited charge (after ref. 27); (b) time-integrated charge-collection measurements for 3 MeV α-particle irradiation; 100 fC deposited charge (after ref. 25). For both series of measurements the drain was biased at 2V .

Early charge-collection measurements on GaAs MESFETs revealed that the charge collected at the drain of the transistor could exceed the total charge deposited by the ion.[22,23] Enhancement factors of two to five are typical,[18,22-24] but factors of up to 60 have been reported for α-particle irradiation of GaAs HFETs under certain bias conditions,[25] as is illustrated in Fig. 3b. To date, the mechanisms leading to this charge amplification have been investigated with a variety of techniques, including ion- and laser-induced time-integrated charge-collection studies as a function of various device geometries and bias conditions, ion-, laser-, and electron-beam fast transient measurements, and two-dimensional device simulations.[6,18,22-37] Those investigations lead to a single, significant conclusion that charge-enhancement effects in GaAs FETs are, to a large extent, a consequence of charge deposited in the substrate below the active region of the device. Some of the key results are summarized below. The data of Fig. 3 illustrate the general features that are characteristic of all GaAs FET devices investigated to date. The data of Fig. 3a are for above-band-gap pulsed laser excitation, and those of Fig 3b for α-particle irradiation. Data obtained for both methods exhibit similar characteristics[26,27] but, because carriers can be injected into the device in a well defined and controlled manner without the concomitant damage associated with particle irradiation methods, the pulsed laser has become an invaluable tool for performing detailed interrogations of the charge-collection and SEU mechanisms of III-V semiconductor devices and circuits.[24-27,33,35]

The time-integrated data of Fig. 3b illustrate that the collected charge can become quite large when the device is biased near threshold.[25] For this data set, a charge enhancement factor (CEF) of 10 is observed for the normal operating bias of 0 V, with CEF of greater than 60 at the most positive bias tested. This factor of 60 is, we believe, the largest charge enhancement factor reported for a GaAs FET. The different mechanisms that contribute to the charge enhancement in these devices can be discerned from time-resolved data, such as that of Fig. 3a.

At short times the data of Fig. 3a exhibit a narrow, sharp feature limited by the time resolution of the instrumentation. This feature, depending on the bandwidth of the measuring instrument, sometimes appears as a shoulder, or not at all. Device simulations[30,36] suggest that this contribution is very fast, and independent measurements on the gate, drain, and source reveal that it can be identified with a gate-drain current arising from the prompt collection of holes at the gate and electrons at the drain.[33]

At longer times the transient data reveal a complex temporal profile that increases in magnitude as the gate is biased more positive. For times greater than 400 ps the transient signal decays on a nanosecond time scale (depending on the bias conditions), and can be identified with a channel-modulation effect in which excess holes present in the substrate beneath the active region of the device serve to backgate the channel. This creates a source-drain current pathway that persists as long as the excess hole density is present.[96,100] When the gate bias is made sufficiently negative, the excess charge density has little effect on the source-drain current. The electron and hole density contour plots presented in ref. 30 and 34 are particularly illuminating in this regard.

Variation of the energy of an applied laser pulse is a powerful tool for investigating the mechanisms of charge collection in semiconductor devices.[26,27,33,36] Representative charge-collection transients measured as a function of the injected carrier density for three different GaAs transistors are shown in Fig. 4. These curves illustrate the unique temporal characteristics of the different device types: Fig. 4a shows the charge-collection transients generated for an enhancement-mode GaAs MESFET fabricated in a lightly p-doped substrate; Figs. 4b and 4c show the analogous transients produced for an n-channel HFET and a MESFET, respectively, both fabricated in semi-insulating substrates. For each device type, as the injected carrier density is increased, there develops a feature that is negligible at the lowest pulse energy, but becomes quite prominent at the higher pulse energies. This contribution exhibits a delayed rise and decays on a timescale of a few hundred picoseconds. Results such as these, when taken together with computer simulation results, suggest a parasitic bipolar transistor mechanism in which the electron-hole plasma distorts the fields in the device, resulting in a reduction of the source/substrate barrier. Electrons emitted from the source propagate through the substrate and are collected at the drain electrode. One set of two-dimensional computer simulation results is presented in Fig. 5, which shows a series of simulations performed on a buried p-layer GaAs MESFET as a function of the injected carrier density. In Fig. 5b the raw data are convoluted with a Gaussian function to approximate the distortion imposed by the measurement apparatus; the data of Fig. 5b can be compared directly to the experimental measurements of fig. 4a. The simulation results reproduce well the general features of the experimental data: clearly evident is a delayed feature that decays on a 100 ps timescale and depends sensitively on the injected carrier density.

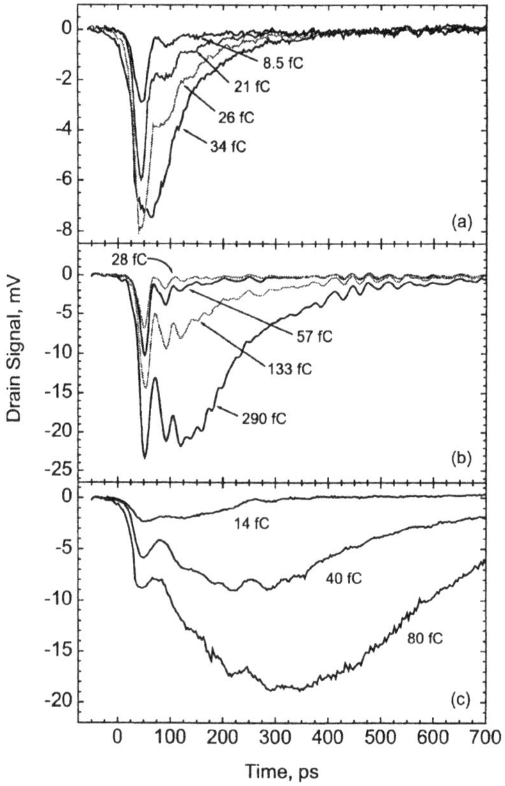

Fig. 4. Typical drain charge-collection transients measured for three types of GaAs field-effect transistors: (a) 1.2 μm gate length enhancement-mode MESFET fabricated by Vitesse Semiconductor [26,29]; $V_D = 2V$, $V_G = -0.3V$; (b) 0.7 μm gate length n-channel HFET fabricated by Honeywell[27]; $V_D = 2V$, $V_G = -0.5V$; and (c) a 1.0 μm gate length enhancement-mode MESFET fabricated by TriQuint Semiconductor[33]; $V_D = 1V$, $V_G = -0.15V$. Measurements performed for nominally 600 nm picosecond pulsed laser excitation focused between the gate and drain for the deposited charge values indicated.

Experimental results available on the charge collection and SEU characteristics of non-GaAs III-V devices and circuits are limited at the present time. To date, charge-collection measurements have been performed on InGaAs/InAlAs HEMTs[38] and HBTs,[39] AlSb/InAs HEMTs,[40] and on resonant interband tunneling diodes (RITDs) in the

AlSb/InAs/GaSb material system.[41] These measurements provide the first insights into the single-event effects performance of these technologies, and provide a basis for future investigations. SEU measurements on III-V ICs other than GaAs are not available at the present time.

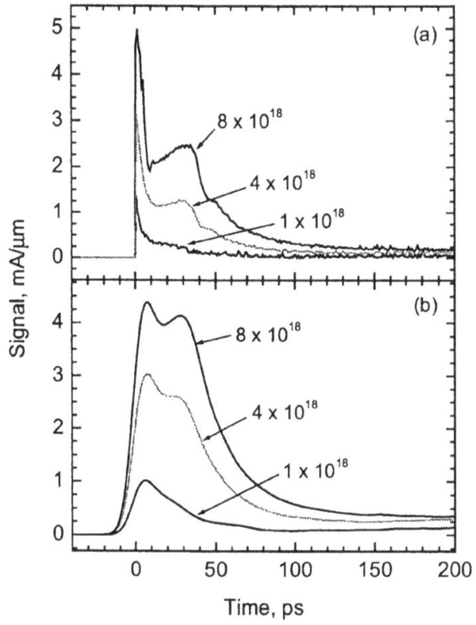

Fig. 5. Drain charge-collection current calculated as a function of laser pulse energy for 1.2 μm gate length GaAs MESFET for 605 nm irradiation: (a) raw simulation data (after ref. 36); (b) data of (a) convoluted with a 9 ps fwhm Gaussian function. The numbers shown indicate the carrier density (in cm^{-3}) at the surface of the device.

Highly functional circuits composed of RITDs and HEMTs in the 6.1 Å lattice constant AlSb/InAs/GaSb material system are an attractive approach for obtaining high-speed low-power circuits.[42] RTDs and HEMTs made with InGaAs and InAlAs lattice matched to InP substrates have been combined to develop low-power static random access memory and logic gates operating at 35 GHz.[43-45] RITDs and HEMTs composed of InAs, GaSb, and AlSb and their ternary and quaternary alloys offer the potential for circuits with lower power dissipation than obtainable in the InP material system. Fig. 6 illustrates the general characteristics of the pulsed laser induced charge-collection transients measured for AlSb/InAs HEMTs (the curves are inverted to be positive valued).[40] Similar results have been reported for InGaAs/InAlAs HEMTs grown on InP substrates.[38] The AlSb/InAs devices exhibit significant charge enhancement factors, and the temporal signature and behavior under variations in the gate bias are similar to those observed for GaAs and InP-based FETs (*cf.*, Fig. 3). Preliminary considerations, therefore, imply charge-enhancement mechanisms analogous to those of GaAs FETs, suggesting that SEE mitigation methodologies analogous to those implemented for GaAs FET-based technologies may be both appropriate and effective in minimizing the charge collection (*vide infra*).

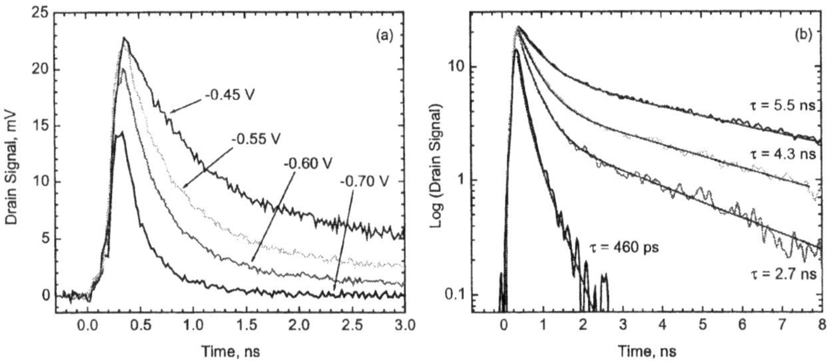

Fig. 6. Typical drain charge-collection transients measured for AlSb/InAs HEMT plotted in linear and logarithmic formats for the gate biases indicated (inverted to be positive valued). In (b) the decaying portion of each curve is fit to a biexponential decay (smooth curves); the time constants for the slower relaxations are indicated, and the time constants for the faster decays are (bottom to top) 140 ps, 330 ps, 420 ps, and 550 ps. 203 fC deposited charge.; $V_D = 0.25$ V.

Device simulations play an important role in sorting out the charge collection processes of III-V devices.[30,32,35-37] The use of spatially localized charge deposition in device simulations permits separation of the different contributions to the charge collection process in a manner that can not be achieved experimentally. In GaAs MESFETs this approach reveals clearly the significant role of substrate carriers in the processes responsible for charge enhancement. Fig. 7a shows a series of charge collection transients for deposition by localized charge tracks (ion-track segments) at various depths beneath the surface of the device calculated for a GaAs MESFET device using a two-dimensional device simulator. The two-dimensional computer simulations were performed utilizing the SILVACO code ATLAS, which solves Poisson's equation and the continuity equations for electrons and holes; additional details are given elsewhere.[37] For

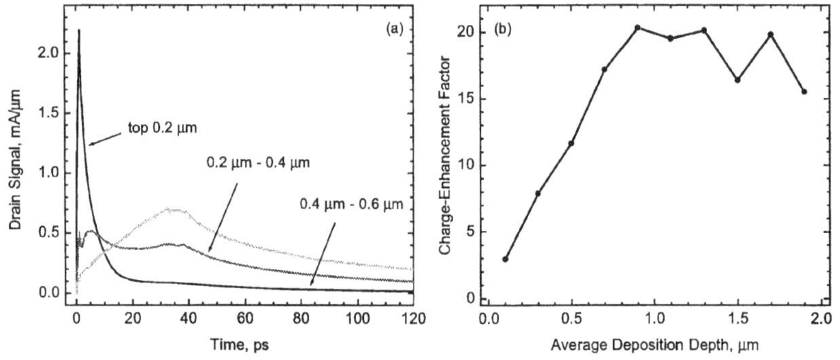

Fig. 7. (a) Drain charge collection transients calculated for excitation by localized charge deposition between the gate and drain at the average depth indicated (after ref. 36); (b) charge collection efficiency for the different localized deposition depths (after ref. 37).

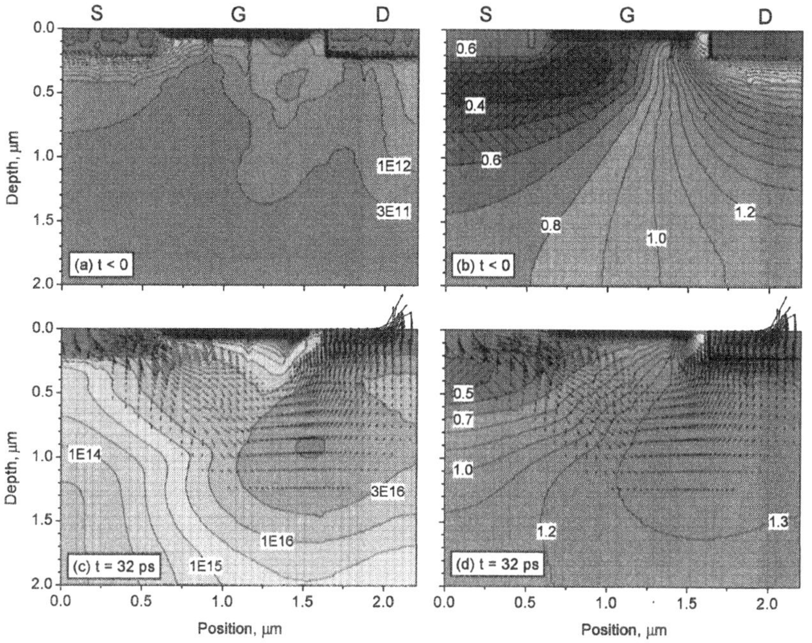

Fig. 8. Electron density (*left*) and constant potential contours (*right*) for GaAs MESFET indicating the source/substrate/drain current pathway and the evolution of the device potential following a localized perturbation centered between the gate and drain 0.9 µm beneath the surface: (a) and (b) prior to the ionization;. and (c) and (d) 32 ps after the ionization event. The arrows indicate the direction and magnitude of electron flow. V_D = 2.5 V, V_G = 0.05V.

shallow deposition, the charge collection process is rapidly completed, and can be identified (primarily) with the conventional collection of electrons and holes by the drain and gate electrodes, respectively. This response can be identified with the sharp features in the data of Figs. 3 and 4. As the deposition depth increases, the charge-collection transient develops a significantly slower character. This slower response arises from source-drain current, and gives rise to charge amplification. The time-integrated data of Fig. 7b reveal the significant charge enhancement factors observed for the localized charge tracks. As is evident, the CEFs increase as the depth of the charge deposition increases, before rolling over in the vicinity of 1 µm.

The two-dimensional electron density and potential contour plots of Fig. 8 illustrate visually some of the primary features of the charge-collection processes of GaAs FETs.[37] Figs. 8a and 8b illustrate the electron density and potential distributions in the device prior to the ionization event (localized ionization track centered at 0.9 µm below the surface). Evident in the unperturbed device is the barrier to electron flow out of the source (blue cross-hatched region). 32 ps after the ionization event (near the peak of the data in Fig. 7a), however, the barrier is reduced significantly (Fig. 8d), leading to a burst of electrons that propagate from the source, deep through the substrate, and to the drain. The source/substrate/drain current pathway evident in the data of Figs. 8c and 8d is responsible for the delayed response in the data of Figs. 3,4,5, and 7, and the very significant charge enhancement that is observed for each case (*cf.*, a CEF of 20 in Fig. 7b).

The results discussed in this section lead to a single, significant conclusion: the charge-collection processes of GaAs FETs are dominated by charge-enhancement effects that are a consequence of charge deposited below the active regions of the device. The direct collection of deposited charge, in the conventional sense, contributes only a small fraction to the total collected charge, and may be largely insignificant in terms of the SEU performance of GaAs field effect transistors.

The data on III-IV technologies other than GaAs are insufficient to draw any firm conclusions regarding the charge-enhancement mechanisms at the present time; the similarity of the non-GaAs experimental data to the GaAs results suggests that similar considerations apply. Recent device simulation results on InAlAs/InGaAs HEMTs, however, reveal the significant role of the InGaAs quantum well in the charge collection/ enhancement process.[59] Those results suggest that the source/drain current flow associated with the charge enhancement process is largely confined to the InGaAs well near the surface of the device, and that the deep source/substrate/drain current pathway that is evident in the case of bulk GaAs MESFETs (*cf.*, Fig. 8) is not an important contributor to the charge collection in InGaAs HEMTs. This result may have ramifications for SEE mitigation approaches in HEMT devices, but additional investigations are required before the significance of this result is clear.

2.2. *SEE mitigation in III-V semiconductor devices*

Various approaches to SEE mitigation have been applied to GaAs circuits.[17,47,32,48-58] Early attempts to harden GaAs systems at the device[52] or circuit[17,48,51,53] level were largely unsuccessful, with the only apparent alternatives involving redundancy at the system or circuit level, and error detection and correction (EDAC) in the system architecture.[51,57] Such methods suffer from reduced performance, increased design complexity, decreased circuit density, and increased power consumption. One circuit-level mitigation scheme has shown some degree of success, however.[58]

Hardening at the device level through a reduction in the collected charge represents the most fundamental approach to circuit hardening, and suffers from none of the shortcomings noted above. Earlier attempts to reduce charge collection at the device level, utilizing potential barriers between the substrate and channel regions, have met with limited success. Those schemes failed to recognize that the primary source of charge collection in GaAs FETs is associated with enhanced source-drain current, and not simply with the efficient collection of charge deposited by the ion.

Device-level mitigation of SEU in III-V FET-based technologies entails minimization of the charge-enhancement processes that dominate the charge collection processes. Significant success using this approach has been achieved through the incorporation of a "low-temperature-grown" GaAs (LT GaAs) buffer layer beneath the active region of the FET .[25,27,34,37,49,50,54-57] The unique characteristic of LT GaAs relevant to this application is its very short carrier lifetime. The LT GaAs material rapidly lowers the concentration of carriers produced by the ionizing event, minimizing the charge-enhancement processes, resulting in a significant (~100x) reduction in the total collected charge.[34] The initial SEU experiments on GaAs HFET ICs with a low-temperature buffer layer revealed that the circuits do not upset for LETs of up to 90 MeVcm^2/mg, rendering them effectively

SEU immune to the natural radiation environment of space.[25,49,50] Fig. 9 illustrates the effect of an LT GaAs buffer layer on the SEE response of GaAs devices. Fig. 9a shows the charge-collection transients measured for 3 MeV-particle excitation of n-channel HFETs fabricated with and without an LT GaAs buffer layer.[25,49] Both devices were biased near pinchoff with $V_G = 0$ V. The transient measured for the conventional (non-LT) HFET shows with clear evidence for the charge-enhancement processes discussed above. The LT GaAs device, in contrast, exhibits a rapid recovery from the perturbation, with the charge-collection transient approaching the baseline within 100 ps of the ionizing event.

Fig. 9b compares charge-sensitive preamplifier measurements (time-integrated charge collection) for α-particle irradiation as a function of the gate bias for LT and non-LT GaAs HFET devices. The dashed line corresponds to a charge-collection efficiency of unity (deposited charge = collected charge). As is evident, the charge-collection efficiency of the LT GaAs HFET is reduced significantly when compared to that of the conventional HFET. These results reveal that the LT GaAs buffer layer results in an approximately 100-fold reduction in the collected charge, consistent with the time-resolved results of Fig. 9a. This 100-fold reduction in collected charge corresponds to a 10^4 reduction in the error rate.

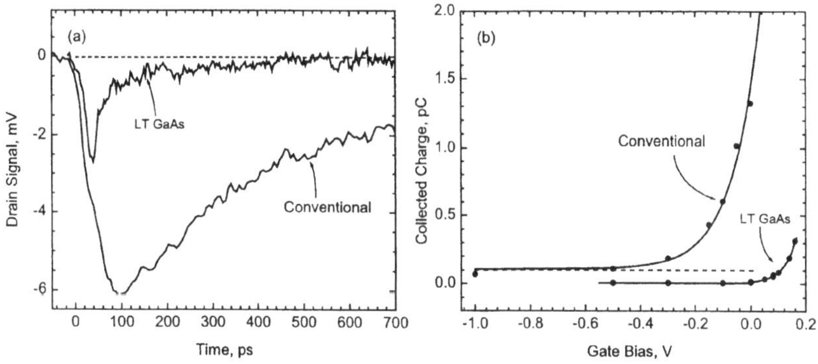

Fig. 9. 3 MeV α-particle induced (a) charge-collection transients and (b) time-integrated collected charge measured for n-channel HFET devices with and without the LT GaAs buffer layer. (a) V_D= 2.0 V; V_S = 0 V; V_G = 0 V for the conventional device and V_D= 4.0 V for the LT device; (b) V_D= 2.0 V; V_S = 0 V. The dashed line in (b) illustrates the charge deposited for the complete ion track (after ref. 25).

The charge-collection data available to date suggest that the SEU susceptibilities of other III-V HEMT-based technologies will be similar to those observed for GaAs circuitry, and it is likely that SEU mitigation will be required for critical operations in high-radiation environments. A device-level mitigation methodology with characteristics analogous to LT GaAs is desirable, especially for very high clock rate circuits for which circuit-level hardening schemes may not be inappropriate and EDAC may be ineffective. The evolution in understanding of the charge-collection mechanisms in GaAs FETs described above, leading up to the development of GaAs FETs fabricated with low-temperature-grown buffer layers, provides a unique example of how such basic-mechanisms research has resulted in a real solution to the severe SEU problem associated with GaAs technology.

3. Summary and Conclusions

Single-event effects are a serious concern for high-speed III-V semiconductor technologies operating in radiation-intense environments. GaAs ICs based on field-effect transistor technology have demonstrated single-event upset sensitivity to protons and very low LET particles. Single-event effects data on other III-V systems are limited, but the general characteristics appear similar to those of GaAs. This primary source of this sensitivity lies in the charge-enhancement processes that occur at the single transistor level. The motivations for utilizing advanced III-V technologies in satellite systems are well established, but it is likely that SEU mitigation will be required for critical operations in high-radiation environments. Of the different SEU mitigation approaches, device-level hardening using short-recombination-time buffer layers appears to be the most effective and generally applicable.

References

1. R. Zuleeg and K. Lehovec, "Radiation effects in GaAs junction field-effect transistors", IEEE Trans. Nuc. Sci. **27** (1980) 1343.
2. D. McMorrow, T.R. Weatherford, S. Buchner, A.R. Knudson, J.S. Melinger, L.H. Tran and A.B. Campbell, "Single event phenomena in GaAs devices and circuits", IEEE Trans. Nucl. Sci. **43** (1996) 628.
3. B.D. Weaver, D. McMorrow, and L.M. Cohn, "Radiation Effects in III-V Semiconductor Electronics", International Journal of High-Speed Electronics and Systems, **13** (2003) 293-326.
4. J.T. Wallmark and S.M. Marcus, "Minimum size and maximum packing density of nonredundant semiconductor devices", Proceedings of the IRE, 286 (1962).
5. D. Binder, E.C. Smith, and A.B. Holman, "Satellite anomalies from galactic cosmic rays", IEEE Trans. Nuc. Sci. **22** (1975) 2675.
6. B.W. Hughlock, G.S. LaRue, and A.H. Johnston, "Single event upset in GaAs E/D MESFET logic", IEEE Trans. Nucl. Sci. **37** (1990) 1894.
7. P. Marshall, C. Dale, and K. LaBel, "Charged particle effects on optoelectronic devices and bit error rate measurements on 400 Mbps fiber based data links", *Proceedings RADECS 93*, 1994 pp. 266-271.
8. C.S. Guenzer, E.A. Wolicki, and R.G. Allas, "Single event upset of dynamic RAM's by neutrons and protons", IEEE Trans. Nuc. Sci. **26** (1979) 5048.
9. R.C. Wyatt, P.J. McNulty, P. Toumbas, P.L. Rothwell, and R.C. Fritz, IEEE Trans. Nuc. Sci. **26** (1979) 4905.
10. P.J. McNulty, G.E. Farrell, R.C. Wyatt, P.L. Rothwell, R.C. Fritz, and N.N. Bradford, "Upset phenomena induced by energetic protons and electrons", IEEE Trans. Nuc. Sci. **27** (1980) 1516-1522.
11. J. Barth, "Modeling space radiation environments", in *1997 IEEE Nuclear and Space Radiation Effects Short Course*, Chapter 1, 1997 pp. I1-I83.
12. J.C. Pickel, "Single event effects rate prediction", IEEE Trans. Nuc. Sci., **43** (1996) 483.
13. E.L. Peterson, J.C. Pickel, J.H. Adams, Jr., and E.C. Smith, "Rate prediction for single event effects–a critique", IEEE Trans. Nuc. Sci. **39** (1992) 1577.
14. M. Simons, "Radiation effects in GaAs integrated circuits: A comparison with silicon", IEEE GaAs Integ. Circ. Symp. Tech. Dig., Oct. 1983, p. 124.
15. P. Shapiro, A.B. Campbell, J.C. Ritter, R. Zuleeg, and J.K.Notthoff, "Single event upset measurements of GaAs E-JFET RAMs", IEEE Trans. Nucl. Sci. **30**, 4610 (1983).
16. R. Zuleeg, J.K. Notthoff, and D.K. Nichols, "SEU of complementary GaAs static RAMS due to heavy ions", IEEE Trans. Nucl. Sci. **31**, 1121 (1984).
17. R. Zuleeg, "Radiation effects in GaAs FET devices", Proc. of the IEEE **77** (1989) 309.

18. B.W. Hughlock, T. Williams, A.H Johnston, and R.E. Plaag, "Ion induced charge collection in GaAs MESFETs and its effect on SEU vulnerability", IEEE Trans. Nucl. Sci. **38** (1991) 1442; B. Hughlock, A. Johnston, T. Williams, and J. Harrang, "A comparison of charge collection Effects Between GaAs MESFETs and III-V HFETs", IEEE Trans. Nucl. Sci. **39** (1992) 1642.
19. T.R. Weatherford, L. Tran, W.J. Stapor, E.L. Peterson, J.B. Langworthy, D. McMorrow, W.G. Abdel-Kader, and P.J. McNulty, "Proton and heavy ion upsets in GaAs MESFET devices", IEEE Trans. Nucl. Sci. **38** (1991) 1450.
20. T.R. Weatherford, A.B. Campbell, D. McMorrow, J.B.Langworthy, W.J. Stapor, A.R. Knudson, E.L. Petersen, D. Wilson, and L. Tran, "Can digital GaAs be used in a space environment?: a look at single event upset in GaAs", in *IEEE GaAs IC Symposium, Technical Digest 1991*, October 20-23, 1991, pp. 79-82.
21. G.L. Hash, J.R. Schwank, M.R. Shaneyfelt, C.E. Sandoval, M.P. Conners, T.J. Sheridan, F.W. Sexton, E.M. Slayton, J.A. Heise, and C.C. Foster, "Proton irradiation effects on advanced digital and microwave III-V components", IEEE Trans. Nuc. Sci. **41**, (1994) 2259-2266.
22. A.B. Campbell, A. Knudson, D. McMorrow, W. Anderson, J. Roussos, S. Espy, S. Buchner, K. Kang, D. Kerns, and S. Kerns, "Ion-induced charge collection in GaAs MESFETs", IEEE Trans. Nucl. Sci. **36** (1989) 2292.
23. Y. Umemoto, N. Matsunaga, and K. Mitsusada, "A bipolar mechanism for alpha particle-induced soft errors in GaAs integrated circuits", IEEE Trans. Elect. Dev. ED-16 (1989) 864.
24. A.R. Knudson, A.B. Campbell, D. McMorrow, S. Buchner, K.Kang, T. Weatherford, V. Srinivas, G.A. Swartzlander, Jr., and Y.J. Chen, "Pulsed laser-induced charge collection in GaAs MESFETs", IEEE Trans. Nucl. Sci., **37** (1992) 1909.
25. D. McMorrow, T.R. Weatherford, A.R. Knudson, S. Buchner, J.S. Melinger, L.H. Tran, A.B. Campbell, P.W. Marshall, C.J. Dale, A. Peczalski, and S. Baier, "Charge collection characteristics of GaAs heterostructure FETs fabricated with a low-temperature grown GaAs buffer layer", IEEE Trans. Nuc. Sci., **43** (1996) 918.
26. D. McMorrow, T. Weatherford, A.R. Knudson, L.H. Tran, J.S. Melinger, and A.B. Campbell, "Single event dynamics of high-performance HBTs and GaAs MESFETs", IEEE Trans. Nucl. Sci. **40** (1993) 1858.
27. D. McMorrow, J.S. Melinger, N. Thantu, A.B. Campbell, T.R. Weatherford, A.R. Knudson, L.H. Tran, and A. Peczalski, "Charge collection mechanisms of heterostructure FETs", IEEE Trans. Nucl. Sci. **41** (1994) 2055.
28. L.D. Flesner, N.M. Davis, and H.H. Wieder, "High-speed response of GaAs metal-semiconductor field effect transistor to electron-beam excitation", J. Appl. Phys. **53**, 3873 (1982); L. Flesner, "Gate charge collection and induced drain current in GaAs FETs", IEEE Trans. Nucl. Sci. **32** (1985) 4110.
29. D. McMorrow, J.S. Melinger, A.R. Knudson, T.R. Weatherford, L.H. Tran, and W.R. Curtice, "Picosecond charge collection dynamics in GaAs MESFETs", IEEE Trans. Nucl. Sci., **39** (1992) 1657.
30. (a) T.R. Weatherford, D. McMorrow, W.R. Curtice, A.R. Knudson, and A.B. Campbell, "Single event induced charge-transport modeling of GaAs MESFETs", IEEE Trans. Nucl. Sci. **40** (1993) 1867; (b) T.R. Weatherford, D. McMorrow, W.R. Curtice, A.B. Campbell, and A.R. Knudson, "Examination of the SEU sensitivity of GaAs MESFETs via 2-D computer simulation and picosecond charge collection experiments", *Proceedings of the Reliability Physics Symposium*, 1993, pp. 357-363.
31. D. McMorrow, A.R. Knudson, and A.B. Campbell, "Fast charge collection in GaAs MESFETs", IEEE Trans. Nucl. Sci., **37** (1990) 1902.
32. T.R. Weatherford, D. McMorrow, A.B. Campbell, and W.R. Curtice, "Significant Reduction in the soft error susceptibility of GaAs field effect transistors with a low-temperature grown GaAs buffer layer", Appl. Phys. Lett. **67** (1995) 703.
33. J.R. Schwank, F.W. Sexton, T.R. Weatherford, D. McMorrow, A.R. Knudson, and J.S. Melinger, "Charge collection in GaAs MESFETs fabricated in semi-insulating substrates", IEEE Trans. Nucl. Sci. **42** (1995) 1585.

34. D. McMorrow, W.R. Curtice, S. Buchner, A.R. Knudson, J.S. Melinger, and A.B. Campbell, "Charge collection characteristics of GaAs MESFETs fabricated with a low-temperature grown GaAs buffer layer: computer simulation", IEEE Trans. Nuc. Sci., **43** (1996) 2904.
35. J.S. Melinger, D. McMorrow, A.B. Campbell, S. Buchner, L.H. Tran, A.R. Knudson, and W.R. Curtice, "Pulsed laser induced single event upset and charge collection measurements as a function of optical penetration depth", J. App. Phys., **84** (1998) 690.
36. D. McMorrow, J.S. Melinger, A.R. Knudson, S. Buchner, L.H. Tran, A.B. Campbell, and W.R. Curtice, "Charge-enhancement mechanisms of GaAs field effect transistors: experiment and simulation", IEEE Trans. Nuc. Sci., **45** (1998) 1494.
37. D. McMorrow, A.R. Knudson, J.S. Melinger, and S. Buchner, "Charge collection efficiency of GaAs field effect transistors fabricated with a low-temperature grown buffer layer: dependence on charge deposition profile", IEEE Trans. Nuc. Sci., **47** (2000) 498.
38. D. McMorrow, J.B. Boos, D. Park, S. Buchner, A.R. Knudson, and J.S. Melinger, "Charge collection dynamics of InP-based high electron mobility transistors (HEMTs)", IEEE Trans. Nuc. Sci., **49** (2002) 1396-1400.
39. T.R. Weatherford, J. Whitaker, K. Jobe, S. Meyer, M. Bustamante, S. Thomas III, and K. Elliott, "Single event induced voltage transients within InP HBT circuits", in *Government Microcircuit Applications Conference (GOMAC 2000), Digest of Papers*, 2000, 479-482.
40. D. McMorrow, J.B. Boos, A.R. Knudson, S. Buchner, M.J. Yang, B.R. Bennett, and J.S. Melinger, "Charge collection characteristics of low-power, ultrahigh speed, metamorphic AlSb/InAs high-electron mobility transistors (HEMTs)", IEEE Trans. Nuc. Sci., **47** (2000) 2662.
41. D. McMorrow, R. Magno, A.S. Bracker, B.R. Bennett, S. Buchner, and J.S. Melinger, "Charge collection dynamics of AlSb-InAs-GaSb resonant interband tunneling diodes (RITDs)", IEEE Trans. Nuc. Sci., **48** (2001) 1973.
42. P. Mazumder, S. Kulkarni, M. Bhattacharya, J.P. Sun, and G.I. Haddad, "Digital circuit applications of resonant tunneling devices", Proc. of the IEEE, **86** (1998) 664.
43. P. Fay, J. Lu, Y. Xu, G.H. Bernstein, A. Gonzalez, P. Mazumder, D.H. Chow and J.N. Schulman, "Digital integrated circuits using InAlAs/InGaAs/InP HEMTs and InAs/AlSb/GaSb RITDs, Electron. Lett. **37**, 758 (2001).
44. J.B. Boos, W. Kruppa, B.R. Bennett, D. Park and H.B. Dietrich, "AlSb/InAs HEMTs with high speed and low power consumption", Compound Semicond. **5**, 36 (1999).
45. J.P.A. van der Wagt, A.C. Seabaugh, and E.A. Beam, III, "RTD/HFET low standby power SRAM gain cell", IEEE Elect. Dev. Lett. **19** (1998) 7.
46. K. Maezawa, H. Matsuzaki, M. Yamamoto, and T. Otsuji, "High-speed and low-power operation of a resonant tunneling logic gate MOBILE", IEEE Elect. Dev. Lett. **19** (1998) 80.
47. P.W. Marshall, C.J. Dale, T.R. Weatherford, M.LaMacchia, and K.A. LaBel, "Particle-induced mitigation of SEU sensitivity in high data rate GaAs HIGFET technologies", IEEE Trans. Nucl. Sci. **42** (1995) 1844.
48. R. Schneiderwind, D. Krening, S. Buchner, K. Kang, and T.R. Weatherford, "Confirmation of SEU ion experiments by laser experiments in radiation-hardened GaAs MESFET combinational logic", IEEE Trans. Nucl. Sci. **39** (1992) 1665.
49. D. McMorrow, T.R. Weatherford, W.R. Curtice, A.R. Knudson, S. Buchner, J.S. Melinger, L.H. Tran, and A.B. Campbell, "Elimination of charge-enhancement effects in GaAs FETs with an LT GaAs buffer layer", IEEE Trans. Nucl. Sci. **42** (1995) 1837.
50. P.W. Marshall, C.J. Dale, T.R. Weatherford, M. Carts, D. McMorrow, A. Peczalski, S. Baier, J. Nohava, and J. Skogen, "Heavy ion SEU immunity of a GaAs Complementary HIGFET circuit fabricated on a low temperature grown buffer layer", IEEE Trans. Nucl. Sci. **42** (1995) 1850.
51. J.R. Hauser, "SEU hardened silicon bipolar and GaAs MESFET SRAM cells using local redundancy techniques", IEEE Trans. Nucl. Sci. **39** (1992) 2.
52. Y. Umemoto, N. Nasuda and K. Mitsusada, "Effects of a buried p-layer on alpha particle immunity of MESFET's fabricated on semi-insulating GaAs substrates", IEEE Electron Dev. Lett. **EDL-7** (1986) 396.

53. F. Barber, "GaAs radiation hardened design techniques", IEEE Proceedings, **76** (1988) 1501.
54. T.R. Weatherford, R. Radice, D. Eskins, J. Devers, D.J. Fouts, P.W. Marshall, C.J. Marshall, H. Dietrich, M. Twigg, and R. Milano, "SEU design considerations for MESFETs on LT GaAs", IEEE Trans. Nuc. Sci., **44** (1997) 2282.
55. T.R. Weatherford, P.W. Marshall, C.J. Marshall, D.J. Fouts, B. Mathes, and M. LaMacchia, "Effects of low-temperature buffer layer thickness and growth temperature on the SEE sensitivity of GaAs HIGFET circuits", IEEE Trans. Nuc. Sci., **44** (1997) 2298.
56. D.J. Fouts, T.R. Weatherford, C.J. Dale, P.W. Marshall, H.B. Dietrich, D. McMorrow, J. Abrokwah, M. Lamacchia, and R. Milano, "Soft-error immune gallium arsenide ICs using COTS designs and foundries", in *Government Microcircuit Applications Conference (GOMAC 1997), Digest of Papers*, 1997, 93.
57. H.T. Nagle, T.K. Miller, and D.F. McAllister, "Fault tolerance for single events", J. Rad. Effects, Research and Engineering, **6-2** (1988) 155.
58. M. LaMacchia, *et al.*, "SEU hardening approach for high speed logic", presented at the 1997 Nuclear and Space Radiation Effects Conference; U.S. Patent number 5,600,260.
59. D. McMorrow, A.R. Knudson, J.B. Boos, D. Park, and J.S. Melinger, "Ionization-induced carrier transport in InAlAs/InGaAs high electron mobility transistors", *presented at* RADECS 2003, and *submitted for publication*, IEEE Trans. Nuc. Sci. (2003).

International Journal of High Speed Electronics and Systems
Vol. 14, No. 2 (2004) 327–339

World Scientific
www.worldscientific.com

INVESTIGATION OF SINGLE-EVENT TRANSIENTS IN FAST INTEGRATED CIRCUITS WITH A PULSED LASER

P. Fouillat[1], V. Pouget[1], D. Lewis[1], S. Buchner[2], D. McMorrow[3]
[1] *IXL, UMR CNRS 5818, Université Bordeaux 1, 33405 Talence, France*
[2] *QSS Group, Inc., Seabrook, MD 20706, USA*
[3] *NRL, Code 6812, Washington, DC 20375, USA*

This paper describes the use of a pulsed laser for studying radiation-induced single-event transients in integrated circuits. The basic failure mechanisms and the fundamentals of the laser testing method are presented. Sample results are presented to illustrate the benefits of using a pulsed laser for studying single-event transients.

Keywords: single-event transient; laser testing.

Introduction

Electronic systems operating in space are exposed to radiation in the form of energetic charged particles, such as protons and heavy ions. As those charged particles pass through the semiconductors and insulators that make up the integrated circuits (ICs), they liberate electrons from the constituent atoms. Under certain conditions, the liberated electrons can disturb the normal operation of the IC, causing a variety of different and potentially harmful effects that have been classified as Single-Event Effects (SEEs). Included among those effects are temporary voltage spikes at internal circuit nodes termed Single-Event Transients (SETs).[1,2] SETs are of great concern because they may lead to a loss of information, to physical failures or, in the worst case, to a total loss of control of the spacecraft. In space, heavy ions responsible for SETs originate from solar flares and cosmic rays or from proton-induced reactions in low earth orbit. More recently, ions resulting from the interaction of atmospheric neutrons with the silicon and boron atoms in electronic circuits themselves have become a threat for modern advanced technologies, even at ground level.

SET testing of ICs is essential in order to evaluate, understand, and mitigate their sensitivity. This paper describes the SET laser testing technique, which is a powerful diagnostic tool for investigating the details of SETs in ICs. The first section describes the basic mechanisms of the SET phenomenon. The second section introduces the fundamentals of the laser testing technique. An example of experimental implementation is presented in the third section. The last section illustrates the power of the laser method for SET testing by describing two typical applications.

1. Basic Mechanisms of a SET

1.1. *Charge generation and collection*

A charged particle moving through a semiconductor deposits a fraction of its energy in the medium by ionizing the atoms of the crystal along its path. The amount of energy received by the medium per unit path length is called linear energy transfer (LET). LET defines the efficiency of the energy transfer and is dependent on both the electrical charge and the energy (i.e. speed) of the particle. It is usually expressed in units of MeV.cm^2/mg (energy loss divided by the material density). In space, most heavy ions from cosmic rays have LETs in silicon ranging from 0.01 to 40 MeV.cm^2/mg. The products of a neutron-induced reaction have LET values less than 10 MeV.cm^2/mg. From a microelectronic point of view, this energy transfer results initially in a strongly localized generation of electron-hole pairs along the path taken by the particle. Another useful unit for LET is given in terms of the amount of electrical charge generated per unit path length. It is based on the fact that it takes an average energy of 3.6 eV for a particle to create an electron-hole pair in silicon. In silicon devices, an LET of 1 MeV.cm^2/mg roughly corresponds to a charge deposition per unit length of 10 fC/μm. The typical range of deposited charge is from 0.1 to 400 fC/μm.

Some fraction of the electron-hole pairs generated by a particle either crossing or passing close to a sensitive junction in a device, such as a transistor, will be collected via drift or diffusion mechanisms, or both, resulting in a parasitic transient current at the associated junction electrode in the device. When the particle actually passes through a sensitive junction, the junction electric field is distributed to some extent along the particle's path due to the very high charge density along the particle's track (cf. Fig. 1a). This so-called funneling phenomenon (shown in Fig. 1b) enhances the charge collection and gives rise to transients with larger amplitudes. Other charge enhancement mechanisms can occur such as the parasitic bipolar effect encountered in III-V or SOI devices[3,4] The transient current following the particle impact can induce different kinds of electrical disturbances (current, charge, potential) in the vicinity of the collecting node.

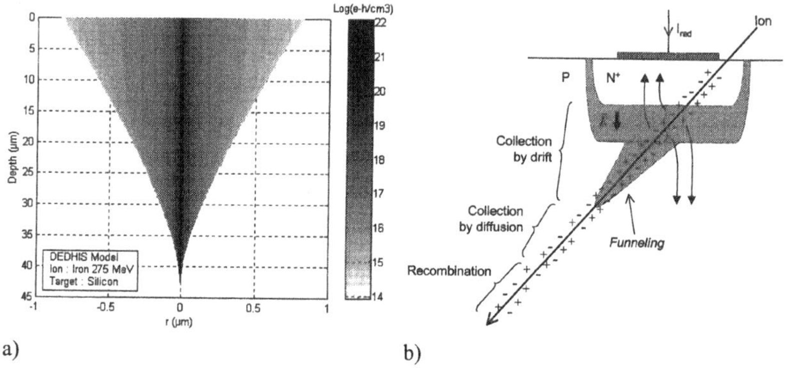

Fig. 1. (a) Carrier densities generated by a 275MeV Fe ion in silicon versus depth and distance from the path, and (b) schematic of the charge collection mechanisms involved when an ion strikes an electronic junction.

The energy transfer from the ion to the lattice is a fast phenomenon that lasts about 1ps. The temporal shape of the collection current depends on the impact location, which, in turn, determines the relative contributions from drift and diffusion. The drift component typically peaks in less than 100ps after the impact, whereas the diffusion current can last for much longer times. The resulting SET will propagate through a digital IC a distance that depends on the capacitance and resistance values of the IC's internal nodes through which the SET propagates. In contrast, SETs generated at internal nodes in analog circuits may have very different pulse shapes by the time they reach the circuit's output, undergoing significant enhancement of both amplitude and temporal extent by the circuit itself.

1.2. *ICs Sensitivity to SET*

SETs disturb the proper functioning of many different kinds of digital and analog circuits. Early observations of SETs were confined to digital architectures where the effect of the deposited energy in a combinational logic gate caused a static bit error once the transient signal reached a sequential logic element, such as a latch, a memory or a register. Clearly, SETs that propagate to and are captured by data storage elements cannot be distinguished from single-event upsets (SEUs) occurring when particles directly strike those same memory elements. Therefore, the overall radiation-induced error rate in digital circuits consists of contributions from both SETs and SEUs.

Interest in the effects of SETs has recently increased because of the claim that the enhanced ability of parasitic signals to propagate through advanced circuits with gate lengths shorter than 0.35 μm and to reach memory elements where they become SEUs[5]. The probability for an SET to be detected when it is generated outside memory elements depends on many parameters[1]. First, the amplitude and temporal extent of the SET determine how far an SET will propagate through a circuit before being filtered by the gates. Second, in a synchronous circuit, the SET must arrive at the memory element's input during a small time interval, termed "window-of-vulnerability[6]," around the clock edge. Third, the SET will only be latched into the memory element if it meets the setup and hold times that are characteristic of the memory element. However, if the SET appears at an asynchronous input, such as "Preset" or "Clear", it need only meet the basic pulse characteristic requirements to potentially generate a bit error or a partial or total reset[7]. The constraints on amplitude and temporal extent are somewhat relaxed in modern technologies because the very low capacitances are not effective in attenuating SETs. Therefore, SETs will propagate unattenuated over long distances though the IC. The increasing operating frequency of modern technologies also means that SETs will have a greater probability of arriving at a latch during the "window of vulnerability," thereby enhancing the probability that they will be captured by a synchronous latch. Both of these effects suggest an overall greater sensitivity to SETs in fast digital electronic circuits. Not only are silicon technologies affected by this phenomenon, but so are very fast digital circuits using GaAs Heterojunction Isolated gate Field Effect Transistors (HIGFETs) or SiGe Heterojunction Bipolar Transistors (HBTs)[8,9].

For the case of SETs in analog circuits, recent publications confirm that SETs can drastically affect the performance of linear devices such as operational amplifiers and comparators, as well as transceivers and opto-couplers[10]. Due to the analog nature of the SET, analog devices will give a significant response whenever the parasitic signal exceeds the noise characteristics. However, because digital circuits usually operate downstream from analog circuits, the key issues determining whether SETs in analog circuits are a problem are similar to those for digital circuits, i.e., the SET's propagation distance from the point of generation in the analog circuit to the point where it disrupts a digital circuit, and the arrival time in the digital circuit relative to the clock edge. Therefore, SETs in analog circuits may cause bit errors in digital circuits whenever these conditions are met.

Understanding the origins of a device's SET sensitivity generally consists of identifying those transistors responsible for the maximum deviation of the output signal for the least amount of deposited charge. This is a relatively cumbersome task for analog circuits due to the complexity of the circuit response to the charge deposition. However, the task is greatly simplified if a pulsed laser is used[11,12], requiring only a few minutes to complete.

1.3. *SET Testing*

Wherever radiation effects are a concern for an application's safety, it is necessary to qualify the devices before using them. Particle accelerators are commonly employed to evaluate device sensitivity to SETs. A typical experiment consists of exposing the device under test (DUT) to a uniform beam of heavy ions and monitoring the DUT's electrical parameters during irradiation. The number (N) of SETs recorded for a known particle fluence (Φ) is recorded and the ratio (N/Φ) defines the device cross-section, which is a measure of the device sensitive area. The experimental results are summarized by a curve representing the cross-section versus LET, from which a LET threshold (below which no SETs are observed) and a saturated cross-section (equal to the maximum sensitive area) can be established. This curve is then used for evaluating the probability of SET occurrence for a given radiation environment.

Evaluating an analog circuit's SET sensitivity is particularly difficult due to the analog nature of the phenomenon, i.e., SETs originating in a single analog device can have a variety of waveforms - positive, negative and bipolar with a variety of amplitudes and pulse widths. Furthermore, analog devices can operate under many different electrical configurations, and the SET sensitivity is known to be affected by the electrical configuration[13,14]. Thus, to fully characterize a bipolar device's SET response, system designers require a large amount of experimental data taken under a variety of bias conditions. Large amounts of data are also necessary for those wishing to model the SET response of analog circuits using device and circuit simulator programs[15]. The data can be used to check the validity of the models, which can, in turn, be used to evaluate the SET response for bias configurations not actually tested.

Device testing with particle accelerators is the most reliable method for simulating a radiation environment. However, this approach is not compatible with industry demands for low-cost and accessible testing. Moreover, these tests cannot provide all the information needed for sensitivity modeling due to their lack of spatial and temporal resolution. To date, microbeams are able to overcome the limitation concerning the spatial resolution[16], but not the temporal one.

The pulsed-laser method for SET testing complements accelerator testing by simultaneously providing high spatial and temporal resolutions at relatively low cost. There are numerous applications for that data[11, 15, 12]. First, it can be used to evaluate the SET sensitivity of devices such as analog devices in configurations for which there are no accelerator test data available. Second, it can be used for rapidly assessing whether SET hardening approaches are effective. Third, it can provide transient waveforms and sensitivities of individual transistors in a circuit, information that is essential for validating device and circuit models used for calculating SETs effects in circuits.

2. SET Laser Testing

2.1. *Principles*

A pulse of laser light focused on a semiconductor can induce a localized transient generation of electron-hole pairs if the photon energy is greater than the semiconductor bandgap energy. Figure 2a presents the generated carrier density as a function of position induced in silicon by a laser pulse of wavelength 800nm and energy 8pJ focused by a 100X microscope objective lens on the silicon surface. The particle-induced generation of electron-hole pairs involves a coulomb interaction, which is intrinsically different from the photoelectric effect responsible for the generation of electron-hole pairs by laser light. Comparison of Figures 1a and 2a reveal the differences in structure between ion-induced and laser-induced charge tracks. Although these figures represent particular cases, they are, nevertheless, representative of the differences in radial distribution and peak density. However, despite the differences in structure between ion-induced and laser-induced tracks , numerous experimental investigations as well as numerical simulations have shown that these two kinds of excitation produce similar transient effects in integrated circuits[17,11].

In order to compare laser-testing results with those of heavy ions it is necessary to define an equivalent laser LET. This definition is based on the rectangular parallelepiped (RPP) model[18], which assumes that the particle-semiconductor interaction can be described by the amount of charge generated in a certain RPP of thickness, d. Assuming that the laser-light- and heavy-ion-induced effects are equivalent when the associated generation rates induce the same number of carriers in the RPP, leads to the definition of L_e, the RPP-equivalent laser LET [19]:

$$L_e = \frac{E_p}{d} \int_{r=0}^{+\infty} \int_{z=0}^{d} \int_{t=-\infty}^{+\infty} G_{las}(r,z,t) 2\pi r dr dz dt \qquad (1)$$

where E_p is 3.6eV in silicon and G_{las} is the laser generation rate. Figure 2b presents the dependence of LET on laser pulse energy for different wavelengths. The non-linear behavior for higher energies is due to the two-photon absorption (TPA) mechanism[20]. Aside from this theoretical approach, a more pragmatic correlation between ion and laser results can be found in the literature[21].

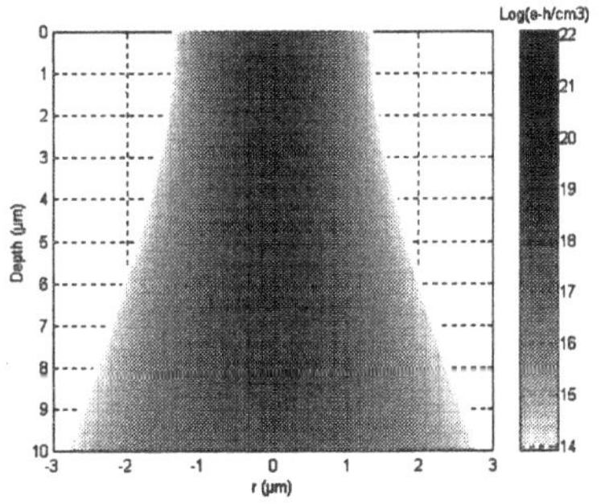

a)

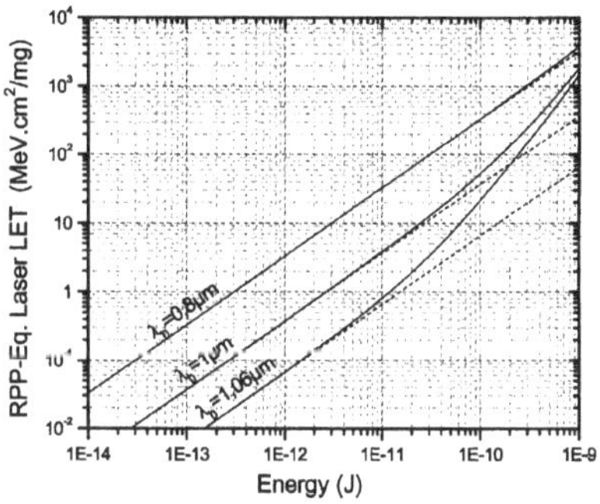

b)

Fig. 2: Carriers density generated by a laser pulse of 8pJ with a wavelength of 800nm in silicon versus depth and distance from the center of the beam (a), and RPP equivalent laser LET versus laser pulse energy (b).

2.2. *Critical evaluation*

Simulating SETs with a laser beam is a convenient laboratory method for investigating the associated failure mechanisms. It provides spatial and temporal information since it is possible to control both the location and time of the laser impact, in contrast to particle accelerator tests, which are usually performed by exposing the whole device to a continuous and almost spatially uniform flux of particles, that strike the device randomly in both space and time.

The RPP-equivalent laser LET can provide a rough estimation of the SEE sensitivity of a device, even if its definition rests on numerous approximations. The attenuation of the laser beam with depth cannot be neglected, in contrast to the longitudinal quasi-uniformity of the ion track. Concerning the radial distributions of the generated carriers, figure 2a shows that the laser-induced carrier distribution has a wider lateral extent and a much lower peak value than the ion track (Fig. 1a). This implies that the amount of charge collected at an electrode of a device by the funneling effect will be lower than in the case of an ion[22]. However, it is important to note that the limitations of the RPP-equivalent laser LET are inherent to the basic RPP model, which is widely used in the exploitation of particle accelerator tests because it provides a simple yet powerful cause-effect relationship. From an experimental point of view, it is clear that several difficulties may arise when trying to perform the RPP-equivalent laser LET calibration. Specifically, an accurate evaluation of the laser energy deposited in the device and a perfect control of the laser spot size and geometry are required. In some cases, these requirements may be hard to fulfill, especially for modern devices with a large number of metallization layers.

Concerning the cross-section measurement, some interesting results were obtained[23,24] but, to date, it is still not clear that the laser method can provide a reliable estimation of the SEE cross-section of a device. Although it is possible to measure a sensitive area by scanning a pulsed laser beam over a device, the correlation between this measurement and the cross-section measured in accelerators is not trivial. Some correction factors to account for the differences in the carrier distributions and the problem of metal layers still have to be defined.

2.3. *Backside testing*

One main limitation when using optical beams for testing microelectronic devices is the opacity of the metal layers. In modern technologies, reaching the semiconductor material for generating carriers with a laser beam is becoming more and more challenging due to the increasing density of metal interconnections. This problem lead to the development of a backside testing technique that involves flipping the device and focusing the beam through the substrate.

For front-side testing of silicon devices, a wavelength of 800nm usually provides a good compromise between penetration depth and generation density. Indeed, the penetration depth of approximately 12μm is sufficient to ensure efficient photo-generation in the active volume of modern devices[17]. For backside testing, a longer penetration depth is needed to reach the active volume through the silicon substrate. The absorption coefficient of silicon rapidly drops for wavelengths longer than 1000nm. In order to induce a sufficient level of photo-generation in the active volume, the best compromise usually lies between 930 and 1000nm depending on the substrate thickness and doping level[11]. Typically, a 1.3 μm illumination beam is well suited for backside visualization of the metal layers at the top of the device.

Laser testing through the substrate requires optical access to the backside as well as a good backside surface quality. Packaged devices have to be opened and thick substrates must be thinned when using laser wavelengths in the absorption domain. This delicate sample preparation is usually performed via micro-mechanical machining, although a new method based on laser ablation[11] should become popular thanks to its ability to open any of the package types used in the microelectronic industry.

Another promising method for backside testing relies on using a wavelength longer than the absorption limit of silicon with pulse energies sufficient for the Two-Photon Absorption (TPA) mechanism to be significant. This way, the beam propagates through the substrate with little or no absorption until it is focused, where the intensity is sufficiently high that free carriers may be generated via TPA. The main advantages of this method are that it does not require substrate thinning and it ensures that the carrier generation is confined to the focus region of the beam[20].

3. Experimental set-up for SET laser testing

Figure 3 presents the main elements of the system developed at IXL, the Microelectronics Laboratory at the University of Bordeaux (France) for SEE testing with a pulsed laser beam. Similar [25,26] or simpler [24,27] systems can be found in the literature.

The laser source is a Ti:Sapphire (Ti:Sa) oscillator pumped by a 10W cw laser. The oscillator delivers 100fs or 1ps pulses at a frequency of 80MHz. The pulse wavelength is tunable in the red-NIR region from 780 to 1000nm. This tunability makes it possible to adjust the penetration depth of the laser light in the semiconductor material for front-side and backside testing.

Because a pulse frequency of 80MHz is too high, in most cases, for the DUT's output to return to a steady state between two consecutive pulses, a pulse picker is used to reduce this frequency. The pulse picker is based on an acousto-optic gate synchronized with the incident pulses. The pulse frequency at the output of the pulse picker is adjustable from single-shot to 4MHz and the pulse triggering can be synchronized with any reference clock with an uncertainty of $\pm$6ns. The pulse energy is controlled by the motorized rotation of a half-wave plate in front of a polarizer. Si and Sr in figure 3 denote incident and reflected energy detectors. The pulse is finally focused down to a spot size of approximately 1μm on the surface of the DUT by a 100x microscope objective. The location of the focused light is visualized with a CCD camera. The customized microscope includes a pulse detector for synchronization (photodiode St in figure 3) and a confocal infrared reflectance measurement system (photodiode Sf in figure 3) for backside imaging. The DUT and its test board are mounted on a 3-axis translation table with 0.1μm resolution.

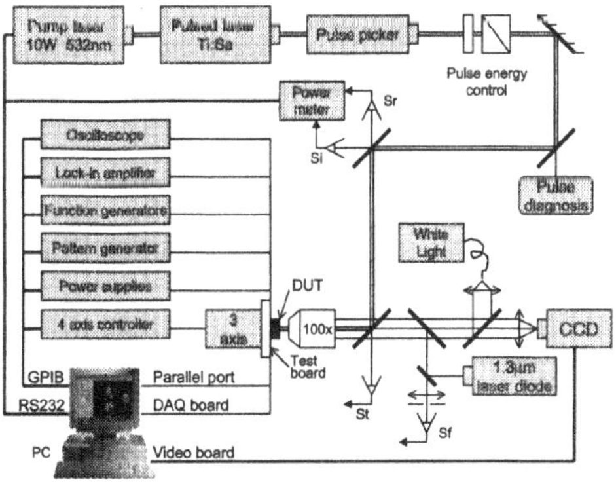

Fig. 3 : Experimental set-up for SET testing with a pulsed laser.

4. Results

4.1. *LM124*

The first application discussed here consists of characterizing the SET sensitivity of the LM124, a quadruple operational amplifier widely used in linear systems. One of the basic system for applications of the LM124 is as a voltage amplifier. The input voltage is maintained constant during the test, which consists of monitoring the transients induced in the output by the laser pulses.

Figures 4a and 4b present experimental results obtained respectively with a particle accelerator at the Nuclear Physics Institute in Orsay (France) and with the pulsed laser set-up at the IXL laboratory in Bordeaux [11,28]. Comparison between the transient responses obtained with boron ions and with laser light pulses having a wavelength of 800 nm and pulse duration of 1 ps shows that the signals are nearlyidentical. This demonstrates the validity of using pulsed laser light to simulate SETs for this circuit with front-side irradiation.

The steepness of the transient's leading edge is typical of SETs. In this particular case, the long tail of the circuit's response is governed primarily by the resistances and capacitances associated with the output node. In fact, the observed signal is directly related to the transfer function between the injection node and the output of the device, since the excitation is virtually a Dirac impulse.

One of the main advantages of the laser method is its capacity to provide SET sensitivity mappings. Among the different parameters that can be extracted, we present in figures 4c and 4d the peak-to-peak amplitude of the transient response versus impact position. These figures were obtained by using front-side and backside irradiation, respectively. The laser pulse energy was adjusted in the backside mode to induce SETs with the same amplitudes. A color map is used to represent the amplitudes, with red points corresponding to the highest values. These mappings allow a detailed analysis of the electrical origin of the sensitivity. In particular, after several experiments, the area most sensitive to SETs was clearly identified as a floating base npn transistor in the vicinity of two transistors of the Darlington amplification stage[11,28,29].

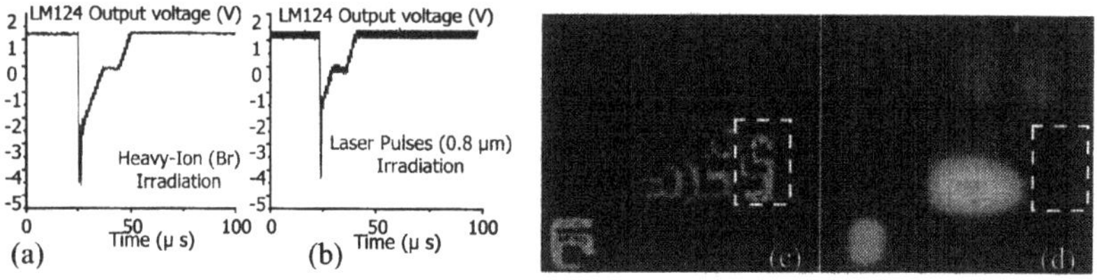

Fig. 4. LM124output transient response with a) ion b)front side laser and LM124 SET mappings for c)front side and d)backside laser testing. LM124 is operating as an non inverter amplifierwith gain 2

In the front-side map, the masking effect of the metal layer clearly affects the apparent shapes of the SET sensitive areas. Sensitive areas appear more clearly in the backside mappings, where each pixel is available for data processing. However, differences may appear in the results of the two testing modes. Whereas the front-side mapping shows an SET sensitive zone (delineated on the front-side map by a dashed line), it is absent in the backside map. We suggest that this phenomenon is due to the existence of a buried layer-substrate junction, which may shunt the generated charges directly to the substrate.

4.2. *GaAs MESFET Circuit Hardening*

GaAs MESFET devices have been shown to be extremely sensitive to SEUs and SETs because of the very small amount of charge needed to produce the effects.[30,31] Therefore, if circuits based on GaAs MESFET technology are to be used in space, their SEU/SET sensitivities must be mitigated. There are two hardening approaches that have been tried. One is a device hardening approach, which involves the use of buffer layers below the active region of the GaAs MESFET where the enhanced recombination reduces the amount of charge that can contribute to SEUs and SETs.[32] A second is a circuit hardening approach in which each bit of data is stored in three separate registers and their outputs are majority voted on each clock cycle before being written back to all three inputs.[33,6] This approach is termed Majority-Vote-and-Scrub (MV&S). Fig. 5 shows a block diagram of the three registers (A, B and C), the voting circuitry and the feedback loop.

To determine whether the MV&S approach to hardening was effective, the part was tested at an accelerator by monitoring the output while exposing the device to a beam of heavy ions. Figure 6 shows the SEU cross-section as a function of frequency for ions having a LET of 10 MeV.cm^2/mg with all "1s" stored in the registers. At very low clock frequencies and high particle fluxes the MV&S technique was not effective because more than one of the three registers could be upset during a single clock cycle. With increasing clock frequency, the error rate decreased, because the data would be rewritten before SEUs occurred in more than one register. At sufficiently high frequencies, it was anticipated that the error rate would approach zero. However, at frequencies above about 50 KHz the error rate once again increased - an unexpected result that could not be explained.

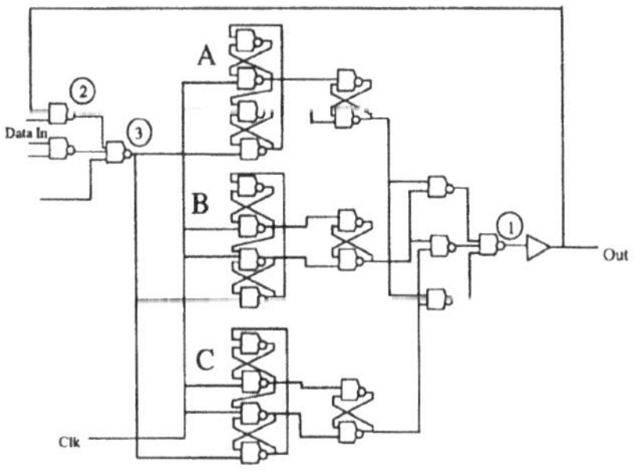

Fig. 5. Circuit diagram for majority-voted register.

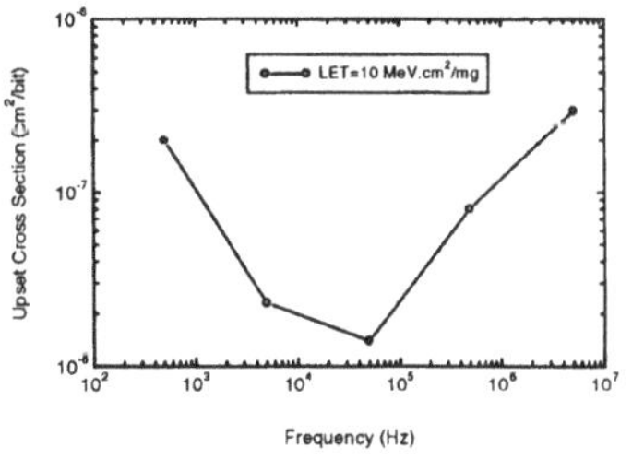

Fig. 6. SEU cross-section versus clock frequency for ions with LET = 10 MeV.cm^2/mg with all the registers containing "1s".

In order to determine the cause of this anomalous frequency-dependent effect, a pulsed laser was used to identify the SET-sensitive nodes in the circuit. SEUs produced in any one of the three registers by the pulsed laser were not observed at the output, confirming the effectiveness of the majority voting circuit. However, when the laser light was focused on nodes 1, 2 and 3 in the feedback circuit, SETs were generated that were

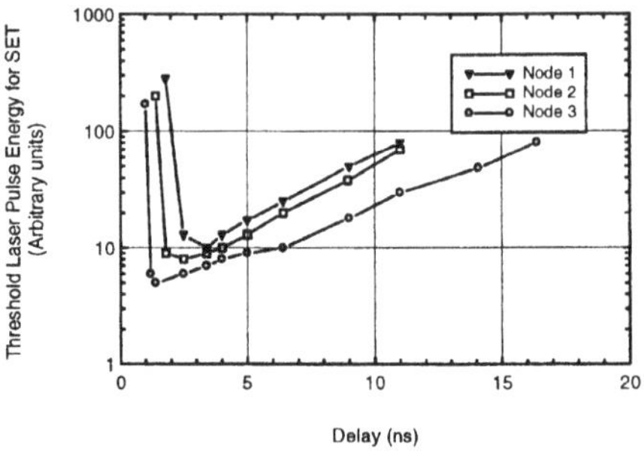

Fig. 7. Threshold laser pulse energy as a function of delay between the firing of the laser and the circuit clock for three different nodes.

loaded into all three registers at the next clock edge. Since all three registers contained incorrect data, there was no way for the majority-vote circuitry to correct the invalid data. These results stress that SETs in digital logic circuits may present a serious problem when they become latched into memory elements, and the pulsed laser is a useful tool for identifying the spatial location of the SET-sensitive nodes.

As previously mentioned, the laser is also capable of providing temporal information on SEUs. Temporal information is important for logic circuits because an SET generated in combinational logic becomes an SEU upon being latched into a register, and that only happens during a short time interval (setup-and-hold time) around the clock edge. An SET that arrives at the register's input after the clock edge will not be stored in the register as an SEU. On the other hand, an SET that arrives at the register's input prior to the clock edge can still be converted to an SEU if its amplitude during setup-and-hold exceeds the required voltage threshold for the register. Large amplitude SETs have larger widths than small amplitude SETs and, therefore, they can arrive earlier at the register's input and still be latched as an SEU. The time interval during which an SET can be latched into a register is proportional to the temporal extent of the SET and has been termed "window-of-vulnerability." The temporal extent of the SET depends on the amount of deposited charge, and, thus, on the ion LET. Therefore, ions with large LETs will have larger windows-of-vulnerability and will be more likely to produce SEUs in logic circuits.

The laser can be used to measure the width of the window-of-vulnerability by using an external signal both to trigger the firing of the laser and to provide a clock signal for the logic circuit. An electronic delay is added to the clock signal for the logic circuit and the amount of laser-light energy needed to produce SETs as a function of the delay is measured. Fig. 7 is a plot of the amount of energy needed to produce SETs as a function of the delay for the three nodes shown in Fig. 5. The least amount of energy is needed if the delay is set for 2 ns because the laser light arrives contemporaneous with the clock edge. For shorter delays, the circuit clock transitions prior to the arrival of the SET generated by the laser light, and no SEUs are captured. As the delay is increased beyond 2 ns, the energy required to produce an SET also increases. The data in Figure 7 (logarithmic y-axis) can be approximated by straight lines indicating that the SETs decay exponentially with time. The figure shows that the width of the window-of-vulnerability for node 3 extends from 1 ns to 6.5 ns when the laser pulse has an energy of 10 units. Increasing the pulse energy increases the width of the window-of-vulnerability. In conclusion, an important consequence of these measurements is that the SEU cross-section in dynamic logic circuits increases with ion LET because of both increased spatial and temporal contributions.

5. Conclusions

The pulsed laser is a very powerful technique for investigating SETs in both digital and analog circuits because it provides both temporal and spatial information on SETs in a relatively rapid and inexpensive way. In spite of the fact that ions and laser light produce free carriers in semiconductors through totally different mechanisms, the SETs generated by both have almost identical characteristics. The pulsed laser is not meant to replace a particle accelerator. Instead, it provides complementary information to enhance our understanding of the mechanisms responsible for SETs. Its use for evaluating the SET response of analog circuits is particularly valuable because, once the SETs generated at a particle accelerator are shown to be identical to those generated by the pulsed laser for a particular bias configuration, the pulsed laser can then be used with confidence to investigate the SET response of that circuit in a multitude of different bias configurations. Finally, the transients produced by the pulsed laser can also be used for validating device and circuit models that are needed to understand the detailed mechanisms responsible for SETs.

References

1. S.P. Buchner, M.P. Baze, "Single-Event Transients in Fast Electronic Circuits", IEEE Nuclear and Space Radiation Effects Conference (NSREC) Short Course, Section V, (2001).

2. T. Turflinger, "Single-Event Effects in Analog and Mixed-Signal Integrated Circuits," *IEEE Trans. Nucl, Sci.*, 43 (1996), p. 594.

3. V. Ferlet-Cavrois, G. Gasiot, C. Marcandella, C. D'Hose, O. Flament, O. Faynot, J. du Port de Pontchara, and C. Raynaud, , "Insights on the Transient Response of Fully and Partially Depleted SOI Technologies under Heavy-Ion and Dose-Rate Irradiations", *IEEE Trans. Nucl. Sci.* 49 (2002) p.2948.

4. A. B. Campbell, A. Knudson, D. Mc Morrow, W. Anderson, J. Roussos, S. Espy, S. Buchner, K. Kang, D. Kerns, S. Kerns, , "Ion-Induced charge collection in GaAs MESFETs", *IEEE Trans. Nucl. Sci.* 36 (1989) p.2292.

5. D.G. Mavis and P.H. Eaton, "SEU and SET mitigation techniques for FPGA circuit and configuration bit storage design," Proceedings MAPLD Conference 2000.

6. S. Buchner, K. Kang, D. Krening, G. Lannan and R. Schneiderwind, "Dependence of the SEU Window of Vulnerability of a Logic Circuit on Magnitude of Deposited Charge," *IEEE Trans. Nucl. Sci.* 40 (1993) p.1853

7. C. Poivey, J.L. Barth, J. McCabe, K.A. LaBel, "A Space Weather Event on the Microwave Anisotropy Probe, Proceedings of Radiation Effects on Component and Systems (RADECS) Workshop, 2002, Padova (Italy).

8. P.W. Marshall, C.J. Dale, and K.A. LaBel, "Space Radiation Effects in High Performance Fiber Optic Data Links for Satellite Data Management," *IEEE Trans. Nucl. Sci.*,43 (1996), p. 645.

9. M.A. Carts, P.W. Marshall, C.J. Marshall, K.A. LaBel, M. Flanegan, and J. Bretthaur, "Single Event Test Methodology and Test Results of Commercial Gigabit per Second Fiber Channel Hardware," *IEEE Trans, Nucl. Sci.*, 44 (1997), p. 1878.

10. R. Koga, S.D. Pinkerton, S.C. Moss et al., "Observations of Single Event Upsets in Analog Microcircuits", *IEEE Trans. Nucl. Sci.*, 40 (1993), p. 1838.

11. D. Lewis, V. Pouget, F. Beaudoin, P. Perdu, H. Lapuyade, P. Fouillat, A. Touboul, "Backside laser testing for SET sensitivity evaluation", *IEEE Trans. Nucl. Sci.*, 48 (2001), p 2193.

12. R. Pease, A. Sternberg, L. Massengill, R. Schrimpf, S. Buchner, M. Savage, J. Titus, T. Turflinger, , "Critical Charge for Single-Event Transients (SETs) in Bipolar Linear Circuits", *IEEE Trans. Nucl. Sci.* 48 (2001) p.1966.

13. A.L. Sternberg, L.W. Massengill, R.D. Schrimpf, Y. Boulghassoul, H.J. Barnaby, S. Buchner, R.L. Pease, J.W. Howard, "Effect of Amplifier Parameters on Single-Event Transients in an Inverting Operationnal Amplifier", *IEEE Trans. Nucl. Sci.*, 49 (2002), p. 1496.

14. R. Koga, S.H. Penzin, K.B. Crawford, W.R. Crain, S.C. Moss, S.D. Pinkerton, S.D. LaLumondiere, and M.C. Maher, "Single Event Upset (SEU) Sensitivity Dependence of Linear Integrated Circuits (ICs) on Bias Conditions," *IEEE Trans. Nucl. Sci.*, 44 (1997), p. 2325.

15. Y. Boulghassoul, L. W. Massengill, A. L. Sternberg, R. L. Pease, S. Buchner, J. W. Howard, D. McMorrow, M. W. Savage and C. Poivey, "Circuit Modeling of the LM124 Operational Amplifier for Analog Single-Event Transient Analysis", *IEEE Trans. Nucl. Sci.* 49 (2002) p.3090.

16. F.W. Sexton, "Microbeam Studies of Single-Event Effects", *IEEE Trans. Nucl. Sci.*, 43 (1996), p. 687.

17. J.S. Melinger, S. Buchner, D. McMorrow, W.J. Stapor, T.R. Weatherford, A.B. Campbell, "Critical Evaluation of the Pulser Laser Method for Single-Event Effects Testing and Fundamental Studies", *IEEE Trans. Nucl. Sci.*, 41 (1994), p. 2574.

18. E. Petersen, "Single-Event Analysis and Prediction", IEEE Nuclear and Space Radiation Effects Conference (NSREC) Short Course, Section III, (1997).

19. V. Pouget, H. Lapuyade, P. Fouillat, D. Lewis, S. Buchner, "Theoretical Investigation on an Equivalent Laser LET", *Microelectronics Reliability*, 41 (2001), p 1513.

20. D. McMorrow, W.T. Lotshaw, J.S. Melinger, S. Buchner, R. Pease, " Sub-bandgap Laser Induced Single Event Effects: Carrier Generation via Two Photon Absorption", *IEEE Trans. Nucl. Sci.* 49 (2002) p.3002.

21. D. Mc Morrow, J. Melinger, S. Buchner, T. Scott, R. Brown, and N. Haddad, "Application of a Pulsed Laser for Evaluation and Optimization of SEU-Hard Designs", *IEEE Trans. Nucl. Sci.* 47 (2000) p.559.

22. A.H. Johnston, "Charge Generation and Collection in p-n Junctions Excited with Pulsed Infrared Lasers", *IEEE Trans. Nucl. Sci.*, 40 (1993), p. 1694.

23. V. Pouget, P. Fouillat, D. Lewis, H. Lapuyade, F. Darracq, A. Touboul , "Laser Cross Section Measurement for the Evaluation of Single-Event Effects in Integrated Circuits", *Microelectronics Reliability*, 40 (2000), p. 1371.

24. F. Darracq, H. Lapuyade, N. Buard, F. Mounsi, B. Foucher, P. Fouillat, M-C. Calvet, R. Dufayel, "Backside SEU Laser Testing for Commercial Off-The-Shelf SRAMs", *IEEE Trans. Nucl. Sci.* 49 (2002) p.2977.

25. S. Buchner, D. McMorrow, J. Melinger, and A.B. Campbell, "Laboratory Tests for Single-Event Effects," *IEEE Trans. Nucl. Sci.*, 43 (1996), p. 678.

26. S.C. Moss, S.D. LaLumondière, J.R. Scarpulla, K.P. MacWilliams, W.R. Crain, R. Koga, "Correlation of Picosecond Laser-Induced Latchup and Energetic Particle-Induced Latchup in CMOS Test Structures", *IEEE Trans. Nucl. Sci.*, 42 (1995), p. 1948.

27. R. Jones, A.M. Chugg, C.M.S. Jones, P.H. Duncan, C.S. Dyer, C. Sanderson, "Comparison Between SRAM SEE Cross-Sections From Ion Beam Testing with those Obtained Using a New Picosecond Pulsed Laser Facility", Proceedings of Radiation Effects on Component and Systems (RADECS) Conference 1999, Fontevraud (France).

28. P. Adell, R.D. Schrimpf , H.J. Barnaby, R. Marec, C. Chatry , P. Calvel , C. Barillot , O. Mion, "Analysis of single event transient in analog circuits", *IEEE Trans. Nucl. Sci.*, 47 (2000), p. 2616.

29. A. Sternberg, L. Massengill, S. Buchner, R. Pease, Y. Boulghassoul, M. Savage, D. Mc Morrow, and R. Weller , "The Role of Parasitic Elements in the Single-Event Transient Response of Linear Circuits", *IEEE Trans. Nucl. Sci.* 49 (2002) p.3115.

30. B.W. Hughlock, G.S. LaRue and A.H. Johnston, "Single Event Upset in GaAs E/D MESFET Logic," *IEEE Trans. Nucl. Sci.* 37 (1990) p.1894.

31. D. Mc Morrow, T. Weatherford, S. Buchner, A. Knudson, J. Melinger, L. Tran, and A. Campbell, "Single-Event Phenomena in GaAs Devices and Circuits", *IEEE Trans. Nucl. Sci.* 43 (1996) p.628.

32. D. McMorrow, W.R. Curtice, S. Buchner, A.R. Knudson, J.S. Melinger, A.B. Campbell, "Charge-Collection Characteristics of GaAs MESFETs Fabricated with a Low-Temperature Grown GaAs Buffer Layer : Computer Simulation", *IEEE Trans. Nucl. Sci.*, 43 (1996), p. 2904.

33. R. Schneiderwind, D. Krening, S. Buchner, K. Kang and T.R. Weatherford, "Laser Confirmation of SEU Experiments in GaAs MESFET Combinatorial Logic," *IEEE Trans. Nucl. Sci.* 39 (1992), p.1665.

International Journal of High Speed Electronics and Systems
Vol. 14, No. 2 (2004) 341–352

World Scientific
www.worldscientific.com

SYSTEM LEVEL SINGLE EVENT UPSET MITIGATION STRATEGIES

W. F. HEIDERGOTT

General Dynamics Decision Systems,
Scottsdale, Arizona, 85257, United States
bill.heidergott@gdds.com

Use of a systems engineering process and the application of techniques and methods of fault tolerant systems are applicable to the development of a mitigation strategy for Single Event Upsets (SEU). Specific methods of fault avoidance, fault masking, detection, containment, and recovery techniques are important elements in the mitigation of single event upsets. Fault avoidance through the use of SEU hardened technology, fault masking using coding and redundancy provisions, and solutions applied at the subsystem and system level are available to the system developer. Validation and verification of SEU mitigation and performance of fault tolerance provisions are essential elements of systems design for operation in energetic particle environments.

Keywords: [Single Event Upset, Soft Error Rate, Fault Tolerant, Fault Avoidance, Fault Masking, Coding, Spatial Redundancy, Temporal Redundancy, Applications Oriented Fault Tolerance, Algorithm Based Fault Tolerance, Fault Containment, System Recovery].

1. Introduction

System design for compatibility with energetic particle environments requires an understanding of the mechanisms responsible for single event effects, an ability to perform assessment of their rate of occurrence, the capability to assess system level consequences of these effects, and a spectrum of mitigation techniques; including component level solutions as well as fault tolerance techniques implemented at the system and subsystem level. Although the energetic charged particle environment of space presents the most extreme environmental exposure, the increased complexity of terrestrial and avionics systems and the susceptibility of high-availability systems to faults induced by the traverse of terrestrial alpha particles and atmospheric neutrons, require the use of a mitigation strategy for Single Event Upsets (SEU) (Soft Error Rate (SER) in the parlance of terrestrial systems). A fault tolerant system design embodies the principles of fault avoidance, fault masking, detection of compromised system operation, containment of error propagation, and recovery to normal system operations. [1]

2. Systems Engineering for Energetic Particle Environment Compatibility

The criticality or risk of a single event upset causing operational impact to a system establishes the need for setting design requirements for single event effects management and mitigation. Analysis of system level SEU criticality based on functional impact and the determination of appropriate mitigation techniques form the basis of a system engineering approach to the design of a fault tolerant system. Evaluation of alternative single event effects mitigation and fault tolerance provisions addressing cost, power, size, weight, performance, reliability, and single event effects mitigation effectiveness must be considered. Understanding of the underlying mechanisms responsible for the occurrence of single event upset, developing an adequate description of component SEU characteristics, establishing energetic particle fluence models suitable for predicting rates of SEU occurrence, and the use of a technique such as Single Event Effects Criticality Analysis to assess effects on system operation are required to establish design requirements for mitigation or tolerance provisions.[2]

System requirements and specifications must address mission phases, performance, availability requirements, operational modes, and external support capability that relate to the fault tolerance characteristics of the system. Support provisions from external agents such as ground operational intervention, including periodic or on-demand activities in support of diagnosis and recovery actions are defined, and the acceptability of different modes of operation including reduced capabilities and degraded levels of service, safe shutdown, and emergency modes of operation must be defined. System architecture and partitioning defines error containment boundaries, establishes the system-wide strategy for coordinated error detection and recovery, and the redundancy provisions that support fault tolerance are determined. Establishing a fault tolerance hierarchy defines detection and recovery functions for individual local subsystems, shared functions over a group of subsystems, and system-wide or global functions.

3. Fault Tolerant Systems Strategies

The general subject of fault tolerant systems addresses numerous causes of faults, including incorrect specifications, design errors, undetected manufacturing defects, human operator actions, component damage or failure, and interaction with the operating environment. A fault may be classified by its nature, duration, and extent; the duration of a fault may be transient, intermittent or permanent. The context in which fault tolerance techniques are discussed herein addresses transient faults introduced by single event upset effects due to the energetic particle environment. The language of fault tolerant systems identifies a *fault* as the initiating event (in this case an energetic particle induced single event upset or transient); *errors* are the undesired system states caused by the fault. If error detection and recovery do not take place in a timely manner, a *failure* can occur that will be manifested by the inability of the system to provide specified service. *Fault tolerance* is the capability of a system to recover from a fault or error without exhibiting failure and compromised availability. A fault in a system does not necessarily result in an error; a fault may be latent in that it exists but has not resulted in an error; the fault must be sensitized by a particular system state and input conditions to produce an error. The principles of fault tolerant systems include *fault avoidance, fault masking, detection* of erroneous or compromised system operation, *containment* of error propagation, and *recovery* to normal system operations. The ability to detect at the lowest level possible, with error containment within the minimal number of components and subsystems, enables simpler and more effective recovery provisions. [3]

Fault tolerant subsystem design includes error detection and recovery provisions for faults within the subsystem as well as support for interfaces with other subsystems and global detection and recovery provisions. Detection and recovery may be either concurrent with normal system operation or preemptive in which case normal system operation is suspended until completion of the recovery activities. Correction provisions may be real-time; recovery operations may be either backward (rollback to a previous error-free state) or forward, constructing a valid error free new state from existing protected information. The recovery sequence includes the identification and removal of errors, restoration of valid subsystem states, and validation of the completion of successful recovery operation. It must be demonstrated that adequate testability is provided to detection and correction/recovery provisions, ensuring that manufacturing or component defects do not exist in systems which compromise the operation or performance of fault tolerance provisions. Furthermore, circuitry added to enable such testability must be examined for the propensity to introduce new fault conditions or to exacerbate the rate of fault occurrence. [4]

System integration and validation and verification activities must include the ability to establish that the design supports the qualitative goals and the quantitative requirements established for fault tolerance. Methods and provisions must be utilized in subsystem evaluation and for various levels of system integration to verify the performance of detection, correction, and recovery provisions. Qualitative evaluation of fault tolerant design provisions includes the use of formal and heuristic methods, test, and experimentation to demonstrate operability of detection and recovery provisions. Quantitative evaluation includes the use of fault injection to obtain information about fault coverage, execution time performance of recovery provisions, and the effects which faults present to system operations when detection and recovery provisions are not totally effective.

3.1 *Fault avoidance provisions*

Fault avoidance addresses endeavors that seek to reduce or eliminate the occurrence of faults through reduction in the severity of the energetic particle environment or the mitigation of circuit response to the traverse of energetic particles. Reduction in the severity of the environment includes consideration of alternative applications conditions, such as orbit altitude and inclination in space systems, and the use of shielding provisions to reduce the severity of the environment when translated to sensitive component locations. [5-7] System operations considerations may include fault avoidance achieved by not performing critical tasks in segments of the orbit in which single event effects are more probable due to elevated particle flux levels. The population of atmospheric neutrons as a function of latitude and altitude is understood and can be utilized in considering fault avoidance in avionics systems. [8] Reduction in the severity of the alpha particle environment can be achieved through screening and selection of materials used in device manufacture; including low alpha emitting solder and plastic encapsulant materials. Due to the limited energy and range of alpha particles emitted by packaging materials, the use of shielding provisions such as thin polyimide material deposited onto the die surface can result in significant reduction in the population of alpha particles reaching sensitive device structures. Due to their energy and resultant range, shielding is much less effective in reducing the severity of the energetic particle environment for atmospheric neutrons, trapped and solar protons, and galactic cosmic ray heavy ions.

Another class of fault avoidance provisions addresses the mitigation of single event upset through reduction in charge generation and collection, and mitigation of circuit response to collected charge. Through direct ionization or generation of recoil products, the traverse of energetic particles produces charge that is collected on circuit nodes. The ability to attenuate the generation of charge and to reduce its collection serves to decrease SEU rates. Techniques directed at reducing collected charge typically rely on the ability to truncate the funnel dimensions, to reduce the carrier lifetime in the region of the sensitive volume, and the use of silicon-on-insulator device isolation instead of reverse biased junction isolation. Commercial manufacturers have demonstrated success in the reduction of alpha particle induced soft error rate through the use of retrograde well profile; the increasing carrier concentration with increased distance into the well region results in reduced carrier lifetime, thus reducing the charge collected from the region below the sensitive volume. [9]

Techniques to mitigate circuit response include provisions to increase critical charge, to suppress response by limiting circuit bandwidth such that no response is generated by the short collected charge transient pulse, and to provide circuit design provisions which block the propagation of pulses in memory cell feedback paths. Commercial SRAM manufacturers have utilized stacked capacitor structures to increase cell capacitance, single event upset hardened devices use high valued polysilicon resistors in the feedback path of the SRAM cell which works with the cell capacitance to prevent the cell from responding to the short duration of charge collection. Numerous unique and novel circuit designs have been developed for memory cells, latches, and registers; all based on techniques to reduce the bandwidth of the cell to achieve immunity to the transient caused by collected charge or to provide redundant storage or blocking provision to prevent upset. [10-13] Although effective in improving cell single event upset characteristics, the increased cell size due to added complexity and the reduced bandwidth are contrary to the achievement of high-speed operation.

3.2 *Fault masking and redundancy*

The basic techniques in the application of redundancy as a fault masking provision are informational (redundant data structures), spatial (redundant hardware), and temporal (redundant sequential operations). The fault detection and correction capabilities of redundant structures are good; recovery may be either simple or highly complex depending upon the nature of the function performed by each of the structures. The increase in hardware content in informational redundancy is typically less than spatial modular redundancy, the increases in hardware content result in increased size, weight, power, and cost; and added components and interconnect adversely affect overall reliability.

Information redundancy and *encoding* provisions are designed such that in the presence of errors, the information is internally inconsistent, enabling error detection and correction (EDAC) implementations. Error detection and correction coding theory is the most widely developed mechanism for detection and recovery in digital systems. Error detection and correction codes can vary widely in detection and correction capabilities, code efficiency, and complexity of encoding and decoding circuitry. Cyclic redundancy checks, other cyclic codes, and convolutional coding schemes are used to detect errors in serial data transfer interfaces and storage media; such codes are generally classified as either block codes or convolutional codes. The encoder for a *block code* divides the information sequence into message blocks of k information bits each and transforms each into an n-bit codeword by appending n-k check bits; block codes are used for data storage systems as well as communications channels. The encoder for a *convolutional code* also accepts k-bit blocks of information and produces an encoded sequence of n-symbol blocks. However, each encoded block depends not only on the present k-bit input block, but also on the m previous message blocks. Convolutional codes are typically utilized for communications channels and other serial data transfer applications.

Hamming codes are the first class of linear block codes that were devised for error correction; these codes and their variations have been widely used in digital communications and data storage systems. The most common block code used in memory systems is the single-error-correcting, double-error-detecting (SEC-DED) Hamming code, the version providing single correction and double bit detection for 64 information bits requiring 8 check bits. More powerful codes may be constructed by using appropriate generating polynomials; a Hamming distance six code can be utilized to provide the ability to correct all double bit errors in 84 bit words using 15 check bits. Shortened Hamming codes are created by deleting columns from the generator matrix, producing a set of codes with reduced detection and correction capabilities. For a code to be useful for high-speed memory applications, its structure must permit rapid parallel encoding and decoding operations. Hsiao developed SEC-DEC codes by examining the properties of shortened Hamming codes to identify specific codes that enable fast encoding and decoding processes. The Bose, Chaudhuri, and Hocquenghem (BCH) codes form a large class of powerful random error correcting cyclic codes. Cyclic codes are an important subclass of linear codes that are attractive because of inherent algebraic structure that enables fast and efficient encoding and decoding implementations. The BCH code is a generalization of the Hamming codes that enable the capability to provide multiple error correction. Among the non-binary BCH codes, the most important subclass are the Reed-Solomon (RS) codes. [14-17]

In memory systems, Single Event Functional Interrupt (SEFI) occurring in memory devices produces conditions in which multiple errors may be produced by a single fault and must be addressed by the fault tolerance provisions. [18,19] Reed-Solomon codes are the general class of byte-oriented codes; however, the code length constraints of RS codes prevent codes having 64 and 128 bit information lengths in usable byte definitions. [20] Kaneda and Fujiwara have developed a class of codes having arbitrary code and byte length. Utilization of EDAC techniques in high-speed processor memory systems typically employs either a hardware or software provision that enables all locations within the currently used address space to be subjected to periodic verification. This provision requires periodic read access of all memory locations and examination by the EDAC protection to prevent the accumulation of single errors from exceeding the capability of the code to perform error correction. This procedure is usually termed "washing" or "scrubbing" the memory subsystem, and is usually performed under hardware control by an Application Specific Integrated Circuit (ASIC) or other standard product device used to implement the detection and correction code logic.

The complex nature of coding precludes its use in most computational operations. Simple arithmetic functions permit the use of parity through multistage adder circuits, and *arithmetic codes* such as AN and *residue codes* exist. AN is a non-separate code in which the added information is not distinct from the initial information, permitting fault detection in addition operations. Other residue codes are separate in nature; inverse residue codes are capable of detecting multiple errors that are unidirectional in nature, and bi-residue codes with two separate encoding levels provide error correction capability. The use of residue coding schemes for fault tolerance has received very limited application due to the high overhead requirement for pre-multiplying input variables, limited correction capabilities, and lack of capability for general computational operations. [21]

The most common form of *spatial or modular* redundancy *is triple modular redundancy (TMR)*, in which a fault in an individual module is corrected by the action of a voter through the majority consensus, or two-out-of-three, voting rules. The individual modules may consist of basic functions such as memory elements, D-flip-flops, or latches, and more complex logic blocks such as state machines or highly complex logic functions; including complete processing functional blocks, or entire subsystems. [22]

The notion of modular redundancy can be extended to the general sense of *n*-modular redundancy (*n*MR) with associated enhancements to the fault tolerance capabilities and expense associated with their implementation. For instance, a *quad redundant* implementation utilizing four modules provides the ability to tolerate errors in two modules, if the voter has the capability to discern the characteristics or signatures of the failure and to provide a plurality vote capability. Using a voter that is capable of responding to the initial failure by removing the voting rights of the failed module and imposing majority voting rules, a second SEU event could be tolerated by the remaining TMR structure. This would be the case if during or before recovery from the initial failure a second SEU event would occur in one of the modules. This capability to change the nature of the modular voting scheme is termed *dynamic* redundancy. In general, dynamic redundancy consists of three actions, *detection* of error, *location* of the error, and *reconfiguration* of the structure to perform in an optimum manner. If there exists a highly probable condition in which failures in both modules would produce identical module output conditions, the redundant module could potentially contain a *complementary* function instead of an identical copy of the primary module. In this case the *dual* of the primary module is utilized in the design of the redundant module. Such complementary logic may be used with *dual-rail codes* in performing error detection.

Use of complex off-the-shelf circuits such as microprocessors, microcontrollers, and complex application specific circuits (ASICs), may require fault tolerance using modular redundancy as a result of the inability to modify their internal design to incorporate fault tolerance provisions at a lower level. In the case of TMR, the ability to detect a single fault is excellent, and the ability to deliver the most probable correct result based upon the two-out-of-three vote is also reliable. The element of difficulty in complex circuits using redundancy of this type is the ability to recover the faulty processor to a known-good condition so that its outputs may again be utilized in detection and correction operations. Often the recovery operation in the faulty processor requires recovery of data and control states and resynchronization processes that cannot be performed concurrent with normal processor operations. The faulty processor must be removed from the redundancy configuration until the recovery operation is performed, resulting in susceptibility of a second fault compromising operation of the system during this interval. To eliminate this vulnerability as well as to provide detection and correction capability in the case of intermittent or permanent faults in individual processors, some architectures use a fourth processor in a quad-redundant configuration to assure the performance of the fault tolerance provisions.

An alternative to spatial redundancy is *temporal* redundancy in which the same hardware or software elements are used in consecutive operations, using diversity in time to provide results which may be compared using similar comparator/detection, and correction techniques discussed above for spatial redundancy. The technique is dependent upon the requirement for known error-free conditions on the input signals or variables used in the operations; either independent detection and correction must be provided to inputs, or the use of *alternating logic* techniques or *complementing functions* must be utilized to provide tolerance to input signal or variable errors. Alternating logic utilizes a class of boolean functions which are self dual, i.e. they satisfy the property $f(\bar{x}_1, \bar{x}_2, \ldots, \bar{x}_n) = \bar{f}(x_1, x_2, \ldots, x_n)$. Recomputing with shifted operands (RESO) is applicable to certain problems in which the shifting of inputs forms a complementing function that produces known relationship in outputs that may be utilized in detection and correction.

In older technologies single event transient pulses would not propagate along a path of combinatorial logic gates and the limited clocking frequency reduced the probability of sampling an incorrect logic state during the presence of a transient. The bandwidth of advanced technologies supports propagation of such pulses, and high clock frequencies provide ample opportunity for clocking of errant results into memory elements. The use of static spatial redundancy as a fault tolerance technique for single event upsets in memory elements (latches, flip-flops, and RAM), along with temporal redundancy as a mitigation provision against single event transients in combinatorial logic has been proposed for advanced technology devices.[23] The combined scheme uses temporal triple mode redundancy by sampling the output of combinatorial logic at three different times, storing the individual result in three different latch or flip-flop elements, and majority voting the latched result. By ensuring that the separation between clock edges is greater than the duration of a transient pulse emerging from the combinatorial logic, the transient can corrupt only one of the three values. Majority voting of static triple mode spatial redundancy ensures that upset of individual latches or flip-flops will also be corrected through action of the majority vote logic.

3.3 System error detection, containment and recovery

Whereas the fault detection capabilities of hardware based schemes such as modular redundancy are good, fault detection is the most difficult aspect of fault tolerance to achieve in applications oriented fault tolerance using algorithms, or software based fault detection, due to the elusive nature of the signature of most faults. A fault that results in a fail-stop consequence enables obvious opportunity for detection; most conditions of compromised system state produce signatures that are difficult to discern. Once detection is accomplished, recovery may be as complete and efficient as necessary to restore normal system operation. [24]

The ability to apply software fault tolerance techniques relies on specific properties of the problems being addressed by processor operations and are embodied in the *constraint predicate*. Three defined predicate subclasses formulate the constraint predicate: *progress*, *feasibility*, and *consistency*. The *progress* predicate utilizes the notion that there exist steps in the sequence of process operations and that the decomposition of the process into a finite number of operations blocks provides opportunity for testability at these intermediate points in the process. The *feasibility* predicate implies constraints that are apparent from the nature of the problem on which the processor is operating. Contained within this notion is the property of boundary conditions as constraints from which to generate a good predicate; testable results must be within the solution space of the problem as defined by the boundary conditions. *Consistency* conditions imply the ability to infer validity in intermediate or final results from input variables and previous intermediate or final process results. Consistency tests are a powerful technique; entire constraint predicates have been developed using only consistency conditions.

<u>Error Detection</u>

Application-oriented fault tolerance is a software solution to the problem of error detection and recovery; the software components are called *assertions*. The application of a set of metrics to a problem specification results in software assertions that may be embedded in the program code; assertions having the form "if not ASSERTION then ERROR". The extent of error detection is determined by the perceptiveness of the *assertions* to discern a compromised state of program execution. The recovery capability of the fault tolerance provision is determined by the response that is embodied in the *error* branch path. Applications-oriented fault tolerance works on the principle that testing of a program's intermediate results for conformance to specification ensures that the end result will be within specification, and that if an error does not manifest itself in failure, then the fault is of no consequence or interest. [25]

Several *acceptance test* techniques employed in detecting the presence of faults are *N-version programming, self-checking software, recovery blocks*, and the *watchdog coprocessor. N - version programming* is intended as a provision to detect defects in software design, coding, and integration. Parallel or sequential execution of programs and comparing the results provides opportunity for fault detection, much in the same way as modular redundancy does for hardware fault detection provisions. [26]

The *self-checking software* technique has aspects of functionality, control flow, and data in order to provide error detection. The functional and data aspects examine the reasonableness of the results, and may include an assessment of the input variables in performing acceptance testing of algorithm results. The capability provided by seemingly ad-hoc techniques in assessing the reasonableness of algorithm results is a powerful technique in the detection of computation errors. The control aspect includes checks on the execution flow from entry point to exit points in algorithm blocks, and only valid paths between the algorithm blocks are permitted. One approach to establishing the correctness of high-level control flow uses information (structure labeling) that is embedded in the syntax of the program text. The introduction of path tags to check the validity of the sequencing of blocks and block-tags to verify that execution of blocks proceeded properly from entry to exit points are examples of such structure labels. Each block contains a unique signature, and upon entry the block-tag is set to the value of the signature. The block-tag is verified upon exit from the routine to confirm that the block was not entered in any manner except the valid entry point. The path-tag is then set to the value of the next block signature, which is checked on entry into all blocks. Similar control checking of program iterative execution loops for illegal entry, completion, and branch related loop termination is utilized.

The *recovery block* approach uses acceptance tests in the form of checksums and other bounds checking on computation results to detect the presence of errors. The recovery block notion in this provision is that subsequent executions of the process using the same or different algorithms provide a recovery path from which acceptable results may be generated. The recovery block is a language construct supporting the incorporation of program redundancy into a fault tolerant program. The syntax of recovery block incorporation may take the form: *ensure T by* B_1 *else by* B_2, . . . *else by* B_n *else error.* T denotes the acceptance test, B_1 denotes the primary try block, and B_n denotes the alternate try blocks. The T acceptance test is a logical expression representing the criterion for determining the acceptability of the execution results of the try blocks. [27]

Another attempt at monitoring the behavior of a system is the use of a *watchdog coprocessor.* Most embedded processor systems use a hardware watchdog timer to detect halts to processor execution or errors in program control flow that are detected through failure of the software execution to reset the timer within the prescribed period of time. The watchdog coprocessor extends the notion from the simple hardware timer to the use of an additional processor to check the results of primary on-line processor elements. An active watchdog may implement interaction with the on-line processor as simple as an "I'm OK" message or heartbeat monitor that the watchdog expects to receive within a prescribed window. If not received within the valid window, the watchdog interprets this result that the on-line processor control flow is disrupted and asserts interrupt or reset of the on-line processor. The watchdog may execute concurrently with the on-line processor, or it may operate off-line and pre-compute results for subsequent acceptance testing. It invokes decisions on the integrity of the system based on assertions about the main process, assuming that faults either disrupt program control flow, corrupt database contents, or produce incorrect numerical results.

One specialized application of information redundancy to computational problems is *Algorithm Based Fault Tolerance* (ABFT). In this approach, some attribute of the function being performed is exploited with the use of information or time redundancy to achieve error detection, correction, or recovery. Most ABFT techniques developed to date address computational problems that exhibit structure and regularity which can be exploited to develop informational redundancy, such as matrix computations, sorting, Fast Fourier Transforms (FFT), QR factorization, singular value decomposition, least squares minimization, and other signal processing applications. Although the term "algorithm based" potentially implies a software implementation, these techniques are directly applicable to hardware-based implementation of the algorithms. [28]

<u>Containment of Error Propagation</u>

To minimize the impact to system operations and the extent of recovery operations, and to enhance the probability of successful system recovery, errors must be confined to the maximum extent possible to the module or subsystem in which the fault occurred. Typically, error containment boundaries are hierarchically defined, with errors confined at the lowest level possible. Containment boundaries can be established by subsystems checking either their own outputs, or by validating all input information. If error detection is activated but error recovery is not supported, the subsystem process is typically halted to prevent error propagation.

Reconfiguration and Recovery

If an unmasked fault has propagated in a system, a recovery period is needed to correct the system. Most recovery schemes restore system operation to a previous correct state or recovery point. A processor is rolled back to a recovery point by restoring the processor state and key variables to a known good condition, invalidating cache memory (which is likely to have been corrupted by error propagation) and forcing cache data to be restored from protected main memory. A checkpoint is a copy of an application's state that is stored in a protected region of the system. When a failure is detected, the application's state is rolled back to the saved previous checkpoint and execution resumed at that point. Backward error recovery can be defined as the capability of a system to return to a consistent state that existed before it failed; a checkpoint is then defined as a consistent state from which the execution can be restarted. [21]

3.4. *Validation and verification of fault tolerance*

Several approaches have been utilized to assess the effects of faults in processor systems and for validating fault handling mechanisms, including analytical modeling, experimental techniques (hardware pin faults, memory corruption, and ion irradiation), simulation modeling (register transfer level, gate level, and op-code level simulation) and fault emulation (memory, bus, and register transfer level). Use of analytical modeling is problematic due to the very large nature of the problem, and the simplifying assumptions that make the analysis tractable are regarded as compromising the usefulness of the technique and the validity of the results. Experimental techniques involve monitoring the behavior of a system while faults are introduced and recording the error detection and correction/recovery performance of the system. This approach requires the use of an accelerating agent to produce sufficient faults to gather statistically valid measurements on the fault tolerant performance of the system. In addition to the ability to quantify fault tolerance, experimental techniques provide the capability to study the consequences of faults that escape detection or are not afforded complete correction or recovery.

Two primary methods have been utilized to perform fault injection: simulation based methods and physical injection provisions. Simulation based methods provide the capability to establish the time of occurrence, location and type of fault, and the transient or permanent characteristics of the fault. Physical injection of faults utilizes charged particles, such as heavy-ion fission fragments from a Californium-252 source, or the use of a heavy ion or proton accelerator to produce single event faults. Most fault injection experiments are not designed around a formalized methodology; designers typically develop customized approaches to validate each system. Complexity of most systems results in the need to perform many experiments to achieve statistical confidence in the validation and verification result. Several simulation based fault injection tools address the need for accelerated fault injection and measurement processes; Messaline, Fiat, Ferrari, Focus, Depend, and React are the ones most discussed in the literature. [29-32]

Validation and verification of system performance utilizes the results of physical and/or simulated fault injection experiments to assess the coverage of faults that produce undesired system response and the ability of detection and recovery provisions to assure proper system operation. These results are utilized in analysis of system performance, often using techniques such as Markov process modeling, to make assertions regarding the fault tolerant performance and availability of the system.

References

1. A.K. Somani and N.H. Viadya, "Understanding Fault Tolerance and Reliability", Computer, pp 45-50, (April 1997).
2. Kenneth LaBel, Michele Gates, Janet Barth, E.G. Stassinopoulos, Allan Johnston and Paul Marshall, "Single Event Criticality Analysis (SEECA)", http://flick.gsfc.nasa.gov/radhome/papers.
3. A. Avizienis, "Toward Systematic Design of Fault-Tolerant Systems", Computer, pp 51-58, (April 1997).
4. Victor P. Nelson, "Fault-Tolerant Computing: Fundamental Concepts", Computer, pp 19-25, (July 1990).
5. Edward Petersen, "Single Event Analysis and Prediction", IEEE Nuclear and Space Radiation Effects Conference Short Course, July 1997.
6. Janet Barth, "Modeling Space Radiation Environments", 1997 NSREC Short Course, July 1997.
7. E.C. Smith, "Effects of Realistic Satellite Shielding on SEE Rates", IEEE Trans. Nucl. Sci., Vol. 41, No. 6, pp 2396-2399, (December 1994).
8. Eugene Normand, "Single-Event Effects in Avionics", IEEE Trans. Nucl. Sci., Vol. 43, No. 2, pp 461-474, (April 1996).
9. "Mitsubishi Electric Develops High-Frequency Synchronous SRAM with Dramatically Reduced Soft Error Rate", International Solid-State Circuits Conference, February 1999, http://www.mitsubishichips.com/press/releases/fsram_99.html.
10. James Edwin Vinson, "Circuit Reliability of Memory Cells with SEU Protection", IEEE Trans. Nucl. Sci., Vol. 39, No. 6, pp 1671-1678, (December 1992).
11. L. Rockett, "An SEU Hardened CMOS Data Latch Design", IEEE Trans. Nucl. Sci., Vol. 35, No. 6, pp 1682-1687, (December 1992).
12. T. Calin, M. Nicolaidis, R. Velazco, "Upset Hardened Memory Design for Submicron CMOS Technology", IEEE Trans. Nucl. Sci., Vol. 43, No. 6, pp 2874-2878, (December 1996).
13. M. Norely Liu, and Sterling Whitaker, "Low Power SEU Immune CMOS Memory Circuits", IEEE Trans. Nucl. Sci., Vol. 39, No. 6, pp 1679-1684, (December 1992).
14. Shu Lin and Daniel J. Costello, Jr., "Error Control Coding – Fundamentals and Applications", Prentice-Hall, (1983).
15. Eiji Fujiwara and Dhiraj K. Pradhan, "Error-Control Coding in Computers", Computer, (July 1990).
16. Ing-Ray Chen and I.-Ling Yen, "Analysis of Probablistic Error Checking Procedures on Storage Systems", The Computer Journal, Vol. 38, No. 5, (1995).
17. Gerardo Rubino and Bruno Sericola, "Internal Availability Analysis Using Denumerable Markov Processes: Application to Multiprocessor Subject to Breakdowns and Repair",

IEEE Transactions on Computers, pp 286-291, (February 1995).

18. R. Koga, S.H. Penzin, K.B. Crawford, and W.R. Crain, "Single Event Functional Interrupt (SEFI) Sensitivity in Microcircuits", Fourth European Conference on Radiation and its Effects on Components and Systems", pp 311-318, September 1997.
19. K.A. LaBel, P.W. Marshall, J.L. Barth, R.B. Katz, R.A. Reed, H.W. Leidecker, H.S. Kim, and C.J. Marshall, "Anatomy of an In-Flight Anomaly: Investigation of Proton-Induced SEE Test Results for Stacked IBM DRAMs", IEEE Trans. Nucl. Sci., Vol. 45, No. 6, pp 2898-2903 (December 1998).
20. James S. Plank, "A Tutorial on Reed-Solomon Coding for Fault-Tolerance in RAID-like Systems", Software – Practice and Experience", Vol. 27(9), (1997).
21. Michael Peercy and Prithviraj Banerjee, "Fault Tolerant VSLI Systems", Proceedings of the IEEE, pp 745-758, (1993).
22. Natin H. Vaidya, "Comparison of Duplex and Triplex Memory Reliability", IEEE Transactions on Computers, Vol. 45, No. 4, pp 503-507, (1996).
23. D.G. Mavis and P.H. Eaton, "Temporally Redundant Latch for Preventing Single Event Disruptions in Sequential Integrated Circuits", Mission Research Corporation Technical Report P8111.29, October 1998.
24. Janusz Sosnowski, "Transient Fault Tolerance in Digital Systems", IEEE Micro, (1994).
25. Bruce M. McMillin, "Fault Tolerance for Multicomputers: The Application Oriented Paradigm", Department of Computer Science, University of Missouri-Rolla, (1997).
26. Michel A. Banatre and Peter A. Lee, "Hardware and Software Architectures for Fault Tolerance, Experiences and Perspectives".
27. K.H. Kim and Howard O. Welch, "Distributed Execution of Recovery Blocks: An Approach for Uniform Treatment of Hardware and Software Faults in Real-Time Applications", Fault-Tolerance Systems: Techniques and Applications, edited by Hoang Phan, pp 95-105, (1992).
28. Yves Blaquiere, Gabriel Gagne, Yvon Savaria, and Claude Evequoz, "A New Efficient Algorithm-Based SEU Tolerant System Architecture", IEEE Trans. Nucl. Sci., Vol. 42, No. 6, (December 1995).
29. Ghani A. Kanawati, Nasser A. Kanawati, and Jacob A. Abraham, "FERRARI: A Flexible Software-Based Fault and Error Injection System", IEEE Transactions on Computers, pp 248-260, (February 1995).
30. Johan Karlsson, Peter Linden, Peter Dahlgren, Rolf Johansson, and Ulf Gunneflo, "Using Heavy-Ion Radiation to Validate Fault-Handling Mechanisms", IEEE Micro, pp 8-23, (February 1994).
31. Mei-Chen Hseuh, Timothy K. Tsai, and Ravishankar K Iyer, "Fault Injection Techniques and Tools", Computer, pp 75-82, (April 1997).
32. Jeffrey A. Clark and Dhiraj K. Pradhan, "Fault Injection: A Method for Validating Computer-System Dependability", Computer, pp 47-56, (June 1995).

International Journal of High Speed Electronics and Systems
Vol. 14, No. 2 (2004) 353–366

RADIATION-TOLERANT DESIGN FOR HIGH PERFORMANCE MIXED-SIGNAL CIRCUITS

W.T. HOLMAN

Vanderbilt University
Department of Electrical Engineering and Computer Science
Institute for Space and Defense Electronics
Nashville, TN 37235 U.S.A.

Modern semiconductor processes can provide significant intrinsic hardness against radiation effects in digital and analog circuits. Current design techniques using commercial processes for radiation-tolerant integrated circuits are summarized, with an emphasis on their application in high performance mixed-signal circuits and systems. Examples of "radiation hardened by design" (RHBD) methodologies are illustrated for reducing the vulnerability of circuits and components to total dose, single-event, and dose-rate effects.

Keywords: radiation hardened by design; radiation-tolerant integrated circuits; mixed-signal circuit design, silicon-on-insulator processes, silicon-germanium, total dose effects, single event effects, dose rate effects.

1. Introduction

The technical advances and economic growth of the commercial semiconductor industry since the late 1960's have resulted in a fundamental transformation in the design of radiation-tolerant analog, digital, and mixed-signal circuits. The end of the Cold War and the exponential growth of the consumer and business markets have dramatically affected the radiation-hardened IC market. The current demand for radiation-hardened integrated circuits is only a tiny fraction of the overall market for high performance commercial mixed-signal circuits and systems, and government funding for radiation-hardened IC process development and support has been slashed. As a result, most semiconductor manufacturers have abandoned the radiation-hardened IC market in favor of the much larger and more profitable commercial market. Few design engineers today can rely on the availability of specialized radiation-hardened IC processes, and those processes that are still available are generally years behind commercial processes in terms of performance.

This situation has led to the development of *radiation hardened by design* (RHBD) techniques, in which designers utilize existing commercial processes for the fabrication of radiation-tolerant mixed-signal circuits and systems. Modern submicron CMOS processes, silicon-on-insulator (SOI) processes, and BiCMOS processes can provide design engineers with significant (and often serendipitous) tolerance to radiation effects in custom integrated circuits. The RHBD philosophy combines this intrinsic radiation tolerance with appropriate component choices, layout methodologies, and circuit topologies to enhance the radiation tolerance of complex digital and analog systems at the expense of integrated circuit power, area, and/or maximum operating frequency. This paper provides a broad overview of common RHBD methodologies and components for radiation-tolerant integrated circuits, with particular emphasis on the mitigation of single-event and total-dose effects in high performance digital circuits, analog circuits, and mixed-signal systems.

2. Radiation Mechanisms in Mixed-Signal Integrated Circuits

Radiation mechanisms in digital and analog circuits are generally divided into three broad categories: *total dose* effects resulting from oxide charge trapping and/or displacement damage induced by γ, proton, or neutron irradiation, *single-event* effects caused by charged particle strikes, and *dose-rate* (*a.k.a.* prompt dose) effects caused by photocurrent generation throughout the entire circuit from very high dose rates over very short time durations. Total dose effects include increased leakage currents, MOSFET threshold voltage shifts, and BJT current gain degradation (resulting in increased base currents). Single-events effects due to charge generation across p-n junctions can result in soft errors, circuit latchup, or component burnout. Dose-rate effects can result in latchup, burnout, or rail-span collapse with resulting data loss. For a design engineer, the relative importance of each of these mechanisms will depend upon the intended application. Integrated circuits for strategic military applications will emphasize immunity to dose-rate effects. Mixed-signal circuits for high altitude aircraft or low earth orbit (LEO) satellites may only require hardening against single-event effects. Circuits for interplanetary probes will require hardening against both total dose and single-event effects.

The intrinsic radiation response of a given IC process will dominate the design of any RHBD circuit. For example, reduced feature size typically results in increased tolerance to total dose effects. Deep submicron CMOS processes have gate oxide thicknesses on the order of 10 Å or less, resulting in relatively little charge trapping. Low to moderate total dose exposure in a 0.13 μm bulk CMOS process may result in negligible effects in most digital circuits, whereas the same exposure amounts in an older 2 μm CMOS process would result in rapid failure of the same circuits. Furthermore, many mitigation techniques that are appropriate for one type of radiation mechanism will have no influence on other mechanisms, or may even exacerbate their effects. Consequently, there is no optimum design methodology for RHBD integrated circuits. Every new design and every new generation of advanced IC processes may require reassessment and modification of existing RHBD techniques.

3. Process, Component, and Layout Choices for Hardened-by-Design Circuits

The evolution of semiconductor processes has resulted in a growing emphasis in "system on chip" applications where digital, analog, and even RF circuits are combined together on the same substrate. Circuit designers now have access to three categories of high-performance IC processes that are suitable for mixed-signal RHBD applications. Bulk CMOS, silicon-on-insulator (SOI), and BiCMOS processes have distinctly different advantages and disadvantages in terms of economics, mixed-signal performance, and radiation response. Deep submicron bulk CMOS processes are widely available, relatively inexpensive, and can provide good inherent immunity to total dose effects provided leakage current paths are eliminated. SOI processes are intrinsically harder than CMOS against single event and prompt dose effects but will still exhibit total dose leakage currents. On the other hand, neither CMOS nor SOI processes will provide optimum low-noise or high-frequency analog circuit performance due to the lack of bipolar transistors. Modern BiCMOS processes combine deep submicron MOSFETs with high-frequency low-noise bipolar (preferably SiGe) transistors, enabling designers to combine analog, digital, and RF circuits on the same substrate. However, BiCMOS processes suffer from the radiation vulnerabilities of CMOS combined with the total dose degradation of bipolar transistors. In general, the inherent single-event immunity of SOI processes has tended to make them more desirable for many applications, but CMOS and BiCMOS processes also remain popular choices for RHBD integrated circuits due to economic constraints or analog performance requirements.

Regardless of the choice of process, the RHBD philosophy dictates that certain types of "rad soft" components and device layouts should be avoided for any design.[1] Components such as pinched resistors, diffused resistors, MOS capacitors, surface Zener diodes, and lateral PNP transistors are inherently vulnerable to radiation effects and should either be replaced with harder components (e.g. poly-poly or poly-metal capacitors instead of MOS capacitors), hardened with appropriate layout techniques (i.e. using a gated lateral PNP transistor), or simply not used at all.

The layout of a component can dramatically affect its radiation response regardless of the process. In some cases the conscientious use of standard commercial layout techniques will improve radiation tolerance. For example, analog and digital circuit designers usually avoid MOSFET transistor layouts with large gate widths (relative to gate lengths) to prevent switching delays due to signal propagation times along the width of the transistor. Instead, folded MOSFET layouts composed of interconnected "fingers" are utilized to provide multiple contacts along each gate and also reduce drain/source capacitances up to 50% via shared diffusions between adjacent fingers. However, sharing drain and source diffusions also reduces charge-collection effects by reducing p-n junction cross-sectional areas.

In other cases, radiation-hardened components require non-standard or extremely conservative layout techniques when compared to commercial designs. For example, leakage currents resulting from total dose exposure can flow between adjacent n-type diffusions in most semiconductor processes. Even in deep submicron IC processes with very thin gate oxides, thicker field oxides will accumulate sufficient trapped charge to create leakage paths as total dose increases. In standard NMOS transistor layouts, this leakage current will flow from drain to source at the edges where the gate polysilicon overlaps the field oxide. As shown in Figure 1, the use of edgeless or re-entrant NMOS layouts ensures that source and drain diffusions are strictly separated by polysilicon over thin oxide, thereby eliminating this leakage path at the cost of increased intrinsic capacitance and layout area.[2] Similarly, leakage currents between n-type diffusions of adjacent NMOS transistors can be controlled by the aggressive use of p-type channel stops or guard rings (structures already available in bulk processes) to interrupt any potential leakage path. The use of guard rings also provides the additional benefit of preventing latchup due to ion strikes or photocurrent effects in parasitic bipolar structures. However, the extensive use of channel stops and guard rings will require additional layout area for RHBD circuits as compared to standard commercial circuit designs.[2]

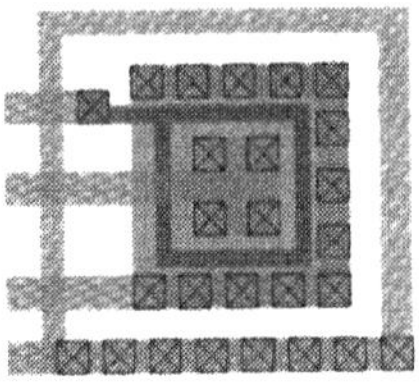

Fig. 1. An edgeless NFET with a rectangular gate layout that eliminates leakage paths between source and drain.

When compared to CMOS and BiCMOS processes, SOI processes provide the advantage of dielectric isolation for individual transistors and the elimination of leakage paths between n-diffusions in adjacent transistors. SOI processes require no guard rings and are inherently immune to CMOS-style latchup mechanisms. This same dielectric isolation also dramatically reduces susceptibility to single event upsets and prompt dose photocurrents by reducing the available p-n junction area for charge collection. In addition, the SOI buried oxide layer limits charge collection in a thin, dielectrically-bounded silicon layer. However, edgeless NMOS transistor layouts and/or body ties are still necessary to eliminate drain-to-source leakage within individual transistors. Body ties are also useful in partially depleted SOI processes to inhibit "single device" latchup due to the parasitic bipolar device present in the source-bulk-drain structure of the SOI MOSFET.[3,4] Finally, SOI MOSFETs can be vulnerable to back-channel leakage, i.e. leakage at the bottom of the channel due to accumulated charge in the buried oxide layer beneath the transistor.[5] Designers of RHBD circuits must ensure that back-channel leakage effects have been characterized or compensated for, as a commercial semiconductor process is typically not created with total dose hardness in mind.

4. Total Dose Hardening

The importance of total dose effects to mixed-signal circuit designers has diminished somewhat as integrated circuit processes have evolved over the past three decades, leading to thinner process layers with fewer defects. For example, SiO_2 layers will tend to trap holes as high-energy photons create hole-electron pairs, and the higher-mobility electrons diffuse out of the oxide. Additional charge traps will also form at defects in silicon-SiO_2 interfaces. As a result, MOSFET threshold voltages will typically exhibit a negative shift with increasing total dose. In the early seventies MOSFET gate oxides were thick and of relatively poor quality, resulting in significant charge trapping and very large threshold voltage (V_T) shifts. In a total dose environment NFETs would quickly become depletion mode devices and be impossible to turn off, while PFETs would become impossible to turn on as the magnitude of V_{TP} grew larger than the magnitude of the power supply voltage.

In contrast, relatively little charge trapping occurs in the very thin, high-quality gate oxides of modern submicron CMOS processes. It has therefore become practical to create digital cell libraries for bulk CMOS technologies that are tolerant of total doses up to several hundred krad [Si].[2,6] These cell libraries utilize edgeless NMOS layouts and guard rings to minimize leakage currents while ignoring the small shifts in V_T. The use of NAND cells over NOR cells may also be preferable in RHBD digital libraries, as the overall performance of series-connected NFETs and parallel-connected PFETs in NAND arrays will be less affected by small changes in V_T.[2]

In analog and mixed-signal circuits the situation is somewhat different. In deep submicron processes, MOSFET parameters such as transconductance and intrinsic noise are generally not strongly affected by cumulative total dose.[7] Some transconductance degradation and intrinsic noise increases will be observed in individual transistors, but in modern processes total doses as high as 30 Mrad [$Si0_2$] may be tolerated.[8] However, even small parametric changes may lead to failure in circuits such as high-resolution analog-to-digital converters, as these changes can be multiplied by large voltage gains.[9] Unlike digital circuits, analog circuits can also be dramatically affected by shifts in internal bias points due to increasing total dose.

In analog circuits where precise component matching is required, relatively small differences in MOSFET threshold voltage shifts can result in circuit failures.[10] Because total dose effects are bias-dependent, two matched transistors will degrade at different rates if biased at different average voltages. When multiplied by a large voltage gain, increasing radiation-induced offsets between MOSFETs can cause failures in differential amplifiers, comparators, and bias circuits. For example, the cascode amplifier circuit shown in Figure 2(a) uses transistors M2 and M6 to establish a bias current, but the two transistors have different gate-source voltages. In a radiation environment the threshold voltages of M2 and M6 will steadily diverge until input transistor M1 is driven into non-saturated operation. The solution (Figure 2(b)) is to replace M6 with transistors M7 through M11 that operate at identical V_{GS} values, thereby significantly hardening the entire circuit to changes in V_T.[10]

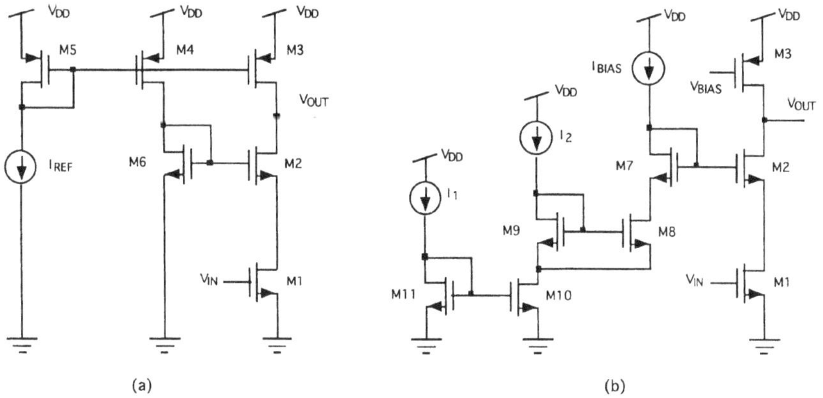

Fig. 2. (a) A standard CMOS bias circuit configuration versus (b) an RHBD configuration with matched V_{GS} values.

More sophisticated circuit techniques can also be utilized to maintain the functionality of high-performance analog circuits.[10] Switched capacitor circuits can be used to cancel or balance radiation-induced offset voltages in amplifiers and comparators. Charge pump circuits and bulk-source biasing circuits can be used to compensate for changes in threshold voltages in processes where transistor wells can be individually isolated. However, it is important to realize that these techniques usually compensate for the symptoms of total dose degradation at the system or circuit level, rather than reversing or preventing degradation at the component level. As a result, an analog circuit may operate normally until a critical threshold of total dose exposure is reached, at which point the circuit will fail catastrophically as voltage headroom or gain falls below a required minimum value.

Increased total dose in bipolar junction transistors results in increased noise and degraded β (or increased base current). As minority carrier devices, BJTs are also susceptible to displacement damage from proton or neutron irradiation. Circuit techniques such as base current compensation and cascoding can used to compensate for some of these effects at the cost of increased circuit area or decreased voltage headroom.[1] However, recent experimental results have shown that SiGe heterojunction bipolar transistors can provide far superior total dose tolerance when compared to conventional silicon BJTs.[11] Analog and RF circuits such as bandgap references and voltage controlled oscillators fabricated in a commercial 0.25 μm SiGe BiCMOS process have demonstrated minimal performance degradation at proton fluences as high as 5×17 p/cm^2.

Some advanced bipolar processes also exhibit enhanced low dose radiation sensitivity (ELDRS) effects, where a given dose at a very low dose rate has a greater cumulative effect than the same dose at higher dose rates.[12] Consequently, a bipolar integrated circuit that is tested at typical laboratory dose rates of 10-100 rad(Si)/s may exhibit far less damage than an identical circuit operating in an actual space environment with a dose rate of 0.01 rad(Si)/s or less. BiCMOS (or bipolar) processes should be carefully characterized for ELDRS effects to accurately predict the performance degradation of mixed-signal circuits in low dose rate environments.

5. Single-Event Effect Hardening

The mitigation of single-event effects (SEEs) has taken on growing importance as feature sizes have decreased. Thirty years ago, typical IC features sizes were on the order of 5 to 10 μm. Today CMOS processes with 0.13 μm gate lengths are commercially available, and manufacturers expect to move to 0.045 μm feature sizes by the year 2010.[13] Hundreds or thousands of transistors operating at frequencies of several GHz can now be placed in the same area once occupied by a single transistor operating at a few MHz, and the same ion energies that had little effect thirty years ago can result in severe voltage transients over multiple circuit nodes today. The use of lightly-doped epitaxial layers in submicron CMOS processes has tended to mitigate single-event effects in digital circuits by reducing charge collection and preventing latchup[14,15], but shrinking features sizes are now forcing commercial manufacturers to address single-event effects as terrestrial radiation sources become a significant source of soft errors in ground-based applications.

The effect of an ion strike on any given node of a mixed-signal circuit is dictated by the basic capacitance equation

$$\Delta q = C \cdot \Delta V \tag{1}$$

where Δq is the collected charge due to an ion strike, C is the effective capacitance of the node, and ΔV is the resulting change in node voltage. The minimum value of C has decreased with the square of shrinking feature size while the range of Δq values has remained relatively constant. As a result, the average nodal ΔV perturbation has increased in modern processes. In addition, IC power supply voltages have steadily decreased along with feature sizes, making the proportional effect of ΔV even greater. Modern circuit topologies that rely upon the temporary storage of charge on a floating capacitor (e.g. DRAM, dynamic logic gates, or switched capacitor circuits) become particularly vulnerable as feature size decreases.[16,17] (It should be noted that total-dose compensation circuits that rely on switched capacitors for error cancellation may also be vulnerable to single-event errors.) In general, circuit-level mitigation techniques for single-event effects must either filter or dissipate the collected charge or provide some form of circuit redundancy to prevent a single event upset from corrupting data.

The operating frequency of a mixed-signal circuit will also determine its susceptibility to single-event effects, as illustrated by the fundamental equation

$$\Delta q = \Delta t \cdot I \tag{2}$$

where Δt is the time needed to dissipate charge from an affected node, and I is the maximum charging or discharging current available to that node. As the clock frequency of a digital circuit increases, the clock period T decreases. As T approaches Δt and then becomes smaller than Δt, the probability of a single event upset propagating through the remainder of the circuit increases. In general, the higher the operating frequency of a circuit, the more likely a single-event effect will result in a significant output transient or soft error, although limited exceptions may occur in some dynamic logic circuits.[16-19]

The principle of *charge dissipation* can be applied to digital or analog circuits to reduce Δt by increasing the maximum charging or discharging current I at critical nodes.[16] Consider the digital inverter shown in Figure 3, where the logic "high" input voltage turns on the NFET and turns off the PFET. Any charge deposited on the equivalent output capacitance C_{OUT} must be discharged through the NFET. By increasing the width-to-length ratios of the inverter transistors, the maximum drain current I_D can be increased so that a net change in capacitor voltage due to an ion strike is dissipated before the subsequent logic stage can propagate the resulting voltage transient as a bit error.[16] Similarly, bias currents in analog circuits can be increased to provide improved charge dissipation and minimize single-event effects in critical analog signal paths. The disadvantages of charge dissipation are the power and area penalties associated with increasing the size and quiescent currents of many of the circuit components.

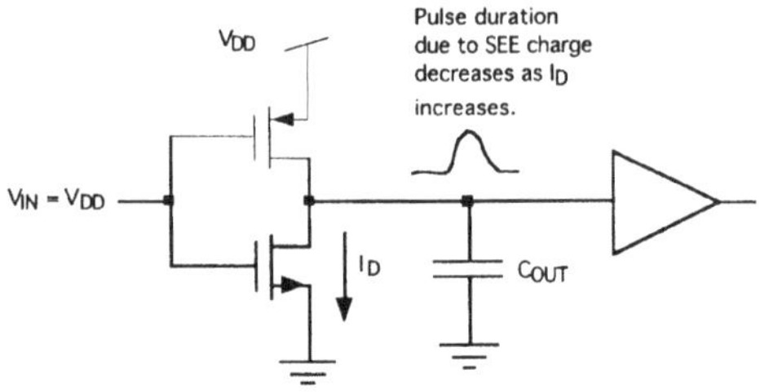

Fig. 3. The principle of charge dissipation as applied to digital circuits.

A related technique for SEE mitigation in mixed-signal circuits is *temporal filtering*.[16] As shown in Figure 4, this method involves adding resistance and/or capacitance at critical circuit nodes to reduce transient response and provide a low-pass filter effect to fast single-event voltage pulses. For example, temporal filtering has been used to harden SRAM cells against soft errors through the addition of resistors in the latch feedback paths.[16] However, the significant area and speed penalties of added resistance and capacitance severely limit the use of temporal filtering in high-performance integrated circuits.

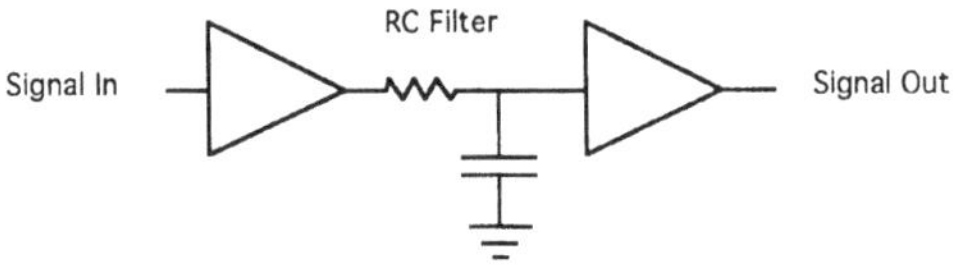

Fig. 4. Temporal (low-pass) filtering to reduce transient response to single-event effects.

In digital circuits, *spatial redundancy* and *temporal redundancy* techniques (or a combination of both) are commonly used to eliminate single-event errors.[16] An example of spatial redundancy is a triple-redundant combinational logic circuit that uses a voting circuit to generate a valid result only if at least two calculations agree. As shown in Figure 5, spatial triple-mode redundancy can be implemented using three physically separated data paths to prevent any single ion strike from upsetting more than one set of logic circuits. Triple-redundant circuits provide excellent SEE immunity at the expense of circuit area, provided the voting circuit is also hardened against errors.

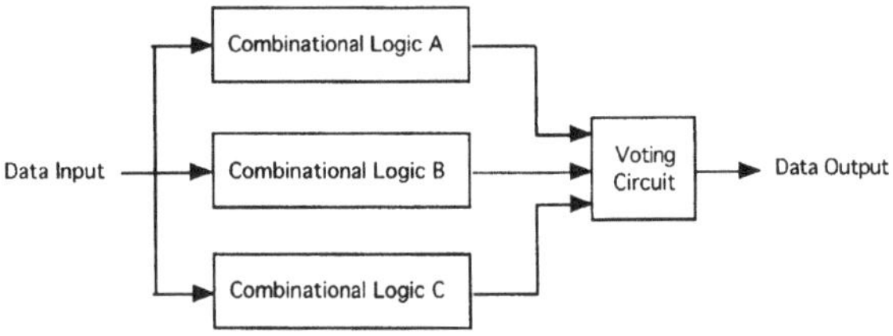

Fig. 5. Spatial triple-mode redundancy in RHBD digital circuits.

Spatial redundancy can also be applied at the gate level through the use of additional transistors, data storage elements, or data paths to ensure that two or more simultaneous SEEs must occur to propagate or store a soft error in a combinational circuit or data latch. The DICE (dual interlocked storage cell)[20] design is a popular example of an SEU-hardened data latch. Spatial redundancy can be applied to CMOS or SOI digital circuits, although some gate-level techniques require that individual MOSFET bulks be isolated from each other.[16,21] Regardless of the application or process, spatially redundant circuits will inevitably require significant area and/or performance penalties. On the other hand, spatial redundancy has no inherent maximum operating frequency.

Just as spatial redundancy can separate identical data signals in space, temporal redundancy separates identical data signals in time in order to filter out single event upsets. Figure 6 shows an SEU-resistant data latch with three separate data paths, a voting circuit, and a 2-input multiplexer connected in a feedback loop.[18] Two of three data paths to the voting circuit incorporate time delays of ΔT and $2\Delta T$, respectively, where ΔT is equal to the time required to dissipate the largest expected charge of an ion strike.

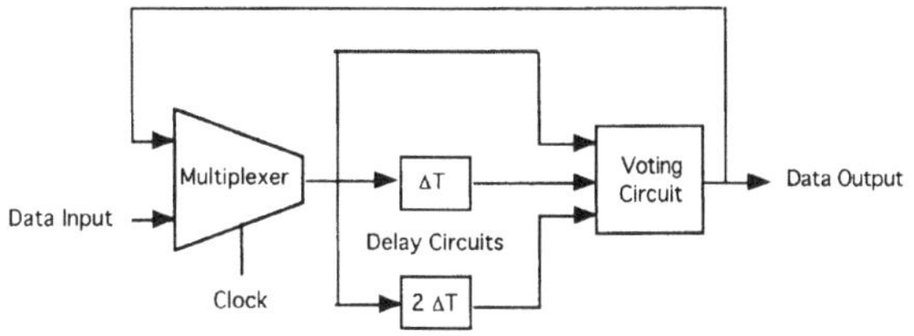

Fig. 6. Triple-path temporal redundancy in an SEU-hardened latch.

The main penalty of temporal redundancy is the inherent latency of waiting $2\Delta T$ seconds to complete a calculation. At operating frequencies below 50 MHz this latency is relatively small, but as clock speeds increase the delay quickly becomes a significant percentage of the overall latency of the circuit.[18] Given a $2\Delta T$ value of 200 ps, temporal redundancy techniques will put an upper limit of 2.5 GHz on the maximum operating frequency of any individual signal path or data latch. Operating frequencies in excess of 3 GHz are already common in high-performance microprocessors, so temporal redundancy may prove less useful as even faster commercial processes become available.

Analog and RF circuits provide very different challenges for SEE hardening when compared to digital circuits. Spatial and temporal redundancy techniques are unsuitable for removing transients from analog/RF signal paths, as no practical means of constructing a voting circuit to compare analog outputs has been demonstrated. Furthermore, short-channel effects dictate that most analog MOS transistors cannot take advantage of deep submicron gate lengths.[22] Charge dissipation or temporal filtering techniques combined with appropriate component layouts may be the only practical alternatives. In addition, the choice of circuit topology can dramatically affect an analog circuit's response to single event transients. For example, cascode circuit configurations can increase voltage gain and small-signal output resistance at the expense of reduced voltage overhead. However, that reduced voltage overhead may result in much longer recovery times if transistors are more easily forced into undesired operating regions by an ion strike.[1] It should also be noted that some mixed-signal circuit topologies are inherently immune to single-event effects. For example, oversampled data converter topologies such as sigma-delta A/D converters are resistant to single-event effects in the analog front end, since any SETs will simply be converted into out-of-band noise by the oversampling process.[23]

At the system level, RHBD circuit designers should keep in mind that few circuits or signal paths are equally affected by single-event effects in digital or analog circuits. Some circuit nodes will be SEE-insensitive due to their intrinsic capacitance or resistance, and some signal paths may be blocked or attenuated in such a way that SEEs do not propagate and become latched as data errors.[16] On the other hand, single-event effects in global signals such as clocks, bias voltages, and reference voltages can create errors throughout the entire system. A careful analysis to identify the most critical nodes or signal paths, and to selectively harden them, can result in significant improvements in radiation hardness while minimizing layout, power, and speed penalties for the overall system.

Another concern for the designer is the occurrence of single-event latchup in SOI integrated circuits. SOI processes are not subject to the destructive power supply rail-to-rail latchup effects of CMOS and BiCMOS processes, but they can suffer from localized latchup effects (i.e. single event snapback) where a low resistance path forms between the drain and source of a single SOI MOSFET.[3,4] Although the current levels are generally not immediately destructive, device burnout may eventually occur if the latchup is not reversed.[27] In such cases the only remedy may be to cycle the power of the affected circuit to restore the transistor to normal operation. Consequently, a radiation-tolerant SOI circuit may require the inclusion of a robust power reset function[24] to ensure that a single-event latchup does not render a remotely operated satellite or sensor permanently inoperative. It should also be noted that the use of body ties can reduce or eliminate single event snapback, but at the cost of layout area.

6. Dose-Rate Effect Hardening

Of the three categories of radiation effects, relatively little information is available in the open literature concerning the modeling and mitigation of dose-rate effects due to the strategic defense applications of circuits that typically encounter this phenomenon. Furthermore, the design of dose-rate hardened circuits generally requires good simulation tools and device models in order to obtain satisfactory model-to-hardware correlation. However, some general design principles are commonly utilized to prevent certain types of circuit failure in dose-rate environments. For example, *rail span collapse* is a dose-rate failure mechanism caused by the simultaneous generation of photocurrents throughout an integrated circuit.[25] As shown in Figure 7, these currents result in large transient voltage drops across the power supply and ground rails. At the very ends of the rails the voltage drops can be so large that the power supply voltage falls below V_T over a portion of the circuit, leading to complete signal or data loss. Rail span collapse can be avoided by widening and distributing power supply rails so as to minimize these voltage drops over any given portion of the integrated circuit.

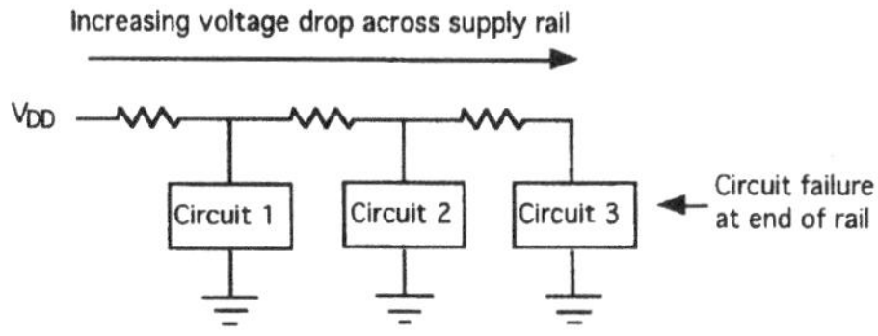

Fig. 7. Rail span collapse results from voltage drops caused by prompt dose photocurrents throughout an IC.

Even if rail span collapse is avoided, dose-rate photocurrents can still result in large transient voltage pulses at circuit nodes connected to reverse-biased *p-n* junctions. In digital circuits such pulses may be large enough to change the values of stored bits in data registers. In many cases the technique of *photocurrent compensation* is applied to such critical circuit nodes. By balancing the total photocurrent entering a node with the total photocurrent exiting a node, the net perturbation at the node is minimized, but at the cost of increased circuit area and increased transient currents flowing between the power supply rails. Figure 8 illustrates a photocurrent compensation technique using dummy devices created using bipolar transistors.[26] If transistor Q1 and the dummy transistor have identical geometries, the induced collector-base photocurrent of Q1 will be equal to the collector-base photocurrent of the dummy transistor, resulting in a net current change of zero for the base node of Q1.

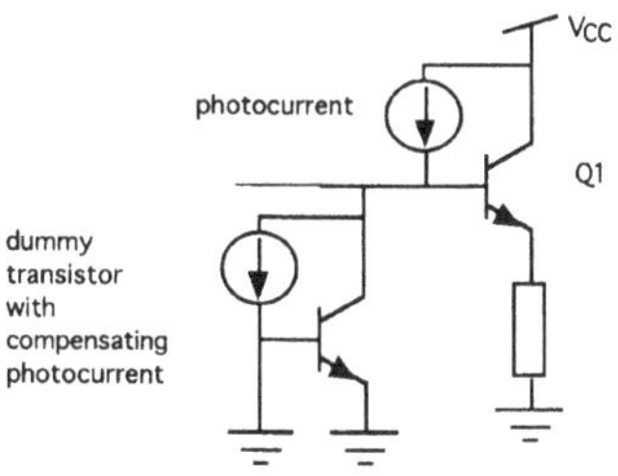

Fig. 8. Dummy transistors can be used to provide photocurrent compensation.

Photocurrent compensation can also be accomplished in CMOS and partially depleted SOI processes by carefully balancing the cross-sectional areas of the reverse-biased drain-bulk and source-bulk junctions for all NFET and PFET transistors connected to a given node. Again, the goal of this technique is to balance all currents entering or exiting a node, and thereby minimize the voltage transient. The drawback of this approach is that it may require undersizing PFET width-to-length ratios compared to NFET width-to-length ratios, resulting in unbalanced transient rise and fall times.

7. Conclusion

Unless current trends in semiconductor manufacturing and government funding change dramatically, the philosophy of "radiation hardening by design" will almost certainly grow in importance for mixed-signal integrated circuit design as IC feature sizes move to 0.09 μm and below. RHBD circuits inevitably require penalties in power, area, or speed when compared to commercial applications, but the use of commercial IC processes will allow design engineers to maintain acceptable performance differentials rather than be left behind by the relentless march of Moore's Law. The overall performance/cost ratio of an RHBD circuit will not match that of a commercial design, but an RHBD circuit fabricated in a commercial IC process can still provide improved speed and functionality in a smaller die area when compared to a circuit fabricated using a specialized radiation-hardened IC process that is one or more generations behind commercial processes in terms of feature size and performance.[6]

Furthermore, help may be on the horizon for RHBD designers in terms of the mitigation of single-event effects. As feature sizes shrink, semiconductor manufacturers must inevitably resolve the issue of soft errors resulting from terrestrial radiation sources. Given the area and speed penalties of spatial and temporal redundancy, manufacturers will have an enormous financial incentive to either intrinsically harden commercial processes against single-event effects or develop new circuit techniques to mitigate their effects with minimal performance penalties. Consequently, the primary focus of future RHBD research may shift to dose-rate hardening over the next decade, with significant total-dose and single-event hardening intrinsically provided by commercial processes and/or automatically incorporated into analog and digital circuits using standard computer-aided design tools.

In the longer term (20 to 30 years) the future of RHBD methodologies is not so clear. Eventually digital components will shrink to molecular or atomic scale, and individual proton, neutron, or ion strikes may not degrade, but rather destroy individual transistors or entire subcircuits. Barring the introduction of new shielding or packaging technologies, some form of dynamic self-repair or reconfiguration of complex digital circuits will be needed. Again, the need to improve manufacturing yields and prevent operational failures will probably force commercial manufacturers to take the lead in solving these problems.

In addition, there is currently no indication that analog and RF circuits will be able to take advantage of the same components and processes that digital circuits will eventually use. The need for high transconductance, low 1/f noise, and high output resistance already puts severe limits on the minimum gate lengths of MOS transistors used in analog circuits.[22] Even today, an analog circuit with a few hundred components may require more chip area than a digital circuit with millions of transistors. Analog and RF circuits may not be compatible or even possible in digital processes employing proposed exotic components such as fullerene molecules, quantum dots, or biological proteins as computational elements. Consequently, future RHBD techniques may evolve in two divergent directions, with analog and digital designers required to work in completely different IC technologies and employ fundamentally different RHBD techniques to implement mixed-signal systems.

References

1. J. Swonger, "Design and simulation of hardened integrated circuits," 1999 IEEE Nuclear and Space Radiation Effects Conference Short Course Notebook.
2. D.R. Alexander, D.G. Mavis, C.P. Brothers, and J.R. Chavez, "Design issues for radiation

tolerant microcircuits for space", 1996 IEEE Nuclear and Space Radiation Effects Conference Short Course Notebook.
3. O. Musseau, "Single-event effects in SOI technologies and devices," *IEEE Transactions on Nuclear Science*, vol. 43, no. 2, pp. 603 –613, 1996.
4. S.E. Kerns, L.W. Massengill, D.V. Kerns, M.L. Alles, T.W. Houston, H. Lu, and L.R. Hite, "Model for CMOS/SOI single event vulnerability," *IEEE Transactions on Nuclear Science*, vol. 36, no. 6, p. 2305-2310, 1989.
5. J.R. Schwank, M.R. Shaneyfelt, P.E. Dodd, J.A. Burns, C.L. Keast, and P.W. Wyatt, "New insights into fully-depleted SOI transistor response after total-dose irradiation," *IEEE Transactions on Nuclear Science*, vol. 47, no. 3, pp. 604-612, 2000.
6. R.C. Lacoe, J.V. Osborn, R. Koga, S. Brown, and D.C. Mayer, "Application of hardness-by-design methodology to radiation-tolerant ASIC technologies," *IEEE Transactions on Nuclear Science*, vol. 47, no. 6, pp. 2334-2341, 2000.
7. M. Manghisoni, L. Ratti, V. Re, and V. Speziali, "Radiation hardness perspectives for the design of analog detector readout circuits in the 0.18-μm CMOS generation," *IEEE Transactions on Nuclear Science*, vol. 49, no. 6, pp. 2902-2909, 2002.
8. G. Anelli, M. Campbell, M. Delmastro, F. Faccio, S. Floria, A. Giraldo, E. Heijne, P. Jarron, K. Kloukinas, A. Marchioro, P. Moreira, and W. Snoeys, "Radiation tolerant VLSI circuits in standard deep submicron CMOS technologies for the LHC experiments: practical design aspects," *IEEE Transactions on Nuclear Science*, vol. 46, no. 6, pp. 1690-1696, 1999.
9. C.I. Lee, B.G. Rax, and A.H. Johnston, "Total ionizing dose effects on high resolution (12-/14-bit) analog-to-digital converters," *IEEE Transactions on Nuclear Science*, vol. 41, no. 6, pp. 2459-2466, 1994.
10. C.F. Edwards, W. Redman-White, M. Bracey, B.M. Tenbroek, M.S.L. Lee, and M.J. Uren, "A multibit ΣΔ modulator in floating-body SOS/SOI CMOS for extreme radiation environments," IEEE Journal of Solid-State Circuits, vol. 34, no. 7, pp. 937-947, 1999.
11. J.D. Cressler, M.C. Hamilton, R. Krithivasan, H. Ainspan, R. Groves, G. Niu, S. Zhang, Z. Jin, C.J. Marshall, P.W. Marshall, H.S. Kim, R.A. Reed, M.J. Palmer, A.J. Joseph, and D.L. Harame, "Proton radiation response of SiGe HBT analog and RF circuits and passives," *IEEE Transactions on Nuclear Science*, vol. 48, no. 6, pp. 2238-2243, 2001.
12. D.M. Fleetwood, S.L. Kosier, R.N. Nowlin, R.D. Schrimpf, R.A. Reber, Jr., M. DeLaus, P.S. Winokur, A. Wei, W.E. Combs, and R.L. Pease, "Physical mechanisms contributing to enhanced bipolar gain degradation at low dose rates," *IEEE Transactions on Nuclear Science*, vol. 41, no. 6, pp. 1871-1883, 1994.
13. Semiconductor Industry Association, International Technology Roadmap for Semiconductors, 2002.
14. A.H. Johnston, "Radiation effects in advanced microelectronics technologies," *IEEE Transactions on Nuclear Science*, vol. 45, no. 3, pp. 1339-1354, 1998.
15. P.E. Dodd, F.W. Sexton, and P.S. Winokur, "Three-dimensional simulation of charge collection and multiple-bit upset in Si devices," *IEEE Transactions on Nuclear Science*, vol. 41, no. 6, pp. 2005-2017, 1994.
16. S.P. Buchner and M.P. Baze, "Single event transients in fast circuits", 2001 IEEE Nuclear and Space Radiation Effects Conference Short Course Notebook.
17. G.R. Agrawal, L.W. Massengill, and K. Gulati, "A proposed SEU tolerant dynamic

random access memory (DRAM) cell," *IEEE Transactions on Nuclear Science*, vol. 41, no. 6, pp. 2035-2042, 1994.

18. D.G. Mavis and P.H. Eaton, "Soft error rate mitigation techniques for modern microcircuits," 40th Annual International Reliability Physics Symposium Proceedings, 2002, pp. 216-225.
19. N. Seifert, X. Zhu, and L.W. Massengill, "Impact of scaling on soft-error rates in commercial microprocessors," *IEEE Transactions on Nuclear Science*, vol. 49, no. 6, pp. 3100-3106, 2002.
20. T. Calin, M. Nicolaidis, and R. Velazco, "Upset hardened memory design for submicron CMOS technology", *IEEE Transactions on Nuclear Science*, vol. 43, no. 6, pp. 2874-2878, 1996.
21. M.P. Baze, S.P. Buchner, and D. McMorrow, "A digital CMOS design technique for SEU hardening," *IEEE Transactions on Nuclear Science*, vol. 47, no. 6, pp. 2603-2608, 2000.
22. **CMOS Analog Circuit Design**, P.E. Allen and D.R. Holberg, Oxford University Press, 2nd edition, 2002.
23. F.P. Cortes, F.P., L. Carro, A. Girardi, and A. Suzim, "A ΣΔ A/D converter insensitive to SEU effects," Proceedings of the Eighth IEEE International On-Line Testing Workshop, 2002, pp. 89-93.
24. P.J. Layton, D.R. Czajkowski, J.C. Marshall, H.F.D. Anthony, and R.W. Boss, "Single event latchup protection of integrated circuits," Fourth European Conference on Radiation and Its Effects on Components and Systems (RADECS 1997), pp. 327-331.
25. L.W. Massengill and S.E. Diehl-Nagle, "Transient radiation upset simulations of CMOS memory circuits," *IEEE Transactions on Nuclear Science*, vol. 31, no. 6, pp. 1337-1343, 1984.
26. Y. Deval, H. Lapuyade, P. Fouillat, H. Barnaby, F. Darracq, R. Briand, D. Lewis, and R.D. Schrimpf, "Evaluation of a design methodology dedicated to dose-rate-hardened linear integrated circuits," *IEEE Transactions on Nuclear Science*, vol. 49, no. 3, pp. 1468-1473, 2002.
27. P.E. Dodd, M.R. Shaneyfelt, D.S. Walsh, J.R. Schwank, G.L. Hash, R.A. Loemker, B.L. Draper, and P. S. Winokur, "Single-event upset and snapback in silicon-on-insulator devices and integrated circuits," *IEEE Transactions on Nuclear Science*, vol. 47, no. 6, pp. 2165-2174, 2000.

International Journal of High Speed Electronics and Systems
Vol. 14, No. 2 (2004) 367–378

A TOTAL-DOSE HARDENING-BY-DESIGN APPROACH FOR HIGH-SPEED MIXED-SIGNAL CMOS INTEGRATED CIRCUITS

NATHAN NOWLIN, JOHN BAILEY, BOB TURFLER, DAVE ALEXANDER
Mission Research Corporation, Microelectronics Division
5001 Indian School Road, NE
Albuquerque, NM 87110 USA
nowlin@mrcmicroe.com

This paper describes design choices and tradeoffs made when designing total-dose hardness into an advanced CMOS integrated circuit. Closed geometry transistors are described and compared, emphasizing their radiation tolerant performance. Speed and area tradeoffs incurred in circuit design when using such closed geometry transistors are illustrated in the design of an advanced IEEE 1394 cable physical layer mixed-signal interface chip.

Keywords: Total-dose, CMOS design, radiation hardening, closed-geometry transistors, mixed-signal.

1. Introduction

Total ionizing dose can result in degradation in microcircuit performance (increased propagation delay times, lower maximum clock rates, reduced output drive, lower noise margins, increased leakage, and catastrophic failure). Radiation hardened parts can be achieved by identifying the mechanisms responsible for radiation effects in microcircuits and implementing electrical and layout design solutions to minimize them. These solutions must be consistent with typical semiconductor processing so that resulting microcircuits can be fabricated in commercial foundries. This approach trades off area and circuit density for radiation hardness.

In a typical CMOS fabrication process, as illustrated in Figure 1, the transition between the isolation or field oxide and the thin, gate oxide produces a parasitic transistor that is susceptible to total ionizing dose effects.[1] The silicon dioxide in this region

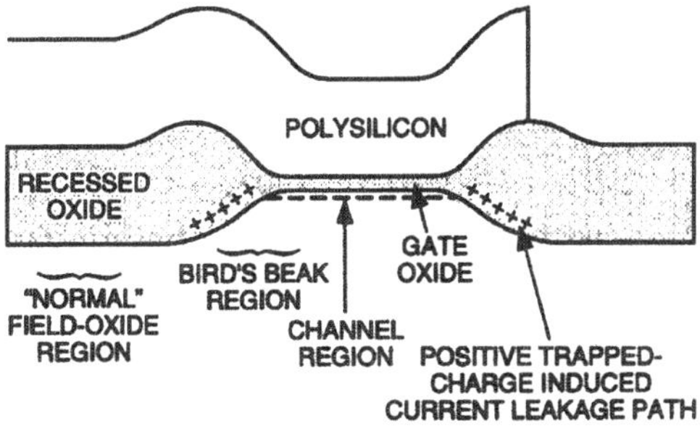

Figure 1. MOS transistor cross section (McLean and Oldham, 1987).[1] The current leakage path is perpendicular to the plane of the page.

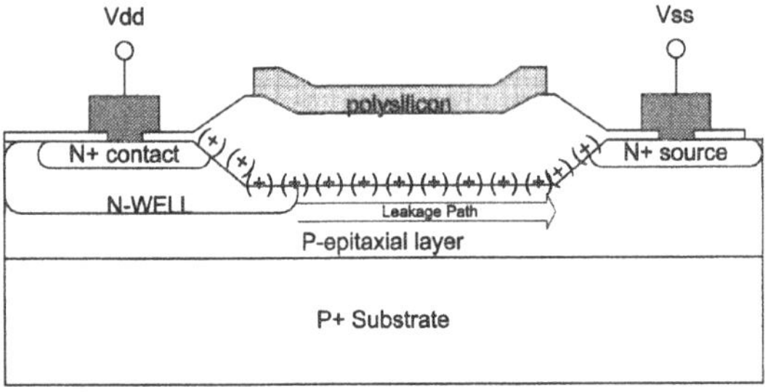

Figure 2. N-well to n[+] source leakage path (Mavis and Alexander, 1997).[2] The positive charges induced in the field oxide by total ionizing radiation are indicated by (+)-symbols.

(known as the bird's beak) is under mechanical stress produced by the dynamics of the oxide growth process and the transition from thin to thick oxide. In this transition region, the oxide is of variable thickness and experiences a relatively high electric field from the combination of polysilicon gate bias and the fringing fields from the source-to-drain bias. When this region is exposed to ionizing radiation, significant hole trapping occurs so that edge leakage currents will appear at the device terminals. At increasing doses, more of the edge parasitic transistor will become involved in the conduction path, and the leakage will quickly rise to become roughly equivalent to the on-state current of the intrinsic transistor. Edge leakage effects are typically manifested as a rapid increase in supply current with increasing total dose, as the parasitic edge transistors shunt the source-to-drain current around the intrinsic transistor. They may also produce functional failure as the composite N-channel becomes permanently "on".

Another total-ionizing-dose-induced leakage mechanism is associated with the field oxide (FOX) that provides isolation between adjacent transistors. An illustration of the structure of the parasitic FOX transistor is shown in Figure 2.[2] The leakage paths are formed when positive charge is trapped in the field oxide and inverts the surface of the P-type material.[3] Two types of leakage paths are possible. The first is between the n-well (typically connected to the positive power supply, V_{DD}), and an n^+ source (typically connected to ground or the negative supply, V_{SS}). The leakage path connects V_{DD} to V_{SS}, producing a change in supply current with increasing radiation. Second, field oxide leakage paths can also span the n^+ source/drain regions between adjacent N-channel transistors. This will also increase the V_{DD} to V_{SS} leakage current.[2]

Typical designs of integrated circuits (ICs) hardened against total-dose radiation use closed-geometry NMOS transistors and guard rings to eliminate the radiation-induced leakage currents.[2,4-10] While the guard rings also aid in hardening ICs against heavy-ion-induced single-event latchup, other circuit techniques, outside the scope of this paper, are sometimes employed to further harden designs against single-event effects.[12,13] These radiation hardening layout and design techniques introduce design penalties in area and speed.[6] This paper compares the total-dose tolerance of closed-geometry transistors and illustrates the speed and area tradeoffs encountered in advanced CMOS processes. The design of an IEEE 1394 cable physical layer mixed-signal interface chip using closed-geometry transistors is described and further illuminates some of the performance tradeoffs encountered in hardening by design.

2. Closed Geometry Transistors

There are two basic layout configurations for closed-geometry transistors. The first design is called variously annular, re-entrant, or edgeless. Sometimes, it is referred to by the shape of the gate, as a circular, square, rectangular, or octagonal gate transistor.[8,9] A typical structure is shown in Figure 3 and illustrates how the gate polysilicon layer forms a ring around the transistor drain node. Conventional "two-edged" NMOS transistors are susceptible to radiation-induced leakage currents along the active diffusion edges between the source and drain.[1,11] This design has no active diffusion edges overlapped by polysilicon that separates the source from the drain (as implied in Figure 1), and it is therefore free of such leakage currents. The selection of source or drain as the internal, enclosed node is left as a design choice. The usual choice is to place the drain on the inside, since this minimizes the drain area and thus the drain-to-substrate capacitance. However, the electric field density has a 1/r-like dependence.[8,9] Therefore, the selection of drain as the internal node can introduce excessively high fields and be a potential reliability problem.[7] For this reason, when the process design rules allow, 90-degree inside corners are usually avoided, and these annular transistors will have the corners beveled at 45 degrees, as shown. This layout also has limitations in terms of minimum effective width W (limited by contact size and space requirements), maximum channel length L (L and W are not independent, due to the radial nature of conduction), and current handling capability of the enclosed contacts (adding more contacts increases W which increases drain current which may in turn require more contacts). Note too, this device is characterized by an effective device width defined by a geometric combination of inner and outer perimeters (effective radii).[8,9]

The second type of closed geometry transistor may be called ringed-source and is illustrated in Figure 4. In this case, rather than being radial, the channel is more nearly transverse as in a standard two-edged transistor. Here the polysilicon overlaps the active diffusion around the source to eliminate the radiation-induced source-to-drain leakage path (the n^+ source diffusion is now not adjacent to the field oxide edge). Now W can be made smaller than in the annular case, there is no limit for L, and contacts can be added as needed to carry the current without increasing W (by extending the source away from the channel). However, there is now a large parasitic MOS gate-to-source capacitance due to the excess polysilicon that encloses the source. This parasitic gate still inverts the surface, but it does not contribute significantly to drain current. This excess polysilicon can also have significant resistance associated with it, especially in large devices. Again, the designer has the option of enclosing either the drain or the source, and sometimes for compactness, stacked transistors may alternate between enclosed drains and enclosed sources. In this case, when the drain is enclosed, the parasitic gate-to-drain capacitance of the excess polysilicon (absent in the annular transistor) appears as a Miller capacitance when referred to the input. The combination of Miller capacitance (excess parasitic capacitance multiplied by the gain of the transistor) and the excess polysilicon resistance in large transistors can introduce a significant RC delay into a circuit. For this reason, the usual choice is to enclose the source. In addition, these transistors are best used as small width (W) devices.

A feature common to both designs is the ring of p^+ diffusion that encloses each transistor. This ring has an important dual role, as a channel stop and a guard ring. As a channel stop, it isolates adjacent NMOS transistors, interrupting the possible inversion layer that can be induced by the total-dose in p^- regions under the field or shallow trench oxides (Figure 2). It also eliminates the n-well-to-n^+-source leakage path of Figure 2. As

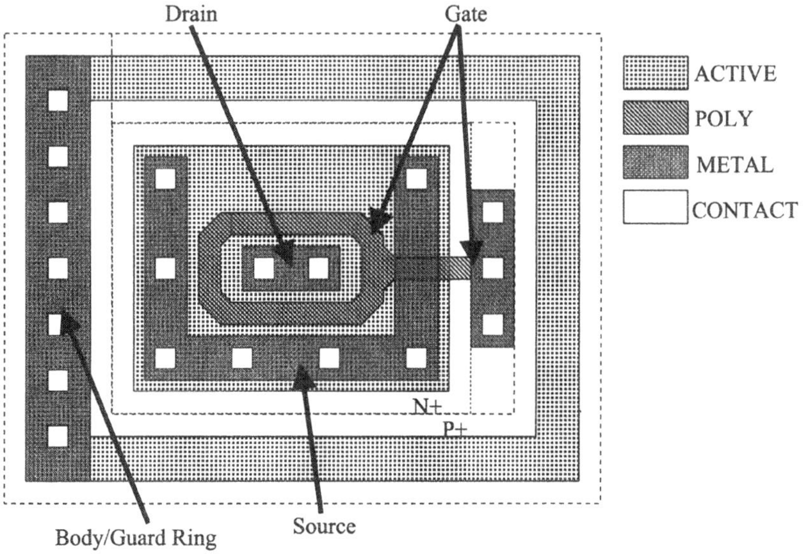

Figure 3. Annular Closed-Geometry NMOS Transistor Layout.

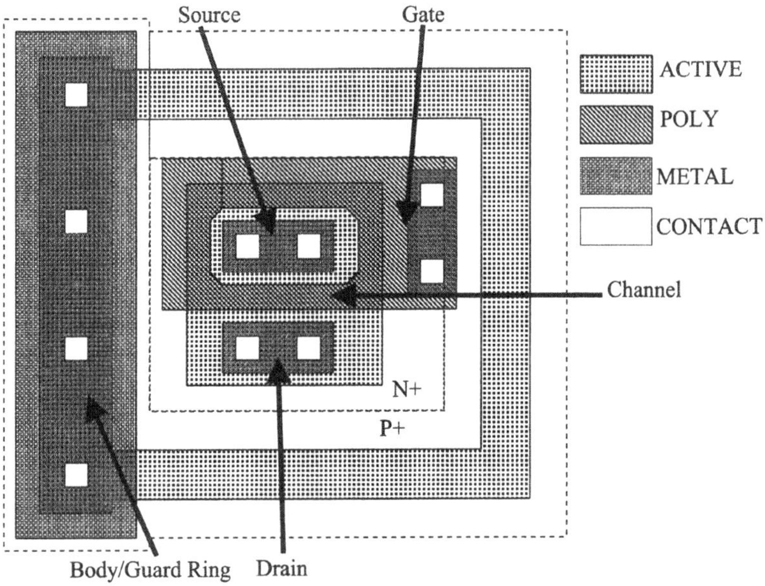

Figure 4. Ringed-Source NMOS Transistor Layout.

a guard ring, it maintains the well or substrate bias (depending on process) through the low resistance p^+ diffusion. This prevents latch-up of the four-layer structures that exist between adjacent NMOS and PMOS transistors. Design hardening approaches may also enclose PMOS transistors in n^+ guard rings to maintain uniform bias in the n^- regions and further protect against latch-up.

We have designed, fabricated, and tested sets of annular and ringed-source transistors in several different modern commercial foundry processes ranging from 0.6μm to 0.18μm feature sizes. These transistors have been tested for their total-dose radiation tolerance, and in all processes, we have observed hardness levels significantly above conventional two-edged NMOS transistors. Examples of the total-dose hardness of these transistors are illustrated in Figure 5 for the Taiwan Semiconductor Manufacturing Company (TSMC) 0.18μm process. The devices, fabricated in early 2002, were irradiated with 50keV x-rays at a dose rate of 165rad(SiO_2)/s. The drain current as a function of gate voltage (drain voltage constant at 0.1V) is shown for pre-irradiation and post-irradiation measurements of (a) a conventional two-edged NMOS (W/L=1.2/0.2), (b) an annular transistor (W/L≈3.8/0.2), and (c) a ringed-source transistor (W/L≈1.2/0.2). The widths of the closed-geometry transistors are approximated as the average of the inside and outside edge perimeters of the channels and represent the minimum achievable transistor dimensions in the process technology. Notice the minimum annular transistor has a W/L nearly three times larger than the minimum ringed-source transistor. This is confirmed by observing the saturation drain current in Figure 5(b) is nearly three times larger than in Figure 5(c). Also, the approximation for W of the ringed-source transistor is reasonably accurate since the pre-irradiation saturation drain current is nearly the same as that for the two-edged transistor. This also validates the assumption that the excess (ring) polysilicon does not contribute to drain current.

The effects of total-dose are clearly seen in the two-edged transistor (Figure 5(a)) where significant leakage current is observed at 50krad(SiO_2) and the device is completely shorted out at 400krad(SiO_2). The closed-geometry transistors show no increased leakage current even after 400krad(SiO_2). Other authors have reported total-dose results of two-edged transistors in the commercial TSMC 0.18μm process showing higher levels of hardness.[14] These other data are taken from lots run at different times in the process history of a commercial process that is not controlled for its total-ionizing-dose hardness levels. This illustrates an additional advantage of design hardening. Radiation hardening by design depends only weakly on the natural or intrinsic process hardness, which may vary significantly with process changes or random variations.

The use of these transistors in the design of a digital library design element, such as a two-input NOR gate, is illustrated in Figure 6. Four gate designs are shown using (from left to right) annular transistors, ringed-source transistors, and conventional two-edged transistors for both a 0.18μm and a 0.25μm process. The annular-transistor gate design conservatively uses a guard ring around the PMOS, but this is eliminated in the ring-source design to achieve more aggressive design scaling. In addition, the annular design does not employ 45-degree inside gate corners so that it can be easily used in processes where such poly beveling is forbidden by design rule. While these 0.18μm Hardened-By-Design (HBD) gates are 25-50% larger in area than the standard (two-edged) cell in the same technology, they are smaller in area than the standard cell in the 0.25μm technology. Commercial designs in a given process can therefore be realized as design-hardened versions in about the same silicon area using the next-generation process technology.

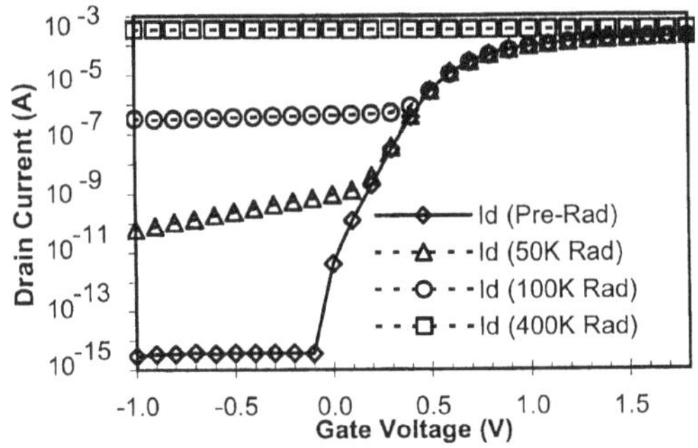

(a)

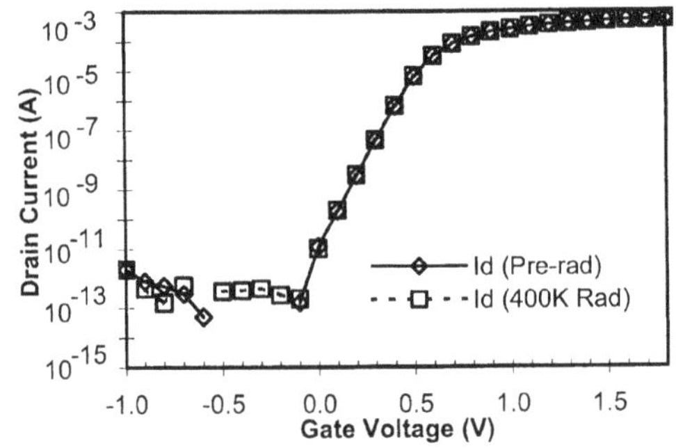

(b)

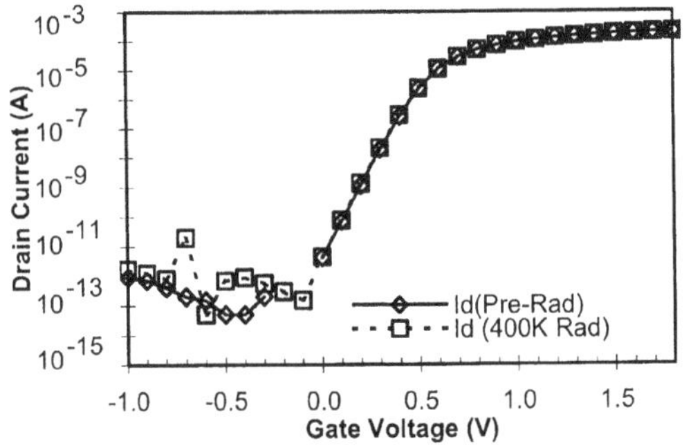

(c)

Figure 5. Total-dose response of a conventional, two-edged transistor (a), an annular transistor (b), and a ringed-source transistor (c). Bias conditions during irradiation: VG=VDD=1.8V; VS=VD=VB=VSS=GND. 50keV x-rays at 165rad(SiO_2)/s. Doses are given in krad(SiO_2).

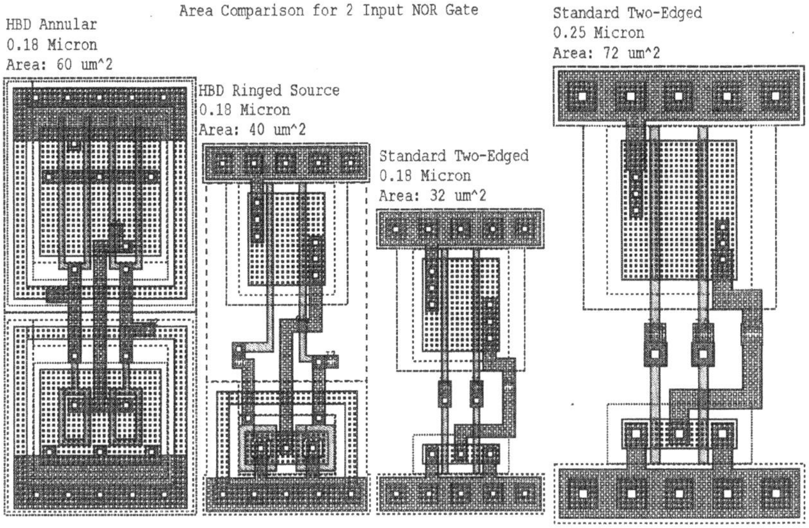

Figure 6. HBD Area Comparison for a Two-Input NOR Gate.

3. Circuit Application—Functional Description

We have recently used these transistors together with Dual Interlocked storage CEll (DICE) latches[12] to design a total-dose tolerant and SEU (Single Event Upset) immune version of an IEEE 1394 Cable PHysical Layer (CPHY) interface chip. The design was actually the mixed-signal portion, designated APHY (for the Analog functions of the PHysical Layer). We used the annular transistors in the high-speed digital portions of the chip (latches, registers, encode and decode logic). For the digital block, area was not critical, but speed was (the ringed-source transistors switch slower than the annular transistors due to their added capacitance). The analog portions of the design were implemented using the ringed-source transistors to minimize area. Our design was a three-port implementation operating at 100MB/s. This was a 5V part designed to withstand total ionizing dose radiation in excess of 300krad(SiO_2). It was fabricated through MOSIS in the AMI Semiconductor (AMIS) C5N 0.5μm commercial process. The design was achieved in a 3.6mm×4.9mm area with 96 I/O pads. A die photograph of the chip is shown in Figure 7.

A functional block diagram of the APHY chip is shown in Figure 8. The Analog Block interfaces directly with the channels (TPA/TPB) on the physical cable and uses comparators and transceivers to sense and transmit the encoded data stream. These analog circuits are differential drivers for sending data on the cable, differential receivers for receiving data on the cable, a sub-circuit for generating cable bias current, and comparators for arbitration signals and cable connect detect. The Phase Locked Loop (PLL) uses a 24.576MHz crystal input to generate the internal 98.304MHz clock. Digital

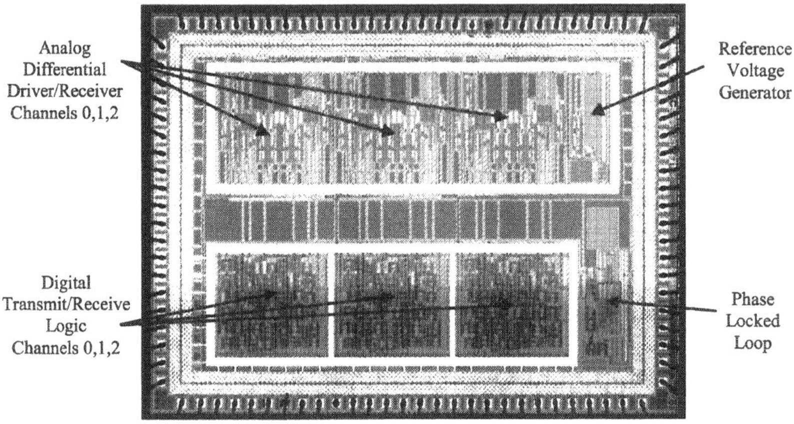

Figure 7. 1394 APHY Die Photograph.

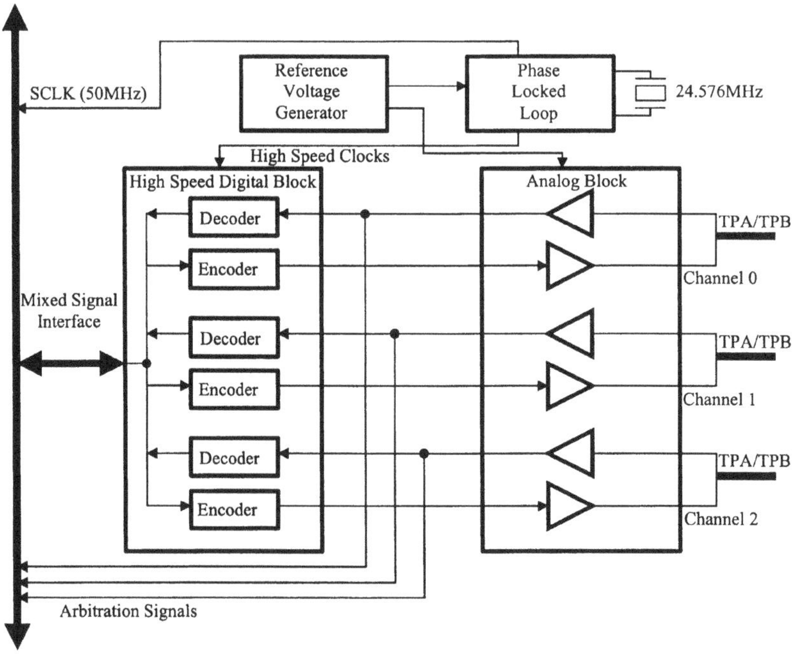

Figure 8. IEEE 1394 Analog PHYsical layer (APHY) Functional Block Diagram.

logic is then used to create the internal 49.152MHz clock and a SCLK output for the DPHY (the Digital portion of the CPHY). The reference voltage generator feeds the reference voltage to the PLL and the differential drivers and cable bias generator in the analog block. The High-Speed Digital Block decodes the received data or encodes the data to be transmitted. Registers latch the data and store it for communication to the Mixed Signal Interface. The Encoder encodes the 2-bit digital data stream from the DPHY to create the digital inputs for the analog block. Next, the Decoder sub-circuit receives data from the analog block, decodes it and synchronizes the data to the internal 49.152MHz clock (3 instances, one for each of the three ports).

4. Performance of HBD APHY Circuit

The key performance challenge in the design of the APHY circuit was in the phase-locked loop. Achieving the target voltage-controlled oscillator (VCO) maximum frequency was strongly dependent on the type of transistor used and which node was enclosed. Figure 9 shows the circuit schematic for the VCO. The NMOS stacks of MN5-MN7 and MN6-MN8 were initially realized using ringed-drain transistors for layout convenience (the drain was enclosed by the polysilicon, rather than the source). These transistors were drawn large (W=75μm for MN7 and MN8 and W=150μm for MN5 and MN6) to reduce single-event-upset effects in the VCO by increasing the node capacitance (this increases the critical charge required in an ion strike to upset or change the node voltage state).[4,13] Figure 10 shows simulation results for three different cases of VCO performance. The top trace shows the result when standard device models are used for the ringed-source (ringed-drain) transistors in the VCO. The maximum oscillator frequency was 177MHz in this case. However, when the parasitic components (R and C) of the excess polysilicon are included as suggested by the model of Figure 11, the maximum VCO frequency drops to 90MHz. Measurements of the VCO with this design also showed a 90MHz maximum oscillator frequency. This was observed by noting the maximum frequency at which the PLL would maintain lock. The model of Figure 11 includes the polysilicon resistance in series with the excess gate capacitance, modeled as the voltage dependent capacitance of a MOS capacitor between the gate and the source. The parameters for these parasitic elements are computed on the basis of the active area under the polysilicon, thus the parasitic elements will have values differing from the elements of the active transistor. Now, when the parasitic MOS capacitor appears in the drain as in MN5, MN6, MN7 or MN8, for example, the RC delay of these parasitic components is significant. Redesigning the VCO circuit to replace the ringed-drain transistors with annular transistors (without the parasitic Miller capacitance) restored the maximum oscillator frequency to 162MHz (even with the RC parasitic components included in the models of all other ringed-source transistors). This is shown in the bottom trace of Figure 10. These results illustrate the potential speed tradeoffs that may be encountered in total-dose hardening a design and suggest the feasibility of attaining both the speed and radiation tolerance requirements in an advanced IC.

5. Conclusions

We have demonstrated the feasibility of realizing advanced, high-speed digital and analog functions in a monolithic integrated circuit using radiation-hardened-by-design techniques. The area penalties were not prohibitive and speed penalties were readily overcome by an appropriate application of closed-geometry transistors. It was seen that

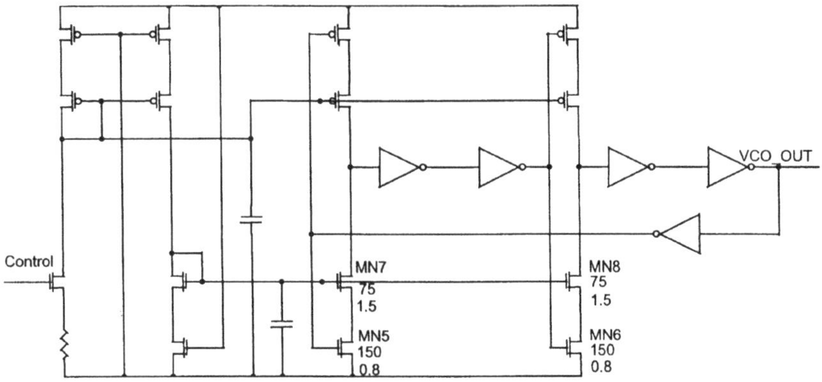

Figure 9. Schematic of VCO.

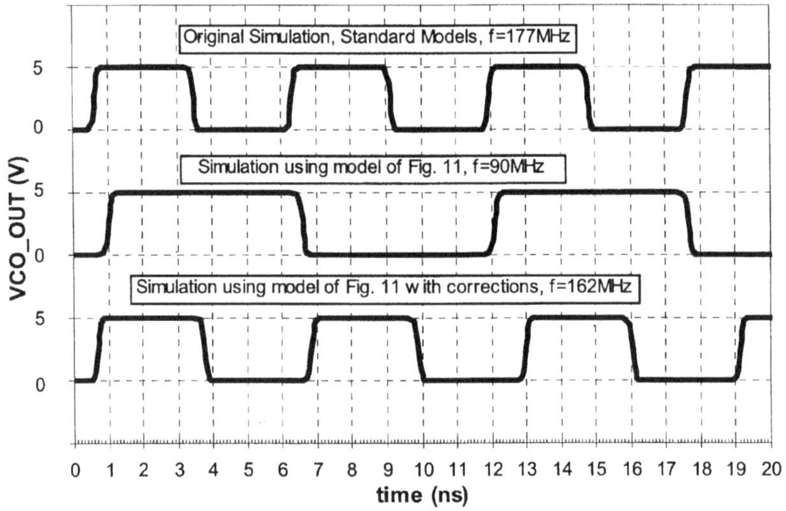

Figure 10. Simulated VCO Output Voltage showing impact of hardened by design techniques.

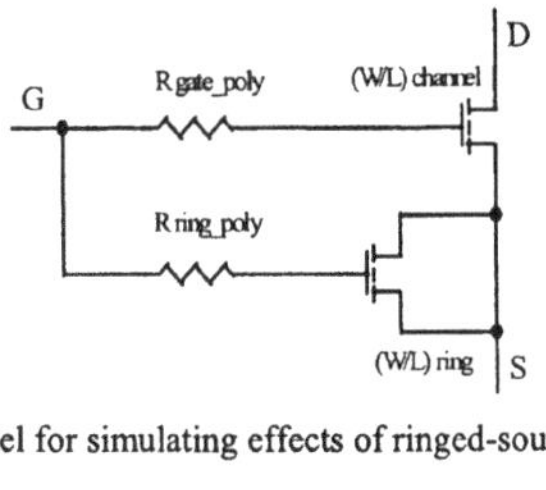

Figure 11. Possible model for simulating effects of ringed-source parasitic components.

parasitic polysilicon resistance could play a significant role in degrading circuit performance, especially if coupled with the Miller-effect multiplication of the parasitic gate-to-drain capacitance.

Acknowledgments

This work was supported through a subcontract with Honeywell under Navy Contract N00173-99-C-2025 and by the Defense Threat Reduction Agency through contract N00164-97-D-0013/0043. Special thanks go to Ken Wolfram for his technical guidance and Jake Tausch for his design and test support. The authors also thank Dave Mavis for his original development of the ringed-source transistor concept under Air Force contract F29601-96-D-0043.

References

1. F.B. McLean and T.R. Oldham, "Basic Mechanisms of Radiation Effects in Electronic Materials and Devices," Harry Diamond Laboratories, Technical Report HDL-TR-2129, 1987.

2. D.G. Mavis, D.R. Alexander, "Employing Radiation Hardness by Design Techniques with Commercial Integrated Circuit Processes", Digital Avionics Systems Conference, 1997, pp. 2.1-15-2.1-22.

3. H.E. Boesch, Jr. and T.L. Taylor, "Charge and Interface State Generation in Field Oxides," IEEE Trans. Nucl. Sci., Dec 1984, p. 1273.

4. D.R. Alexander, D.G. Mavis, C.P. Brothers, J.R. Chavez, "Design Issues for Radiation Tolerant Microcircuits in Space", 1996 NSREC Short Course, Indian Wells, Ca.

5. D.R. Alexander, D.G. Mavis, R.M. Turfler, J.R. Bailey, M. Williams, "A Scaleable HBD Cell Library for Radiation Tolerant ASICs for Space Applications", GOMAC 2001.

6. R.C. Lacoe, J.V. Osborn, R. Koga, S. Brown, D.C. Mayer, "Application of Hardness-By-Design Methodology to Radiation Tolerant ASIC Technologies", IEEE Trans. Nucl. Sci., Dec. 2000, pp. 2334-2341.

7. R.C. Lacoe, D. Mayer, E. King, "Reliable and Hardened-by-Design (RHBD) Components for Space", presented at 2002 Design Hardening Workshop, Albuquerque, NM.

8. G. Anelli, M. Campbell, M. Delmastro, F. Faccio, S. Florian, A. Giraldo, E. Heijne, P. Jarron, K. Kloukinas, A. Marchioro, P. Moreira, W. Snoeys, "Radiation Tolerant VLSI Circuits in Standard Deep Submicron CMOS Technologies for the LHC Experiments: Practical Design Aspects", IEEE Trans. Nucl. Sci., Dec. 1999, pp. 1690-1696.

9. A. Giraldo, "Evaluation of Deep Submicron Technologies with Radiation Tolerant Layout for Electronics in LHC Environments", Ph.D. Thesis at the University of Padova, Italy, December 1998.

10. P.V. Dressendorfer, "Radiation-Hardening Technology", in T.P. Ma, P.V. Dressendorfer, Ionizing Radiation Effects in MOS Devices and Circuits, John Wiley & Sons: 1989, pp. 383-387.

11. J.R. Adams, F.N. Coppage, "Field Oxide Inversion Effects in Irradiated CMOS Devices", IEEE Trans. Nucl. Sci., Dec. 1976, pp. 1604-1609.

12. T Calin, M. Nicolaidis, R. Velazco, "Upset Hardened Memory Design for Submicron CMOS Technology", IEEE Trans. Nucl. Sci., Dec. 1996, pp. 2874-2878.

13. S.E. Kerns, "Transient-Ionization and Single-Event Phenomena", in T.P. Ma, P.V. Dressendorfer, Ionizing Radiation Effects in MOS Devices and Circuits, John Wiley & Sons: 1989, pp. 508-519.

14. R. Lacoe, D. Mayer, "The Effects of Total Ionizing Dose Irradiation on CMOS Technology and the Use of Design Techniques to Mitigate Total Dose Effects", presented at Reliability Physics Symposium, 2002. Available at http://www.irps.org/02-40th/Special_Topic/st_1.pdf.

International Journal of High Speed Electronics and Systems
Vol. 14, No. 2 (2004) 379–399

RADIATION ISSUES IN THE NEW GENERATION OF HIGH ENERGY PHYSICS EXPERIMENTS

F.FACCIO
CERN, 1211 Geneva 23, Switzerland
Federico.Faccio@cern.ch

Abstract—With the construction of the Large Hadron Collider at the European Center for Nuclear Research (CERN), the radiation levels at large High Energy Physics (HEP) experiments are significantly increased with respect to past experience. The approach the HEP community is using to ensure radiation tolerance of the electronics installed in these new generation experiments is described. Particular attention is devoted to developments that led to original work: the estimate of the SEU rate in the complex LHC radiation environment and the use of hardness by design techniques to achieve radiation hardness of ASICs in a commercial CMOS technology.

Keywords—radiation effects, enclosed layout transistors, radiation tolerance, SEU estimate

1. Introduction

The increasing use of embedded electronics components together with the increase in particle energy and beam intensity render the new generation of High Energy Physics (HEP) experiments considerably more vulnerable to radiation effects. In this respect, as in many others related to size and complexity, the most demanding example in the HEP community is the Large Hadron Collider (LHC) with its giant experiments. The LHC is the new particle collider that is being constructed at the European Center for Nuclear Research (CERN) located in Geneva, Switzerland. Scheduled to become operational in 2007 this circular machine with a radius of 4.3 km will accelerate bunches of protons up to 7 TeV and collide them head-on in the center of large detector systems built with different technologies to measure the tracks and energies of the various collision products.

Particle collisions at such energy are expected to produce momentarily conditions that allow physicists to progress in the understanding of the deepest structure of matter, but this happens with such a low probability that a very large number of collisions is required to observe a few such interesting events. The vast majority of such collisions will not be interesting to study, but will anyway produce particles that will travel through the detectors surrounding the collision point. The job of the detectors, with some intelligence provided by a large amount of integrated electronics, is to enable the physicists to selectively observe and study only those very rare collisions enlarging man's understanding of the universe. The other collisions will merely produce a radiation background posing a hazard for the reliability of the experiments.

Compared with previous HEP experiments, the novelty resides both in the intensity of the background, due to the very high collision rate, and in the massive use of electronic components inside the detector system, where they are exposed directly to the radiation background. This on-detector electronics is necessary to select potentially interesting collision events and only transfer meaningful detector data (eliminating most of the background) for analysis off-detector. In practice, the devices sensitive to radiation effects are larger in number and types, and the radiation background is considerably more intense, which results in unprecedented radiation concerns for the HEP community.

This paper aims at summarizing these radiation issues. Focus will be on the two largest LHC experiments, but similar problems (though on a smaller scale) have been encountered in other HEP experiments around the world. In particular, original or unusual developments will be highlighted in which either the requirements or the solutions chosen differ from those typical in the space and nuclear radiation effects communities.

2. The Radiation Environment

2.1. The LHC experiments

Four HEP experiments will be positioned around the LHC ring, corresponding to four particle collision points. Each experiment is actually a large detector system made up of several sub-detector systems. The most challenging in terms of radiation effects are the two general-purpose systems called ATLAS (A Toroidal LHC ApparatuS) and CMS (Compact Muon Solenoid), designed to study p-p collision events. Bunches of protons will circulate in opposite directions in the LHC ring and will be focused at the center of these two large detectors where they will cross at a frequency of 40 MHz. An average of 20 p-p collisions will occur at every bunch crossing. Each of the detectors is built to cover all the solid angle around the collision point, and designed in layers surrounding it, each layer having a different detection function. This is shown in Figure 1 for the CMS experiment. The cylindrical detector layers, called the "barrel" portion of the experiment, surround the collision point. The barrel is closed by detector end-caps in the forward region (at low angle with respect to the beam pipe).

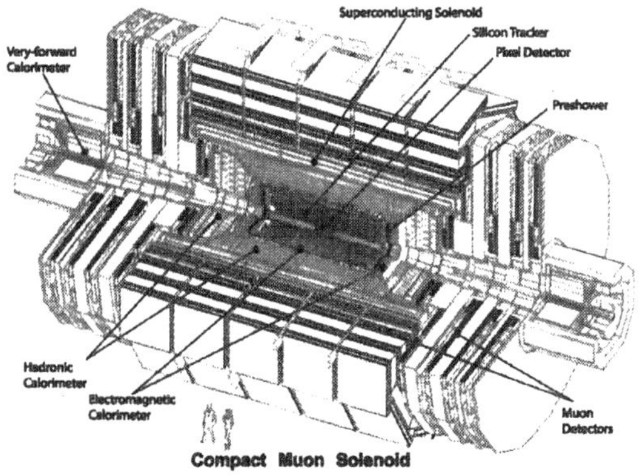

Figure 1: View of the CMS experiment

The innermost sub-detector, called the "tracker", is used to reconstruct the trajectory of charged particles coming from the interaction point. Particle tracks are naturally denser close to the collision point, therefore requiring detectors able to spatially resolve the tracks with very high precision. This task is performed by pixel detectors, where each sensitive element might be as small as 500x50 μm and the readout electronics is bump-bonded to the sensor with the same small pitch. This close to the collision point, the radiation environment is at its most intense, and radiation hardness of both detectors and electronics is indispensable. To limit the radiation-induced damage to the silicon detectors, all the equipment must be kept constantly at low temperature (around -10°C), imposing strict constraints on the power budget of the electronics.[1] Because of the electronic functions required, the limited power and space budget and the radiation levels, all electronics installed in this sub-detector is custom-developed (Application Specific Integrated Circuits, ASICs).

The "Electromagnetic CALorimeter" (ECAL) is the sub-detector that particles meet after the tracker in their flight away from the collision point. Its purpose is to measure the energy of electrons and photons with high precision. Space constraints are more relaxed than in the tracker, and also the radiation background is less severe. The "Hadronic CALorimeter" (HCAL) is the following layer, and its aim is to measure the energy of hadrons (such as protons, neutron and pions) produced by the p-p collisions. Finally, the "Muon Detector" measures the trajectory and energy of the muons, particles interacting so weakly that they can cross all the other detector layers. The whole experiment is immersed in an intense magnetic field (2 to 4 T) produced by a complex and massive system of magnets.

The experiments are located in large underground caverns (all the LHC machine sits at a depth varying from 70 to 150 m), and a very large fraction of the control, power and data acquisition electronics is installed inside the cavern and exposed to a radiation field.

2.2. *The radiation environment*

The radiation environment of ATLAS and CMS is determined by the p-p collisions at their center. The main difference between the two detectors is the presence of the massive return yoke for the magnetic field that in CMS is embedded in the muon detector system and effectively shields the muon detectors electronics and the experimental cavern from particles produced in the calorimeters. As a consequence, the CMS radiation levels in these regions are about an order of magnitude lower.

The inner tracker is exposed to a primary particle flux from the interaction region – mainly composed of pions – and to neutron albedo emerging from the calorimeters. Already in the outer tracker, the charged hadron flux is significantly reduced and neutrons emitted from the surrounding ECAL dominate the particle environment. Neutrons, with practically isotropic distribution, actually dominate the environment also of all other regions of the experiment.

The charged hadron energy distribution covers the range between about 1 MeV and 10 GeV. The spectrum is peaked at around 200-300 MeV and remains almost invariant in shape from the tracker to the outer periphery of the experimental cavern, the absolute fluxes changing meanwhile by 6 orders of magnitude. The neutron distribution has instead a substantial flat low-energy component typically only 1 order of magnitude lower than the peak at 1 MeV. This contributes significantly to displacement damage effects. The neutron spectrum has another peak at around 60-100 MeV, and extends at high energy up to about 1-10 GeV depending on the location.

Table 1 summarizes the radiation levels foreseen in different regions of ATLAS as cumulative doses and fluences over the 10 years expected lifetime of the LHC. These numbers have been obtained with sophisticated simulation tools developed specifically for this purpose. In each region, the field is characterized by three parameters: TID level, fluence of neutrons above 100 keV and fluence of all hadrons above 20 MeV. They are reference levels for ionization, displacement damage and Single Event Effects respectively. All numbers in the table are expressed as maximum levels in the corresponding region of the experiment.

2.3. *Main radiation issues*

Due to the absence of heavy ions in the experiments' radiation environment, some issues typical of space applications only apply marginally to the two main LHC experiments. For instance, Single Event Latchup (SEL) or Single Event Gate Rupture (SEGR) are observed with very low frequency – if at all – in proton irradiation tests. Therefore, these effects have very low probability in the hadron-dominated LHC particle environment. On the other hand, the abundance of neutrons with energy higher than 100 keV renders displacement damage a much more relevant concern than for space applications.

To illustrate the radiation issues and how to address them, it is appropriate to distinguish two regions of the experiments. In the inner region, namely the tracker and some parts of the ECAL (the forward end-cap), the TID levels are so high that it is unthinkable to use commercial electronics. In this region, all electronics is implemented by custom-developed ASICs, manufactured in technologies and with techniques that enable them to stand tens of Mrad of total dose. In this case, all radiation concerns are tackled at the development stage: appropriate technology or design techniques are chosen to achieve TID and displacement damage hardness, to prevent destructive SEEs and to minimize the rate and/or impact of soft errors such as Single Event Upset, Transient or Functional Interrupt (SEU, SET or SEFI). This approach will be illustrated in more detail in section 4.

In the outer region of the experiment and in the experimental cavern most of the electronics functions are implemented with commercial devices, which have no guarantee on radiation effects. Overall, the TID level in this outer region varies between less than 1 krad (most often, in particular for CMS) to more than 25 krad. If in the latter case both cumulative and SE effects have to be carefully addressed, the major concern for the former (about 1 krad) environment is certainly represented by SEEs in general, and in particular by the Burnout (SEB) of power devices. Not only is this a destructive event, hence with permanent effects, but also it affects devices often distributing power to relatively large systems, consequently their failure has a large impact. SEB has been observed frequently in proton and neutron tests [2, 3], and will certainly threaten power devices in the LHC experiments. The problems encountered in employing Commercial Off The Shelf (COTS) electronics in a radiation environment are well known in the space community. The approach suggested in the HEP community to deal with these problems [4] is not substantially different from what is known already, therefore it will not be addressed by this paper. Instead, the next section will describe the estimation of the SEU rate in the LHC hadron-dominated particle environment, a problem that is common to all the regions of the experiments and that has led to the development of an original computational method.

Table 1: Foreseen radiation levels in the ATLAS experiment, integrated over the scheduled LHC lifetime of 10 years.

Detector region	*TID [rad]*	*Neutrons 1 MeV equiv. [cm^{-2}]*	*Ch. hadrons E>21 MeV [cm^{-2}]*
Innermost tracker (barrel)	$2.7\ 10^7$	$4.7\ 10^{14}$	$5.4\ 10^{14}$
Outer tracker (barrel)	$5.1\ 10^6$	$1.9\ 10^{14}$	$9.3\ 10^{13}$
ECAL (barrel)	$1.1\ 10^4$	$3.1\ 10^{12}$	$6.0\ 10^{11}$
HCAL (barrel)	$1.0\ 10^3$	$4.3\ 10^{11}$	$1.0\ 10^{11}$
Muon detector (forward)	$3.4\ 10^4$	$1.2\ 10^{13}$	$1.9\ 10^{12}$
Experimental cavern	$1.7\ 10^3$	$6.5\ 10^{10}$	$1.5\ 10^{10}$

3. SEU Rate Estimate in the LHC

In the early days of the LHC electronics developments, the HEP community devoted little attention to SEEs, in particular to non-destructive events such as SEUs. This was both for historical reasons (lack of experience in the community) and for the absence of heavy ions in the LHC environment. It was generally believed that the SEU concerns in the space community were largely due to the presence of heavy ions in space. Moreover, the LHC radiation environment was at the time typically only described in terms of TID and 1-MeV neutron equivalent fluence. Detailed information on the energy spectra of particles, a determinant parameter in evaluating the SEE risks, was not generally available. To change this situation, it was necessary to develop a computational method to estimate the SEU rates in the very special LHC radiation environment, and thus draw attention to the non-negligible error rates as well as providing the HEP community with a tool to extrapolate test results available in the literature that normally refer to mono-energetic proton or heavy ion beams. This method has been developed by Huhtinen and Faccio [5], and the main points will be summarized.

In the LHC environment, an energy deposition sufficient to induce SEU is only possible through the interaction of the hadrons with nuclei in the electronics circuit, the hadrons themselves not having enough Linear Energy Transfer (LET, dE/dx) to directly deposit the required charge. Such interaction typically produces a shower of secondary particles and one nuclear recoil, but in some cases there can be several fragments. The recoils normally have low energy, therefore a limited range – implying that they have to be produced locally to be effective in inducing SEU. The difficulty in estimating the error rate resides in the necessity of computing the probability of nuclear interaction, covering all reaction channels, and the amount of ionizing energy deposited by all fragments in the Sensitive Volume (SV, the volume where deposited charge can be collected by the sensitive node and contribute to the SEU). For this purpose, a powerful simulation tool was employed, based on the use of the FLUKA simulation package [6] for the explicit generation of nuclear fragments for neutrons with energy above 20 MeV. For lower energy neutrons, comprehensive neutron cross-section data are available and the fragment generation was made using the information contained in the ENDFB-VI cross-section tabulations. The fragments generated either way were then transported through silicon and, whenever one of them entered a predefined SV, a detailed accounting for its energy loss by ionization was done with a modified version of the TRIM code.[7] Although mainly focused on interaction of hadrons with silicon nuclei, the work also checked for possible contributions from interactions in heavier elements (such as tungsten, quite frequently used in CMOS technologies) and from the interactions of thermal neutrons with boron dopants, finding that neither should dominate the error rates in the LHC.

The result of the simulation is a curve giving the probability per unit fluence to have within the SV an ionizing deposition greater than or equal to any given energy value, such as curve "P" in Figure 2. This probability is therefore expressed in cm^2 and refers to the individual SV simulated. The simulation can be run for any environment, from mono-energetic protons hitting the device along one direction, like in typical SEU laboratory tests, to the complicated mixed hadron environment of the LHC.

Each electronic circuit can have a different sensitivity to SEUs, and the error estimate must somehow contain this information to lead to meaningful results. In the developed method, this is achieved with the introduction of the SV size (area and depth) and of the critical energy E_{crit} for upset. An error can occur only if the energy deposited by the fragments in the SV is larger than E_{crit}. The SV has been chosen for simplicity to be a parallelepiped whose vertical dimension (depth) was varied between 0.5 and 2 μm – finally the depth of 0.5 μm was chosen as it most often lead to the best fit with experimental data. The area of the SV and the critical energy can be obtained for each device when heavy ion irradiation data are available. This information must be combined with the probability curve obtained with the simulation for a unitary area SV (1μm^2) to finally compute the error rate for the device in any environment.

A pictorial image of this process is shown in Figure 2. The curve "S", characterizing the sensitivity of the device, is the Weibull fit to the heavy ions experimental points, whilst the curve "P" is the above described probability from simulation for a SV 0.5 μm deep and with a sensitive area of 1 μm^2. The Weibull function typically describes the sensitivity of the device (cross-section, therefore sensitive area) as a function of the LET of the heavy ions. To be able to plot both curves on the same scale (energy deposited in the SV) it is necessary to translate the heavy ion LET into deposited energy, which can be done assuming the LET is constant across the SV. The energy deposited by the ion is therefore in first approximation the product of the LET by the depth of the SV.

The shaded area intersection of curves P and S is proportional to the cross-section of the device in the desired environment, which can be computed by binning the intersection energy range and performing the following summation over all energy bins:

$$\Sigma = \sum_i P_i \frac{s_{i+1} - s_i}{A} \tag{1}$$

where Σ is the desired cross-section, P_i is the probability at the end of the energy bin (not to under-estimate the sensitivity), s_i is the heavy ion cross-section at the beginning of the energy bin, and A is the area of the SV in the simulation (1 μm$_2$). Multiplying the obtained cross-section Σ by the particle flux (cm^{-2}s^{-1}), the result expresses the error rate (per device per unit time) in the radiation environment of interest.

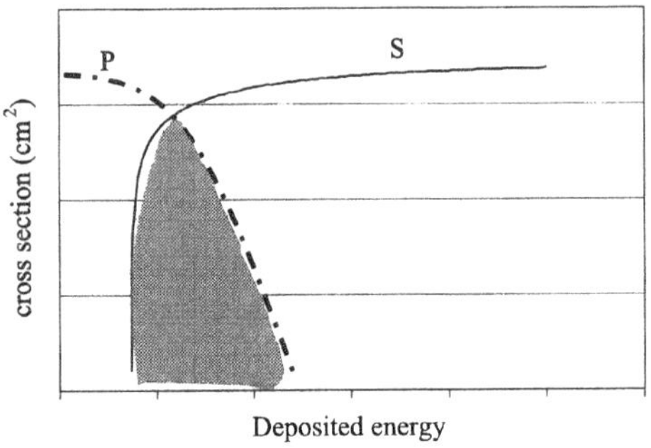

Figure 2: The shaded area is proportional to the cross-section of the device described by the sensitivity curve S (Weibull fit) in the environment described by the probability curve P (from simulation).

The full method has been tested with data available in the literature – mostly published data in the IEEE Transactions on Nuclear Science – for several classes of devices (SRAMs, DRAMs and a microprocessor). For the 18 devices selected, both the heavy ion and the proton cross-section curves as measured in experiments were available. The test consisted in calculating the proton cross-section with the simulation method using the Weibull fit to the heavy ion points as only input, then comparing the result with the available measured proton cross-section. The ratio of the simulation versus experimental results varied between 0.5 and

2.3, a very satisfactory result. The same test was repeated for 4 devices for which also the experimental cross-section for 14 MeV neutrons was available, obtaining in this case a ratio varying between 1 and 4.1. At this low energy of the hadrons, the energy distribution of the produced fragments drops sharply at high energy, and the correctness of the simulation result is much more sensitive to the choice of the sensitive depth.

Having gained confidence in the simulation method, it was possible to compare the SEU cross-section of the same device in different hadron environments: proton, pion and neutron beams from 20 MeV to 10 GeV, isotropic neutron and proton environments from 3 to 200 MeV. The result is summarized in Table 2 for one example device, and several interesting observations can be made:

(i) Neutrons and protons of the same energy produce almost identical upset rates above 20 MeV.

(ii) Simulations for isotropic and monodirectional (beam) irradiation show negligible differences in the upset rate.

(iii) Elastic scattering gives a minor contribution to the SEU rate.

(iv) Above 60 MeV, the upset rate depends little on the hadron energy. This is true for devices with relatively low LET threshold (hence E_{crit}), which are the most sensitive and the large majority. This might be different for devices with LET threshold above about 8-10 $MeVcm^2mg^{-1}$. Figure 3 illustrates how the probability curves for 60 and 200 MeV protons have very similar values for E_{crit} values around 0.2-1.5 MeV (corresponding to 1.7-12.5 $MeVcm^2mg^{-1}$ for a 0.5 μm deep SV). It also appears that the probability curve for the foreseen environment of the CMS tracker is similar to the two proton curves.

(v) For 200 MeV pions we observe twice the upset rate of protons of the same energy. This is due to the higher pion inelastic cross-section for reaction with silicon at energy around 200-300 MeV (delta resonance). Experimental measurements have reported contradictory results in this respect: sometimes identical SEU cross-sections for pions and protons have been measured [8], and sometimes a factor of 5 difference has been reported [9]. The factor of two in our simulation could not therefore be confirmed by experimental results.

Table 2: SEU cross-section, normalized to the 200 MeV proton value (8.4×10^{-7} cm^2), for a sample device in different hadron environments, from reference [5]. The calculation has been made for a 1x1x1 μm SV. The device is the Texas Instruments SMJ44100 4M DRAM, with a LET_{th} of 1.39 $MeVcm^2mg^{-1}$.[10]

Particle	*Direction*	*Total σ*	*Elastic σ*
10 GeV proton	Beam	1.2	0.006
200 MeV proton	Beam	1.0	0.011
200 MeV neutron	Beam	0.97	0.007
200 MeV pion	Beam	2.0	0.012
60 MeV proton	Beam	1.05	0.028
20 MeV proton	Beam	0.45	0.10
20 MeV neutron	Beam	0.46	0.022
200 MeV proton	Isotropic	1.03	0.010
20 MeV neutron	Isotropic	0.43	0.099
14 MeV neutron	Isotropic	0.44	0.041
3.1 MeV neutron	Isotropic	2e-7	---

The above observations can be used to recommend an appropriate qualification test procedure for components to be used in the LHC experiments. The simulation of the full LHC environment has shown that, for all example devices used, the upset rates in all regions of the experiments will be dominated by the contribution of hadrons with energy above 20 MeV. With neutron and protons being equally effective in inducing SEU, and with small energy dependence of the upset rate above 60 MeV, it is clearly recommended to use the widely available mono-energetic protons as preferred test source. Additionally, the monodirectional beams typical of cyclotron proton sources well simulate the almost isotropic environment of the LHC[a]. Therefore, a reasonable estimate of the SEU rate in most locations around the LHC can be obtained by using a 60-200 MeV proton beam to irradiate the candidate devices. Only for inner trackers, where the particle environment will be dominated by pions, this might be too optimistic by a factor of 2-5 at most.

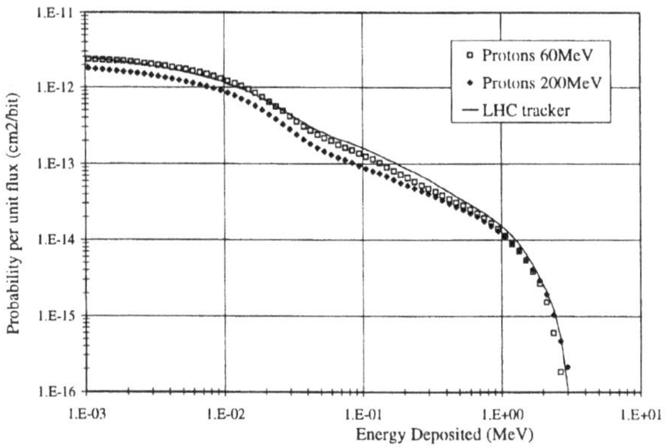

Figure 3: Energy deposition probabilities, for a 1x1x1 μm SV, for protons of 60 and 200 MeV and for the CMS tracker environment. The curves show the probability to have within the SV an ionizing deposition greater or equal to the indicated "Energy Deposition" value.

4. Radiation-Hard ASICs Design

4.1. ASICs design in dedicated rad-hard processes

In the inner region of the LHC experiments, where TID levels are expected to largely exceed 500 krad, the electronics functions have to be performed by radiation-hard ASICs. In the early days of the Research and Development (R&D) for the LHC, the only possible solution seemed to be the use of a dedicated radiation-hard technology. This happened at the beginning of the 90s, when several such technologies were still available both in Europe and in the United States. The R&D efforts concentrated in assessing the use of some of these technologies[11,12,13] for the particular LHC requirements: mixed analog-digital circuits with strict specifications on the noise performance, relatively large volumes (hundreds of wafers) in comparison with the markets these technologies were usually serving. It soon turned out that there were difficulties in following this approach: the required performance in terms of low noise and high radiation

[a] Neutrons, largely dominating the particle environment as from the outer tracker, are originated from interaction of primary hadrons with the detector materials and have therefore practically isotropic flux.

hardness (10 Mrad for the inner tracker) were difficult to be met regularly in the low-volume manufacturing foundries. The resulting low yield further increased the cost per circuit, which was already high for these dedicated radiation-hard processes. The shrinking of the global demand of these technologies as from the first half of the 90s, and the interruption of governmental funds to maintain them, obliged most companies to withdraw access to the processes for ASIC design, or sometimes even to close the manufacturing lines.

Essentially, the only radiation-hard process still in use today for circuit design for the LHC is the BiCMOS DMILL process.[14] This technology was developed in the early 90s and was tailored to satisfy also the LHC requirements. Though several devices planned to be available have been abandoned during the development (CJFETs and PNP bipolar transistors), this SOI technology offers CMOS and NPN bipolar transistors. The process, originally developed at CEA-LETI (Grenoble, France), was industrially transferred to ATMEL (formerly TEMIC) and is used for a number of ASICs amongst which the highest volume demand is for the readout electronics for the ATLAS tracker (about 900 wafers).[15] Although the process is qualified and guaranteed for 10 Mrad and 10^{14} neutrons/cm^2, for some applications with very tight power consumption requirements it was necessary to assess the radiation tolerance of the ASICs for every production lot. Such a time-consuming procedure has been setup and is at present carried out at the wafer level with an X-ray generator (for TID).

4.2. Hardness by design using commercial technologies

Another approach, based on the use of commercial CMOS technologies, has recently become popular in the HEP community and at this time most of the LHC ASICs are designed this way. Research in the 80s demonstrated that radiation effects in laboratory SiO_2 layers decrease quadratically with the oxide thickness, and even more sharply when the thickness is reduced below about 10 nm.[16,17] When scaling down CMOS processes following what is known as Moore's law [18], the gate oxide thickness of the transistors is also decreased. A study of radiation effects in several technology generations [19,20] demonstrated that these commercial-grade gate oxides followed the expected behavior, as shown in Figure 4. It appears that the 5-6 nm gate oxide of the quarter micron technology node that started to become available around 1996-97 is appropriate for multi-Mrad applications.

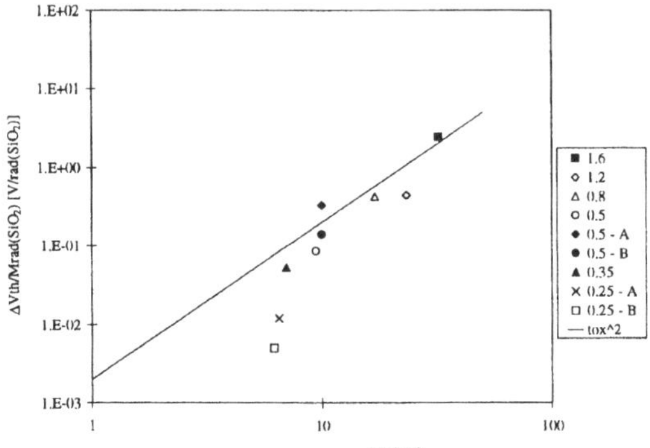

Figure 4: The threshold voltage shift of commercial-grade transistors from several technology generations follows the expected trend, sharply decreasing below the t_{ox}^2 law for a thickness of around 10 nm. Some data are from.[21]

Unfortunately, the lateral and isolation oxide thickness does not need to scale down, and despite the introduction of the Shallow Trench Isolation (STI) process around the 0.25 μm technology node this represents the weak point, in terms of TID tolerance, of modern submicron CMOS.[22] Radiation-induced charge trapping in the thick lateral oxide induces source-drain leakage currents in NMOS transistors.[23] Trapping in the isolation oxide can open leakage current paths between n+ diffusions at different potential (transistor to transistor or Nwell to transistor leakage), as shown in Figure 5. These leakage currents at first increase the current consumption of the still functional circuit, until failure occurs at some point typically in the 10-100 krad TID range.

It is nevertheless possible, with the use of custom layout techniques, to prevent the opening of these leakage current paths.[24] Enlcosed Layout Transistors (ELT, sometimes called "edgeless transistors") are very effective in eliminating any source-drain leakage and were in fact already used in the early days of the CMOS processes to this aim.[25] The absence of any thick oxide between the source and drain diffusion, a consequence of the enclosed layout where the drain (or source) is completely surrounded by the thin gate oxide, is the reason for that. In Figure 6, the comparison between the standard and enclosed layout transistors clearly illustrates the hardness improvement of ELTs. Another layout technique, namely the systematic use of guardrings, serves the purpose of cutting the leakage current paths under the isolation oxide. In this approach, a heavily doped p ring separates all n+ diffusions (Nwell included) at different potential. The charges trapped in the isolation oxide in Figure 5 can invert the surface of the low-doped p substrate, hence opening a conductive channel between adjacent n+ diffusions, but they cannot invert the heavily doped p guardring, which effectively cuts the leakage path.

The combined use of ELTs and guardrings, called "radiation tolerant layout technique", has been known for some time[24], but only with the availability of deep submicron technologies, with naturally radiation tolerant gate oxides, could it enable the development of ASICs capable of surviving in multi-Mrad environments. An R&D effort at CERN [26] has deeply studied this approach and developed the tools necessary for ASICs design in a selected commercial quarter micron technology that is now widely used by the HEP community.

Compared to the use of a dedicated rad-hard technology, the use of a commercial technology brings advantages in terms of low cost, high yield, high speed, low power consumption and high density. The typical commercial foundry has a throughput of a few thousands wafers per week, which is a guarantee for high yield and process reproducibility. Moreover, despite the use of the cumbersome ELTs and guardrings, that considerably decrease the design density, the integration density that can be achieved with this approach is still considerably higher than for dedicated radiation-hard processes, which are typically at least two technology generations late.[27] Overall, the cost difference for the same performance – whenever the same performance can be achieved – is very large.

Since radiation hardness is achieved because of the small gate oxide thickness, which is necessarily very well controlled in these modern processes, the good performance in a radiation environment is expected to be reliably obtained across the full production cycle of any ASIC.

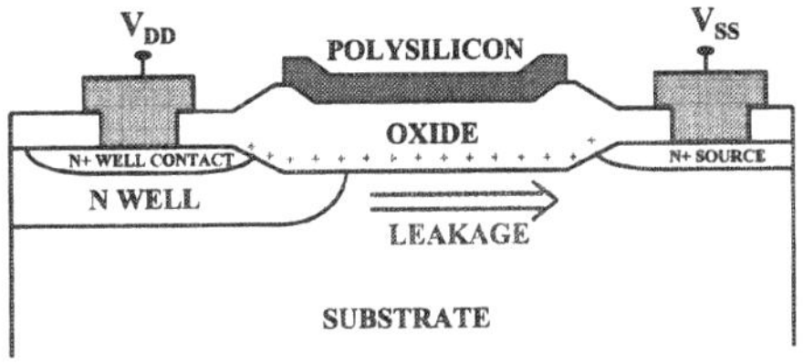

Figure 5: Radiation-induced leakage path under the isolation oxide. In this case, the current flows between an N-well and an n+ diffusion (source or drain of a NMOS transistor). The polysilicon line, in this example used to route a signal as a "level-0" metal, acts as an equivalent gate for the parasitic transistor where the leakage current flows.

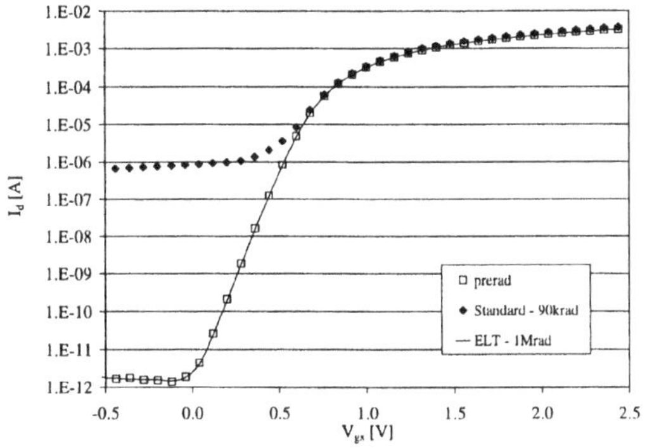

Figure 6: The effectiveness of ELT in eliminating source-drain leakage currents is demonstrated in a 0.25 μm CMOS technology for transistors with W/L = 10/0.25. Drain-source bias is 2.5V, the maximum Vdd in this technology. Devices were biased in the worst-case condition during irradiation.

In the R&D phase, and later in prototype circuits for the LHC, the effectiveness of this approach with the selected 0.25 μm technology has been demonstrated: all of the circuits irradiated to TID levels varying between 10 and 30 Mrad have continued to work with unchanged performance.[28,29,30,31,32,33] These circuits contain a very wide range of analog and digital functions, including low-noise amplifiers, high speed (1.6 Gbit/s) digital optical transmitters, PLLs and DLLs, analog memories, 10-bit ADCs, optical receivers and 60 kgate digital circuits. The degradation in the static parameters of individual transistors summarized in Table 3 is useful to appreciate the magnitude of the radiation effects.

Since the used technology, as all other modern CMOS technologies, is not sensitive to displacement damage for fluences below 10^{15} n/cm^2 (and possibly even well above this level), displacement effects will not be discussed in the following sections.

Table 3: Degradation of the static parameters of NMOS and PMOS transistors in the quarter micron technology. after 30 Mrad(SiO_2) and annealing.[34] Annealing was at 100°C for 168 hours, under worst case bias.

Parameter	*NMOS*	*PMOS*
ΔV_{th}	+ 45 mV	- 55 mV
Mobility degradation	- 6 %	- 2 %

4.3. *Drawbacks of the hardness by design approach*

The use of the radiation tolerant layout technique introduced a number of complications as compared to the standard design flow in a commercial process. In the first place, a model for the effective size (width W and length L) of ELTs had to be developed, since it was not known how to account for the 4 transistor "corners".[35,36] The possible variations of the enclosed shape being almost without limitation, we defined one "standard" shape (see Figure 7) and only model that. Moreover, these peculiar transistors are not source-drain symmetrical, and as a consequence the output conductance is higher when the central electrode is the drain.[27] These characteristics are not integrated in the SPICE simulation model, and the development of a custom model was far beyond the available resources. Therefore, we had to integrate a series of work-arounds in the simulation flow to accurately predict the circuit response when using ELTs.

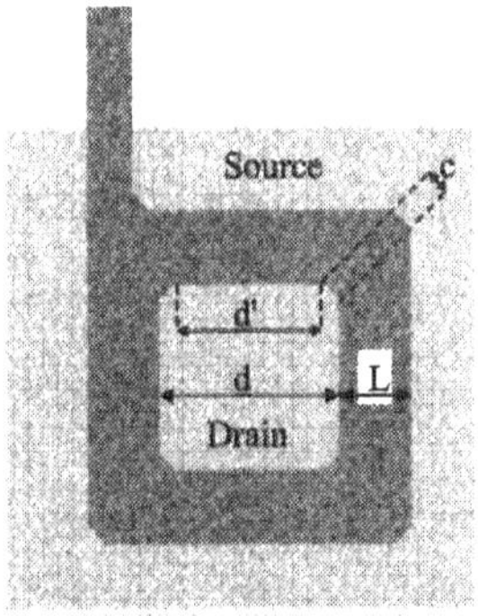

Figure 7: The enclosed shape that was chosen as "standard" and modeled has corners cut such that the length "c" is the same irrespective of the gate length L.

Logic blocks and circuits are normally designed using commercially available libraries, but none existed using ELTs and guardrings. Therefore, we had to build our own "radiation-tolerant LHC library" – and fully characterize it – for that purpose.[37] Because of the limited resources available, the number of gates and I/Os in the library is certainly not comparable with commercial libraries, but it is sufficient for our purposes (the largest digital circuit, a 36 mm^2 chip for the control of the CMS tracker, contained 60 thousand gates).

Finally, there is a large density penalty to pay when using ELTs and guardrings, a factor between 1.5 and 3.5 for the digital cells in our library – possibly larger for complex blocks. This penalty was not critical for most of our circuits, and is the price to pay for radiation hardness. On the other hand, as mentioned above, the alternative approach of using dedicated rad-hard technologies leads – when comparing technologies generations available at a given time – to even lower densities.

4.4. *Noise and matching of ELTs*

Noise performance is of paramount importance in the design of low-noise front-end circuits for particle detectors, because it determines the minimum detectable signal from the sensor. It was therefore essential to study the noise characteristics of the quarter micron process chosen for LHC circuit design, and in particular their evolution with TID.[38] Measurements were performed on NMOS and PMOS transistors in weak, moderate and strong inversion in the frequency range 200 Hz – 30 MHz. The inversion regions more interesting for readout of particle detectors are weak and moderate: the input transistor of readout amplifiers has a large W/L ratio to match the detector capacitance, hence it works very close to weak inversion for maximum g_m/I_d and minimum power for a given transconductance. A typical noise spectrum, for an enclosed NMOS transistor, is shown in Figure 8.

PMOS transistors have a smaller 1/f noise, a result generally common to all technologies, but also the 1/f noise measured on NMOS is acceptable for the low-noise application foreseen, provided a channel length larger than 0.5 μm is chosen.[38] More relevant to our applications is the white noise performance, which has been measured as very good since the excess noise factor – expressing the ratio between the measured and the theoretical noise levels – is very close to 1 for all transistors in moderate and weak inversion. Irradiation up to 100 Mrad with X-rays only increases the white noise by 15% for NMOS and 7% for PMOS (in $V/Hz^{1/2}$). The chosen technology therefore certainly appeared to satisfy, even after irradiation, the low-noise requirements for particle readout electronics.

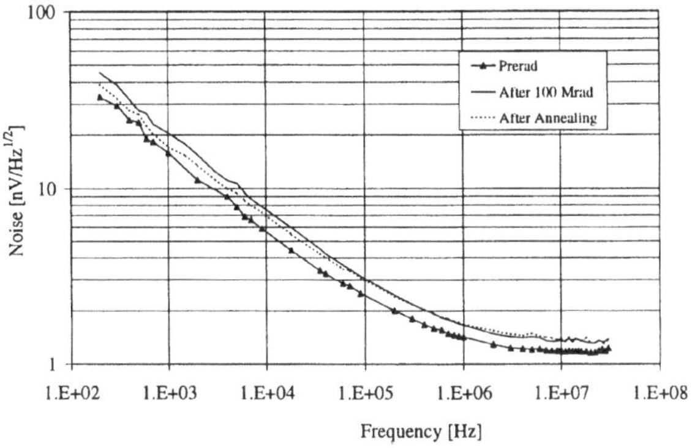

Figure 8: Noise spectrum of an enclosed NMOS transistor with W/L=200/0.5 μm.

Another critical parameter for analog design is the matching of transistors of identical geometry[39,40]: circuit blocks such as differential pairs and current mirrors heavily rely on device matching. For these reasons matching is a subject widely studied, but no literature on irradiation effects on matching was known to us, and certainly the matching of ELTs had never been studied before. A dedicated set of test structures has been designed in the quarter micron technology to explore this field, and though eventually several aspects where not understood, a thorough picture was obtained.[41] The results that are detailed in the following have not been widely published so far.

Matching being a statistical property, it is necessary to measure a large number of transistors to get meaningful results. This is normally done at the wafer level, which complicates the irradiation procedure. To this purpose, we used an X-ray source (with a tungsten target to produce 10 keV fluorescent lines as in reference [42]) and a semi-automatic wafer prober. With this setup, it was possible to keep the transistors under bias during irradiation with the help of a probe card, and to collimate the X-rays on the biased devices only. This procedure was repeated by stepping across all positions in the wafer at room temperature up to a total dose of 1.5 Mrad(SiO_2) and using a dose rate of 17 krad(SiO_2)/min. The typical worst-case bias condition was chosen: all terminals grounded for PMOS transistors, and Vdd potential to the NMOS transistor gate (2.5 V, all other terminals being grounded).[43] After irradiation, the wafer was annealed with devices unbiased at 100°C for a week and matching was measured again.

Matching is measured on transistor pairs, where the two transistors have been laid out carefully to be completely identical – same geometry, same environment, same connections to the terminals. The two transistors of each of our pairs were statically measured to extract the threshold voltage V_{th} and the current factor β. We calculated the difference ΔV_{th} and the relative difference $\Delta\beta/\beta$ for the pair, and repeated that for 90 identical pairs integrated regularly in the wafer, obtaining two distributions. The sigma values of these distributions give respectively the mismatch of the threshold voltage and of the current factors for the transistor pair under study (given size and shape). This procedure was repeated on a series of transistor pairs with different size to study the mismatch dependence on the transistor gate area. These data were plotted, as usually done in matching studies,[40] as a function of the square root of the inverse of the gate area, as shown in Figure 9 for the threshold voltage of regular NMOS transistors. The slope of these curves, A_{Vth} and $A_{\Delta\beta/\beta}$ respectively, is a good indicator of the matching performance. It has been found that A_{Vth} scales with the gate oxide thickness t_{ox}, and in technologies with good matching performance the value of A_{Vth}/t_{ox} is 1 mV·µm/nm, whilst $A_{\Delta\beta/\beta}$ is typically about 1-3 %·µm.[44]

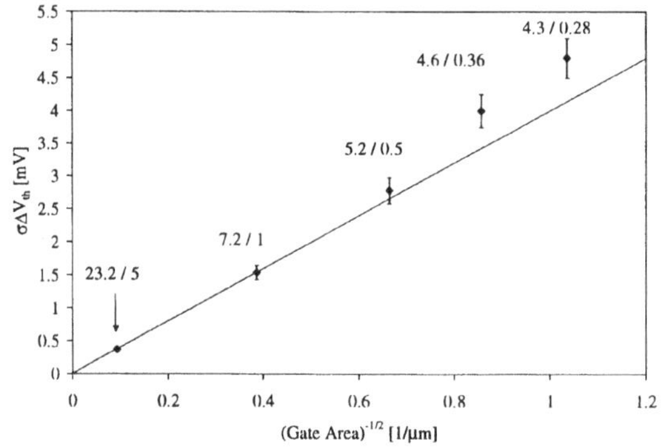

Figure 9: Threshold voltage matching of NMOS transistors with standard layout in the studied quarter micron technology, from reference.[41] The W/L ratio of the measured transistors is indicated.

Before irradiation, for regular NMOS and PMOS transistors in the chosen quarter micron technology the slopes were well within the expected range for a 5.5 nm gate oxide thickness: 4 and 3.7 mV·µm/nm for the threshold voltage of NMOS and PMOS transistors respectively, and about 1 %·µm for the current factor. However, for NMOS ELTs instead the experimental points in the same plots do not follow a straight line, as can be seen in Figure 10 for the threshold voltage in the case in which the enclosed diffusion is used as the transistor drain. The experimental points can be fitted with the following equation (for the threshold voltage, but similarly for the current factor):

$$\sigma_{\Delta V_{th}} = \sqrt{\frac{A_{V_{th}}^2}{WL} + \sigma_0^2} \qquad (2)$$

The values of A_{Vth} and σ_0 used in the fit of Figure 10 are 4.3 mV·µm and 1 mV, the slope being actually very close to the value of 4 measured for standard NMOS transistors. It appears that in the case of ELTs there is an additional mismatch source, independent of the gate area and never observed before in standard transistors. Experiments showed that the relative contribution of the additional source, quantified by σ_0, depends on the size/shape of the drain electrode. This can be seen in Figure 10, where the two circular symbols represent two ELTs with the same gate area as those on the curve fitting equation (2), but with the surface of the inner diffusion (here used as the drain) four times larger. It can also be seen in Figure 11, where the two curves are for the same transistors with drain chosen as the inner and outer diffusion respectively. In the latter case, the value of σ_0 decreases to 0.5 mV. This peculiar behavior has not been understood yet, and is even more difficult to explain considering that the extraction of the parameters has been performed in the linear region of operation of the transistor (low electric field in the drain region).

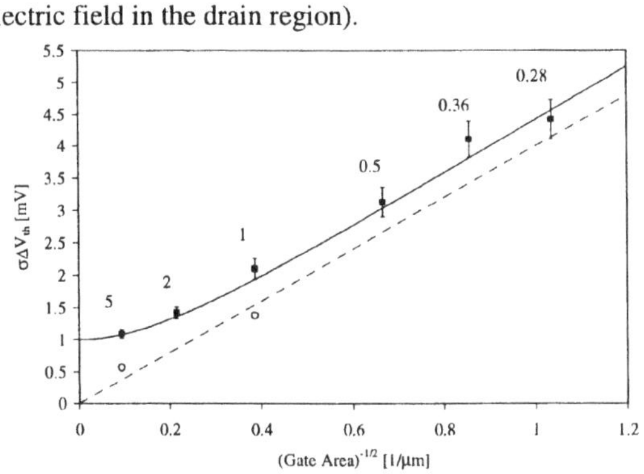

Figure 10: Threshold voltage matching of ELT NMOS transistors with the inner diffusion used as the drain, from reference.[41] The gate length of the measured transistors is indicated close to the points. The white circles represent two transistors with larger area of the inner diffusion, and the error bars are not indicated. The dotted line is the fit for the standard layout transistors.

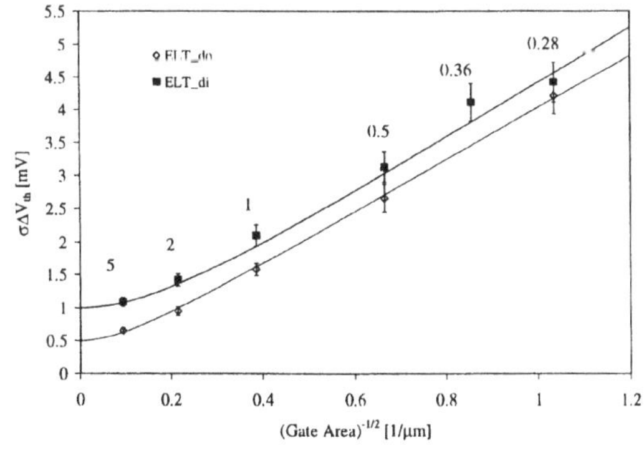

Figure 11: Comparison of the threshold voltage matching of the same ELT pairs with the inner (ELT_di) or the outer (ELT_do) diffusion used as the drain, from reference.[41] The gate length of the measured transistors is indicated close to the points.

While the current factor matching experimental points can be fitted with an equation similar to (2), with $A_{\Delta\beta/\beta} = 1.3\ \%{\cdot}\mu m$ and $\sigma_0 = 0.3\ \%$, the matching in this case does not depend on which diffusion is chosen as drain nor on the size of the inner diffusion.

Irradiation affects in a different manner the threshold voltage matching of PMOS and NMOS transistors, whilst for both devices the current factor matching is unaffected, in agreement with the very low degradation of the transconductance. As shown in Figure 12, the TID of 1.5 Mrad(SiO_2) mildly influences the threshold voltage matching of PMOS transistors. For NMOS transistors, instead, the slope A_{Vth} increases to 5.4 mV·µm, an increase of 45% meaning that another mechanism begins to dominate the mismatch performance (Figure 12). Such a large increase cannot be correlated with any other irradiation effect measured on the same transistors, in terms of static parameters (threshold voltage, subthreshold slope, transconductance) or noise. In practical terms, this means that NMOS differential pairs will degrade their matching (hence their offset) with TID to an extent larger than for PMOS transistors.

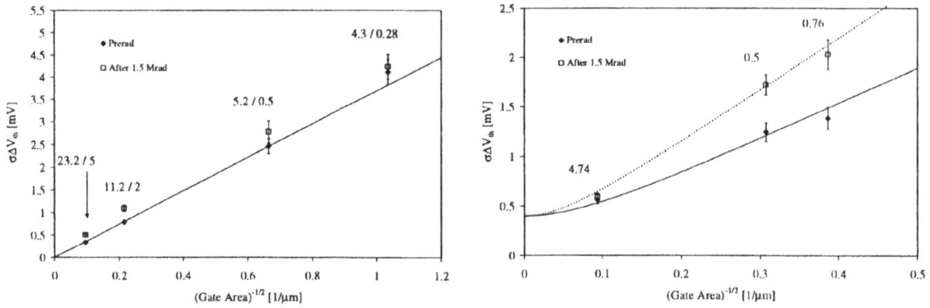

Figure 12: Irradiation up to 1.5 Mrad(SiO_2) mildly influences the threshold voltage matching performance of PMOS transistors (left).[41] This is not the case for NMOS ELT devices (right), where the inner diffusion is used as the drain.[41]

4.5. Single Event Effects

In the LHC experiments, SEE will threaten the reliable performance of ASICs designed with the above-described hardness by design approach. SEGR is not considered to be an issue in the LHC hadron environment, because this effect does not appear to be a threat for modern commercial deep submicron circuits even in the presence of heavy ions.[45,46] The main causes of concern are therefore SEL and SEU.

SEL sensitivity has been shown to decrease in modern deep submicron processes due to the combination of several favorable technological choices such as retrograde wells and STI isolation.[47] Moreover, device simulation has pointed out the effectiveness of guardrings in substantially decreasing SEL susceptibility.[48] Considering that our designs employ both a deep submicron process and guardrings, and because the environment is relatively "benign" due to the absence of heavy ions, we did not expect SEL to be a real concern. In fact, digital test circuits designed using cells from our "radiation tolerant" library have been irradiated at room temperature and nominal supply voltage (2.5 V) with heavy ions of LET up to 89 $MeVcm^2mg^{-1}$, and no latch-up has been observed.[49] Moreover, two prototype ICs for the LHC have been exposed to heavy ions up to an LET of 59 and 110 $MeVcm^2mg^{-1}$ respectively, without any SEL[50,51] Proton and pion irradiations at different energy have also been performed on several prototype ICs as a routine qualification procedure, and no latch-up has ever been reported. All the above contributes to rule out the SEL concern for circuits developed using our hardness by design approach.

Concerning SEU, both the frequency and system impact of this type of single event effect depend too closely on the circuit and system architecture to be able to generalize. Nevertheless, it can be said that the penalty paid in terms of density and power consumption when using ELTs turned out to be an advantage in terms of SEU sensitivity. The use of enclosed geometry for NMOS transistors sets a lower limit to the gate width (about 3.3 μm in our case), since the gate has to completely surround the minimum size drain. In digital cells, the PMOS transistor has to be sized correspondently large to match the driving strength of the NMOS. As a result, both the node capacitance (proportional to the gate area) and the current drive of the transistors (proportional to the aspect ratio W/L) are considerably larger than for the minimum size transistors achievable with standard layout. This has two consequences: on one hand the threshold charge for upset increases, and on the other hand the saturation cross-section, proportional to the area of the sensitive node, also increases. In the hadron-dominated LHC particle environment, with LET of the generated fragments limited to below about 15 $MeVcm^2mg^{-1}$, the increase in the threshold has by far the larger impact, and the error rate decreases.

The simple flip-flop cell (DFF) in our library has an LET threshold for upset close to 14 $MeVcm^2mg^{-1}$,[49] very high with respect to typical values for designs in a quarter micron process. The same trend has been confirmed on an SRAM cell, for which we have measured a LET threshold of 5.4 $MeVcm^2mg^{-1}$.[52] With such high thresholds to start with, and in a particle environment dominated by hadrons, it is possible to slightly modify the design of the cells when it is necessary to decrease the SEU sensitivity. For instance, in the case of the DFF mentioned above, a slight modification of the transistor size or of the capacitive load of the nodes is sufficient to decrease the foreseen error rate in the LHC by a factor of 10.[52] This approach has been followed occasionally for critical elements of some LHC ASICs. Alternatively, designers have looked for cells with equivalent functionality but smaller sensitivity, developed specifically to be SEU tolerant. Several types of such custom cells are available in the literature,[53,54] and most often they have been used to protect the content of critical registers. In other cases, the Triple Modular Redundancy (TMR) approach has been used, where each bit of information is stored in three identical cells, and a voting between the three ensures the correct value is output. Finally, Error Detection And Correction (EDAC) in transmission protocols is very frequently used to decrease the Bit Error Rate of the systems, and efficiently corrects for errors due to both noise and radiation.

5. Conclusions

With the new generation of physics experiments, the HEP community is facing unprecedented radiation concerns. Not only do the foreseen radiation levels range up to tens of Mrad, but the required volumes of radiation tolerant electronics and the complexity of the detector systems make reliable operation a very difficult objective. The particle composition of the radiation environment – dominated by charged and neutral hadrons over a wide energy range – also complicates the estimate of the SEE rate in the LHC experiments.

This paper has summarized the main radiation issues for reliable operation of detector electronics in the LHC experiments, focusing on two original developments: a novel computational method to estimate SEU rates, and a radiation tolerant layout approach for ASIC design.

To reliably estimate SEU rates in an environment composed of hadrons with a wide energy spectrum, it was necessary to develop a computational method based on the use of software packages developed by the HEP community and containing all the information concerning the nuclear interaction of hadrons with silicon nuclei. Although several radical assumptions have to be made, in particular on the size and shape of the volume in which the charge effectively contributing to SEU is deposited, comparison with experimental benchmarks has shown that this method leads to reliable rate predictions. Lessons learned during the development were used to propose a standard radiation qualification protocol – based on the use of mono-energetic proton beams of 60-200 MeV – for the LHC electronics components.

A large number of ASICs in relatively large volumes (for the typical applications in which radiation hardness is required) are needed for the inner parts of the LHC experiments. The use of radiation tolerant layout techniques, coupled with the natural hardness of the thin gate oxide of a quarter micron CMOS technology, is the solution which has most often been chosen by the LHC ASIC designers. Advantages and drawbacks of this approach have been investigated, and all the tools necessary to efficiently design ASICs with this methodology have been developed for the chosen commercial CMOS process: models for the special enclosed layout transistors, a small digital library and all software tools necessary for the typical design flow. All aspects related to precise low-noise analog design, such as noise and matching, have been investigated and eventually the radiation tolerance of all prototypes has proven to be largely sufficient for LHC applications.

The HEP community has devoted so far a considerable effort to solve the problems related to the reliability of the electronics that will be installed in the radiation environment of the LHC experiments. Most of these were not in the field of expertise of the community, and required therefore a long learning phase. Although progress has certainly been made, and technical solutions are at present generally well recognized, a large amount of work remains to be done in producing, testing, qualifying and assembling the electronics components in reliable electronics systems. This task will surely continue to demand a large effort from the community in the few remaining years before the first proton beam will circulate in the LHC tunnel.

Acknowledgements

The work on radiation hardness by design described and richly referenced in this paper has been accomplished over several years by a large number of people in the Microelectronics group of the CERN EP division, under the RD49 research programme. H.P. Tuinhout, from Philips Research Laboratories, has greatly contributed to the matching study that was mainly performed by G.Anelli, CERN. I am grateful to M.Huhtinen, CERN, for his essential collaboration in the estimate of SEU rates described in section 3. I wish to thank F.Anghinolfi, CERN, for the data concerning the ATLAS radiation environment.

References

1. G.Lindström, M.Moll and E.Fretwurst, "Radiation hardness of silicon detectors - a challenge from high-energy physics", Nuclear Instruments and Methods in Physics Research A 426 (1999) 1-15
2. E.Normand, J.Wert, D.Oberg, P.Majewski, P.Voss, S.Wender "Neutron-Induced SingleEvent Burtout in High Voltage Electronics" in *IEEE Trans. on Nuclear Science*, Vol. 44, No.6, pp. 2358-2366, December 1997

3. D.Oberg, J.Wert, E.Normand, P.Majewski, S.Wender "First Observation of Power MOSFET Burnout with High-Energy Neutrons" in *IEEE Trans. on Nuclear Science*, Vol.43, No.6, pp. 2913-2920, December 1996
4. F.Faccio, "COTS for the LHC radiation environment: the rules of the game", in the proceedings of the 6th Workshop on Electronics for LHC Experiments, Cracow, 11-15 September 200, p.50, CERN 2000-010, CERN/LHCC/2000-041, 25 October 2000
5. M.Huhtinen and F.Faccio, "Computational method to estimate Single Event Upset rates in an accelerator environment". Nuclear Instruments and Methods in Physics Research A 450 (2000), 155-172
6. A.Fassò, A.Ferrari, J.Ranft, P.Sala, in: A.Menzione, A.Scribano (Eds), Proceedings of the IV International Conference on Calorimetry in High Energy Physics, La Biodola, Sept. 20-25, 1993, World Scientific, Singapore, 1993, p.493
7. J.F.Ziegler, J.P.Biersack, U.Littmar, The Stopping and Range of Ions in Solids, Vol.1, Pergamon Press, Oxford, 1985
8. S.Duzellier, D.Falguère, M.Tverskoy, E.Ivanov, R.Dufayel, M.C.Calvet, "SEU Induced by Pions in Memories From Different Generations, IEEE Trans. Nucl. Science, Vol.48, No.6, December 2001, p.1960
9. C.J.Gelderloos et al., "Pion-induced soft upsets in 16 Mbit DRAM chips", IEEE Trans. Nucl. Science, Vol.44, No.6, December 1997, p.2237
10. P.Calvel et al., "An Empirical Model for Predicting Proton Induced Upset", IEEE Trans. Nucl. Science, Vol.43, No.6, December 1996, p.2827
11. F.Faccio et al., "Study of device parameters for analog IC design in a 1.2μm CMOS-SOI technology after 10 Mrad", IEEE Trans. Nucl. Science, Vol.39, No.6, December 1992, p.1739
12. S.L.Thomas et al., "Measurements of a radiation hardened process: Harris AVLSIRA", Nuclear Instruments and Methods in Physics Research A 342 (1994) 164-168
13. S.Tedja et al., "Noise Spectral Density Measurements of a Radiation Hardened CMOS Process in the Weak and Moderate Inversion", IEEE Trans. Nucl. Science, Vol.39, No.4, 1992, p.804
14. M.Dentan et al., "Study of a CMOS-JFET-Bipolar radiation hard analog-digital technology suitable for High Energy Physics electronics", IEEE Trans. Nucl. Science, Vol.40, No.6, December 1993, p.1555
15. W.Dabrowski et al., "Design and performance of the ABCD chip for the binary readout of silicon strip detectors in the ATLAS Semiconductor Tracker", IEEE Trans. Nucl. Science, Vol.47, No.6, December 2000, p.1843
16. N.S. Saks, M.G.Ancona and J.A.Modolo, "Radiation effects in MOS capacitors with very thin oxides at 80°K", IEEE Trans. Nucl. Science, Vol.31, p.1249-1255, December 1984
17. N.S. Saks, M.G.Ancona and J.A.Modolo, "Generation of interface states by ionizing radiation in very thin MOS oxides", IEEE Trans. Nucl. Science, Vol.33, p.1185-1190, December 1986
18. G.E.Moore, "Cramming More Components onto Integrated Circuits", Electronics, vol.38, no.8, 1965
19. G.Anelli et al., "Total Dose behavior of submicron and deep submicron CMOS technologies", in the proceedings of the Third Workshop on Electronics for the LHC Experiments, London, September 22-29, 1997, pp.139-143 (CERN/LHCC/97-60, 21 October 1997)
20. W.Snoeys et al., "Layout techniques to enhance the radiation tolerance of standard CMOS technologies demonstrated on a pixel detector readout chip", Nuclear Instruments and Methods in Physics Research A 439 (2000) 349-360
21. J.V.Osborn et al., "Total Dose Hardness of Three Commercial CMOS Microelectronics Foundries", in the proceedings of the Fourth RADECS conference, Cannes, France, 15-19 September 1997, p.265-270
22. M.R.Shaneyfelt et al., "Challenges in Hardening Technologies Using Shallow-Trench Isolation", IEEE Trans. Nucl. Science, Vol.45, No.6, pp.2584-2592, December 1998
23. T.R.Oldham et al., "Post-Irradiation effects in field-oxide isolation structures", IEEE Trans. Nucl. Science, Vol.34, No.6, p.1184-1189, December 1987
24. D.R.Alexander, "Design issues for radiation tolerant microcircuits for space", Short Course of the Nuclear and Space Radiation Effects Conference (NSREC), July 1996
25. A.G.F.Dingwall and R.E.Sticker, "C^2L: A New High-Speed High-Density Bulk CMOS Technology", IEEE Journal of Solid-State Circuits, Vol.12, pp.344-349, August 1977

26. "3rd RD49 Status Report: Study of the Radiation Tolerance of ICs for LHC", LEB Status Report/RD49, CERN/LHCC 2000-003, 13 January 2000
27. G.Anelli et al., "Radiation Tolerant VLSI Circuits in Standard Deep Submicron CMOS Technologies for the LHC Experiments: Practical Design Aspects", IEEE Trans. Nucl. Science, Vol.46, No.6, pp.1690-1696, December 1999
28. M.Campbell et al., "A Pixel Readout Chip for 10-30 Mrad in Standard 0.25μm CMOS", IEEE Trans. Nucl. Science, Vol.46, No.3, pp.156-160, June 1999
29. P.Moreira et al., "G-Link and Gigabit Ethernet Compliant Serializer for LHC Data Transmission", 2000 IEEE Nuclear Science Symposium Conference Record, pp.9.6-9.9. Lyon, October 15-20, 2000
30. F.Faccio, P.Moreira and A.Marchioro, "An 80Mbit/s radiation tolerant Optical Receiver for the CMS optical digital link", in the proceedings of SPIE 4134, San Diego, July 2000
31. G. Anelli et al., "A Large Dynamic Range Radiation-Tolerant Analog Memory in a Quarter-Micron CMOS Technology", IEEE Transactions on Nuclear Science, vol. 48, no. 3, pp. 435-439, June 2001
32. A. Rivetti et al., "A Low-Power 10-bit ADC in a 0.25-mm CMOS: Design Considerations and Test Results", IEEE Transactions on Nuclear Science, vol. 48, no. 4, pp. 1225-1228, August 2001
33. W.Snoeys et al., "Integrated Circuits for Particle Physics Experiments", IEEE Journal of Solid-State Circuits, Vol.35, No.12, pp.2018-2030, December 2000
34. F.Faccio et al., "Total Dose and Single Event Effects (SEE) in a 0.25μm CMOS Technology", in the proceedings of the Fourth Workshop on Electronics for LHC Experiments, Rome, September 21-25, 1998, pp.105-113 (CERN/LHCC/98-36, 30 October 1998)
35. A.Giraldo, A.Paccagnella and A.Minzoni, "Aspect ratio calculation in n-channel MOSFETs with a gate-enclosed layout", Solid-State Electronics, Vol.44, 1st June 2000, pp.981-989
36. A.Giraldo, "Evaluation of Deep Submicron Technologies with Radiation Tolerant Layout for Electronics in LHC Environments", PhD. Thesis at the University of Padova, Italy, December 1998 (http://wwwcdf.pd.infn.it/cdf/sirad/giraldo/tesigiraldo.html)
37. K.Kloukinas, F.Faccio, A.Marchioro and P.Moreira, "Development of a radiation tolerant 2.0V standard cell library using a commercial deep submicron CMOS technology for the LHC experiments", in the proceedings of the Fourth Workshop on Electronics for LHC Experiments, Rome, September 21-25, 1998, pp.574-580 (CERN/LHCC/98-36, 30 October 1998)
38. G.Anelli, F.Faccio, S.Florian and P.Jarron, "Noise characterization of a 0.25 μm CMOS technology for the LHC experiments", Nuclear Instruments and Methods in Physics Research A 457 (2001) 361-368
39. K.R.Lakshmikumar, R.A.Hadaway and M.A.Copeland, "Characterization and Modeling of Mismatch in MOS Transistors for Precision Analog Design", IEEE Journal of Solid-State Circuits, Vol.21, No.6, pp.1057-1066, December 1986
40. M.J.Pelgrom et al., "Matching Properties of MOS Transistors", IEEE Journal of Solid-State Circuits, Vol.24, pp.1433-1440, October 1989
41. G.Anelli, "Conception et caractérisation de circuits intégrés résistants aux radiations pour les détecteurs de particules du LHC en technologies CMOS submicroniques profondes", Ph.D. Thesis at the Politechnic School of Grenoble (INPG), France, December 2000, availeble on the web at the URL: http://rd49.web.cern.ch/RD49/RD49Docs/anelli/these.html
42. L.J.Palkuti and J.J.LePage, "X-ray Wafer Probe for Total Dose Testing", IEEE Trans. Nucl. Science, Vol.29, No.6, pp.1832-1837, December 1982
43 T.P.Ma, P.V.Dressendorfer, Ionizing radiation effects in MOS devices and circuits, John Wiley & Sons, 1989, p.264
44. H.P.Tuinhout, "Matching of MOS Transistors", Lecture at the Course "Deep Submicron: Modeling & Simulation" held at the Ecol Polytechnique Fédérale of Lausanne, Switzerland, October 12-15, 1998
45. F.W.Sexton et al., "Single Event Gate Rupture in Thin Gate Oxides", IEEE Trans. Nucl. Science, Vol.44, No.6, pp.2345-2352, December 1997
46. L.W.Massengill et al., "Heavy-Ion Induced Breakdown in Ultr-Thin Gate Oxides and High-k Dielectrics", IEEE Trans. Nucl. Science, Vol.48, No.6, pp.1904-1912, December 2001
47. A.H.Johnston, "The Influence of VLSI Technology Evolution on Radiation-Induced Latchup in Space Systems", IEEE Trans. Nucl. Science, Vol.43, No.2, P.505, April 1996
48. T.Aoki, "Dynamics of heavy-ion-induced latchup in CMOS structures", IEEE Trans. El. Devices, Vol.35, No.11, p.1885, November 1988

49. F.Faccio et al., "Single Event Effects in Static and Dynamic Registers in a 0.25μm CMOS Technology", IEEE Trans. Nucl. Science, Vol.46, No.6, pp.1434-1439, December 1999

50. E.Noah et al., "Single event upset studies on the CMS tracker APV25 readout chip", Nuclear Instruments and Methods in Physics Research A 492 (2002) 434-450

51. J.J.Van Hunen et al., "Irradiation and SPS Beam Tests of the Alice1LHCb Pixel Chip", in the proceedings of the 7th Workshop on Electronics for the LHC Experiments, Stockholm, 10-14 September 2001, pp.85-89 (CERN 2001-005, CERN/LHC/2001-34, 22 October 2001)

52. F.Faccio et al., "SEU effects in registers and in a Dual-Ported Static RAM designed in a 0.25μm CMOS technology for applications in the LHC", in the proceedings of the Fifth Workshop on Electronics for LHC Experiments, Snowmass, September 20-24, 1999, pp.571-575 (CERN 99-09, CERN/LHCC/99-33, 29 October 1999)

53. R.Velazco et al., "2 CMOS Memory Cells Suitable for the Design of SEU-Tolerant VLSI Circuits", IEEE Trans. Nucl. Science, Vol.41, No.6, p.2229, December 1994

54. T.Calin, M.Nicolaidis, R.Velazco, "Upset Hardened Memory Design for Submicron CMOS Technology", IEEE Trans. Nucl. Science, Vol.43, No.6, p.2874, December 1996

International Journal of High Speed Electronics and Systems
Vol. 14, No. 2 (2004) 401–417

World Scientific
www.worldscientific.com

SPACE RADIATION EFFECTS IN OPTOCOUPLERS

ROBERT A. REED
NASA/Goddard Space Flight Center
Code 561.4
Building. 22, Room 052
Greenbelt, Maryland, 20771, U.S.A.
Robert.A.Reed@nasa.gov

PAUL W. MARSHALL
Consultant
NASA/Goddard Space Flight Center
7655 Hat Creek Road
Brookneal, Virginia, 24528, U.S.A.
PWMarshall@aol.com

KENNETH A. LABEL
NASA/Goddard Space Flight Center
Code 561.4
Building 22, Room 050
Greenbelt, Maryland, 20771, U.S.A.
Kenneth.A.Label@nasa.gov

Assessing the risk of using optocouplers in satellite applications offers challenges that incorporate those of commercial off-the-shelf devices compounded by hybrid module construction techniques. In this paper, we discuss these challenges with regard to radiation-induced effects, e.g., single event transients in the photodiodes and displacement damage effects in light emitting diodes. Radiation-induced effects can depend on several important variables including circuit application, radiation environment, annealing, temperature, and others. Despite the ominous list of test variables above, it is possible to make a conservative estimate of optocoupler performance in the space radiation environment. However, these predictions could result in an overestimation of on-orbit performance by an order of magnitude or more.

Keywords: Optocouplers; space radiation; single event transients; displacement damage.

1.0 Introduction

Optocouplers operate on the principle that light can be emitted, transmitted, and absorbed by a semiconductor to communicate signals optically. The optical signal electrically isolates the microelectronic circuits on either side of the optocoupler—a primary use for optocouplers. **Figure 1** shows a schematic representation of an optocoupler.

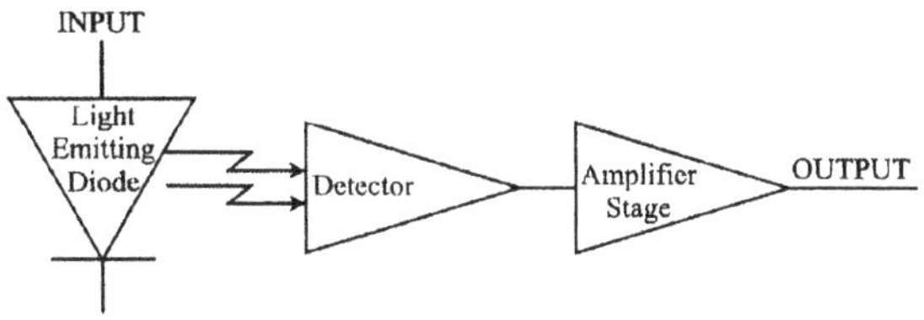

Fig. 1. Representative optocoupler design. The receiver side contains a detector and possibly a simple amplification stage.[7]

Optocouplers use a light emitter—typically a Light-Emitting Diode (LED)—to provide an internal optical signal to a photodetector and amplifier (or phototransistor). There is no direct electrical connection between the LED and output; this results in a very high degree of isolation between the electrical signal that drives the LED and the output of the amplifier. Optocouplers—which implement an optical link and isolation function within a single package—are strictly commercial off-the-shelf (COTS) devices. Past studies have shown that optocoupler performance is adversely affected by both radiation-induced permanent damage and single event effects.[1-12] The components that make up an optocoupler rely on carefully grown, well-defined microscopic structures that can have very low tolerance for slight changes in their characteristics. When a component is exposed to radiation, the radiation transfers some of its energy to the component materials, changing the local material properties. This can have significant effects on component functionality and/or parametrics with the end result depending on the type of radiation and where within the optocoupler the energy deposition occurred.

In this paper we will review the Single Event Transient (SET), Total Ionizing Dose (TID) and Displacement Damage (DD) effects that occur when an optocoupler is exposed to a radiation environment. **Table 1** lists components used to make up optocouplers and the performance degradation mechanisms that are most important for each. We will briefly consider these radiation effects and how these effects may degrade optocoupler performance. Single-event transients in photodetectors and amplification circuitry will be discussed first, then we will describe permanent degradation (TID and DD) of optocoupler parameters.

Table 1. Dominant performance degradation mechanism for each component of an optocoupler.

LED	DD
Photodetector	TID, SET
Amplification Circuit	TID, SET
Coupling Medium	TID

2.0 Optocoupler Types Addressed in This Paper

This paper will focus on radiation-induced effects in high-bandwidth digital signal isolators and current transfer optocouplers. While the degradation mechanisms described below can cause performance reduction or failure in other types of optocouplers, we will not specifically address these in this paper.

Figure 2 shows an x-ray image taken of a typical current transfer optocoupler. The optocoupler contains an LED (white dashed box outline), transmission medium (outlined with black dashed line), and phototransistor (solid white box outline). The LED's function is to convert the electrical signal into an optical format. The role of the transmission medium is to transfer the optical signal from the transmitter to the receiver. The optical receiver's (in this case a phototransistor) purpose is to convert the optical signal back into an electrical signal. The optical signal is detected by the phototransistor, its output current is related to the intensity of the optical signal.

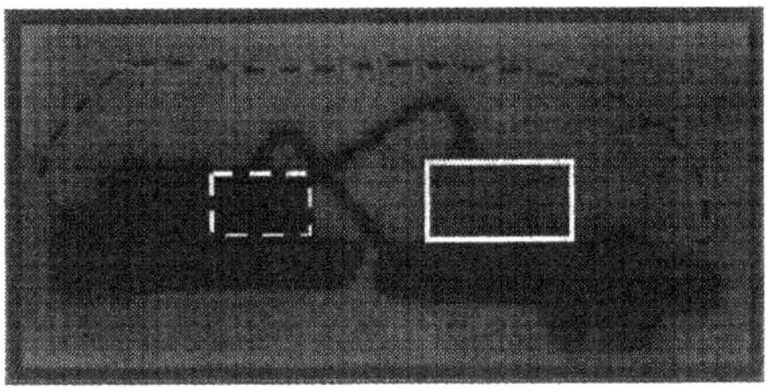

Fig. 2 X-ray image of a simple current transfer optocoupler.

Figure 3 shows a typical design for a high bandwidth digital signal isolator. Like the current transfer optocoupler, these types of couplers have an emitter, transmission medium (not shown) and a receiver. However, in this case the receiver is a photodiode followed by an amplification circuit.

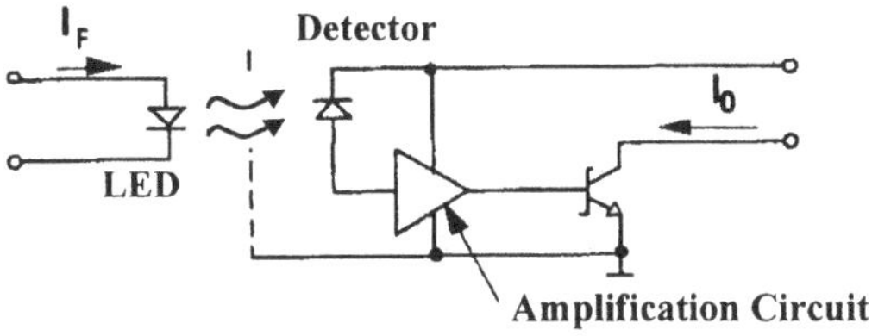

Fig. 3 Typical high bandwidth digital signal isolators design.

Optocouplers, because of their COTS orientation, typically use low-cost design approaches that target high-yield above minimum performance levels. Optocouplers are hybrid devices, and while some vendors are attuned to the specialized needs of the aerospace community, most are not. Traceability of internal components is not the rule, and as a result (as we will see later) device-to-device and lot-to-lot variability in radiation response is common.

3.0 Physical Mechanisms of Particle-Induced Transient Effects

3.1. Introduction

Single event error mechanisms for optocouplers are fundamentally different from conventional microelectronics. The most SET sensitive element is usually the optoelectronic photodetector. The problem is exacerbated by the fact that detector elements are often large area diode structures with diameters of several hundred microns.

Also, an optocoupler's detector diode is commonly incorporated monolithically using conventional bipolar processing methods. The importance of this arises from the fact that the p-n junction found in bipolar processes does not collect charge primarily via drift from a fully depleted intrinsic region. Instead, in bipolar p-n junctions, diffusion from the field-free substrate bulk may be a significant source of both the optically detected signal, as well as the charge deposited by the ionizing particle. Characteristics of optocoupler cross-section measurements consistent with charge collection via diffusion were first noted by LaBel, et al.[6], and this was examined by Reed, et al.[7] and Johnston, et al.[8, 11] with indications that diffusion lengths approaching 50 μm could be possible.

3.2 Single Event Transients

An ionizing particle generates electron-hole pairs along its path as it passes through a material, resulting initially in a line charge distribution with equal numbers of holes and electrons. Because the operation of active optocoupler components is governed by the controlled injection of charge into the depletion layers of p-n junctions, the uncontrolled charge injection resulting from ionization can produce a self-recoverable false-signal on the output of the optocoupler. These false signals are know as SETs. Particles incident on either the photodetector or amplification circuitry can produce SETs, but the photodetector is typically much more sensitive. Proton-induced SETs in high-bandwidth optocouplers (like those in **Figure 3**) were reported first by LaBel, et al. following an investigation of onboard flight anomalies due to single particle effects in optocouplers used on the Hubble Space Telescope (HST).[6]

Protons can generate ionization by either direct means—in which the proton itself generates the charge—or by indirect means—in which the products of a proton-nucleus collision generate the charge. Either of these event types can induce an SET at the output of an optocoupler. For most microelectronics, direct proton ionization generates far too little charge to be an SET concern. However, the high sensitivity of some photodetectors contained in optocouplers means that direct proton ionization cannot be ignored. **Figure 4** shows, in pictorial form, the fact that both direct and indirect ionization can induce an SET at the output of an optocoupler. The long path length through the photodetector at grazing angles plays a critical role for the case of direct ionization, while the location of the reaction and its products plays a critical role for the indirect ionization case. Indirect ionization tends to dominate the device response for protons incident normal to the large diode surface, while direct ionization tends to dominate for grazing angle events. Quantifying how each event type impacts the over all circuit performance depends on the detector geometry, the amplification circuitry, and the sensitivity of the circuitry that is affected by the output of the optocoupler.

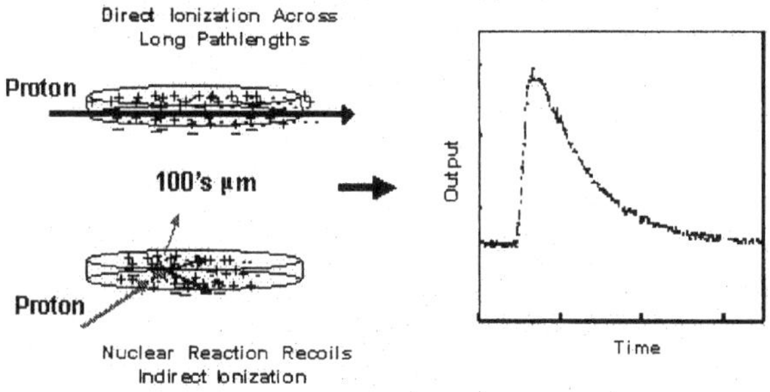

Fig. 4 Protons can generate charge in a device by either direct or indirect ionization. If sufficient ionization is generated, a single-event transient (SET) may result.

Figure 5 plots the cross-section (number of events divided by the proton fluence) for generating an SET on the output of an optocoupler. Except for grazing angles, the cross-section is constant and presumed to be dominated by nuclear elastic and inelastic interaction recoils. The SET cross-section for protons at near-grazing incidence (90 degrees on this chart) is more than an order of magnitude higher than the cross-section for normally incident protons—this despite the fact that the physical device cross-section is actually lower for grazing-incidence protons. The enhancement around grazing angle is due to direct ionization from protons, and suggests a low charge threshold and LET dependence for the error cross-section.

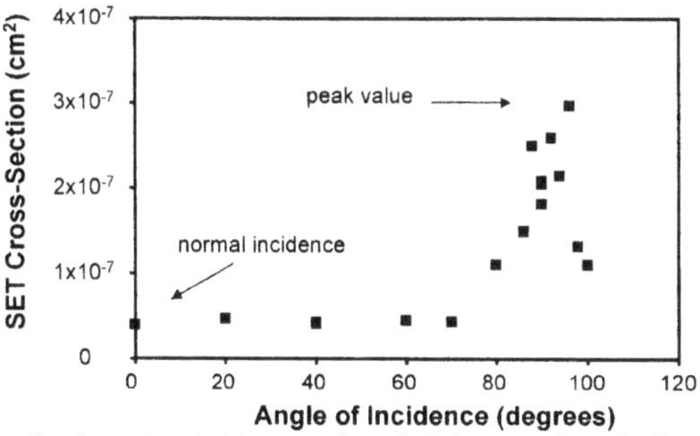

Fig. 5 The SET cross-section for protons incident on a photodiode increases dramatically as the protons approach grazing incidence—an indication that direct ionization is becoming important.[7]

Figure 6 shows an example of the increase in SET cross-section with the incident angle for the Agilent HCPL5231 optocoupler for three different proton energies. At 60 MeV, the angular dependence is very weak, but for the lower energies the effect is stronger. For 42 MeV protons, the cross-section at grazing incidence is a factor of 10 higher than the cross-section at normal incidence. For 31 MeV protons, the grazing incidence cross-section increases by nearly two orders of magnitude. This dependence indicates that transients caused by direct ionization by protons traversing the diameter of the photodiode dominate at grazing incidence.[7,10] Johnston and coworkers explored the possibility of elastic scattering as the cause for the angular dependence.[11] But, by examining the lower proton energy (higher LET) response of 6N134 optocouplers from two manufacturers and showing the cross-section increase at grazing angles could be as much as 3 orders of magnitude at a proton energy of 15 MeV, Johnston et al. agreed with the conclusion that the angular dependence was due to direct ionization.[8]

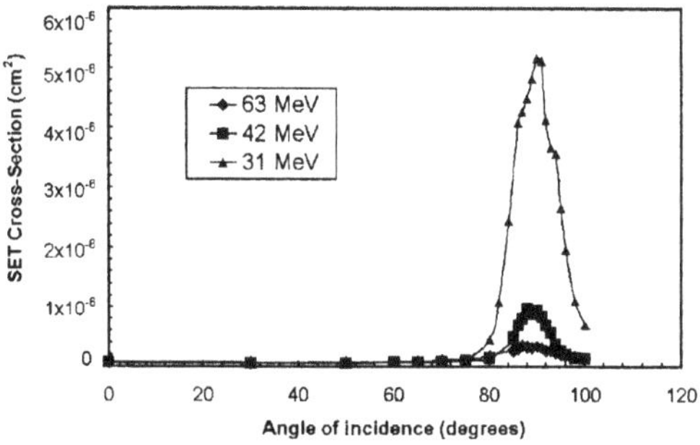

Fig. 6 Effect of incident angle on cross-section of the Agilent HCPL5231 optocoupler for various proton energies.[10]

The initial work by LaBel, et al. correctly identified the sensitivity in high-bandwidth (> ~5 MHz) devices, and described the role of direct ionization from protons in the optocoupler's internal receiver photodetector.[6] The error condition results from deposition of charge in the detector that leads to a transient that propagates during the "off" state of the device. Errors can occur when ions initiate transients with sufficient pulse width and voltage amplitude, provided the optocoupler amplifier stage is of sufficient bandwidth to propagate the signal to the optocoupler output.

3.3 Heavy Ion-Induced Transients

It is generally accepted that any device that is SET sensitive to protons can also be upset by heavy ions. Optocouplers are not an exception to this ideology. However, because of the monolithic construction of the detector circuitry, as the heavy ion Linear Energy Transfer (LET) is increased, other portions of the design can become sensitive. Johnston et al., investigated the response of the HP 6N134 to heavy ions and showed that in addition to the photodiode, amplifier stage circuitry could be susceptible to ion-induced transients, and transients with longer time signatures would result.[11]

3.4 Final Comments on Transient Effects

High-bandwidth optocouplers have demonstrated a significant enhancement in their SET response that can be directly attributed to direct ionization from protons. The degree of enhancement is a strong function of proton LET. At oblique angles, the proton induced error rates are dominated by nuclear elastic and inelastic reaction recoils. The combination of mechanisms poses a significant challenge when calculating space-radiation error rates. In Galactic Cosmic Ray (GCR) dominated environments, the contributions to the gain stages cannot be ignored, and the temporal response of the error may vary.

Also, we note that Reed et al. noticed a significant part-to-part variability in the transient error sensitivity, as indicated in **Figure 7**.[10] Presumably, this follows from the COTS nature of the manufacturing process and part-to-part variability in the gain stages. This has implications for the confidence interval appropriate for a predicted space radiation-induced error rate.

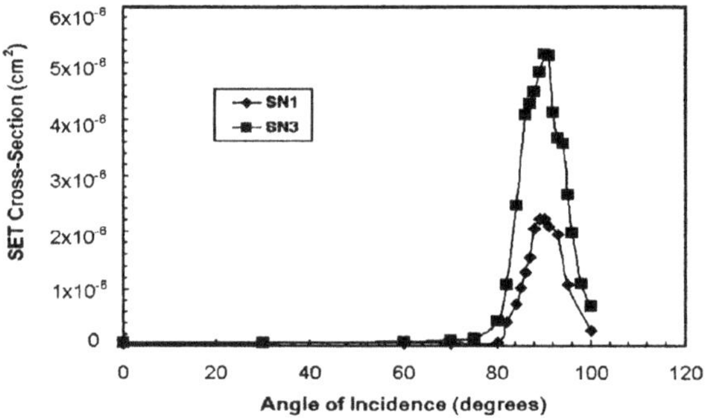

Fig. 7 Example of part-to-part dispersion observed during proton SET characterization of the Agilent HCPL5231.[10]

An assessment of the status of methods and tools to predict space radiation-induced error rates in optocouplers indicates that we are presently limited to factor of 3-5 estimates, and then only with data-intensive characterization efforts. Presently, no single tool exists that is appropriate for the full range of variables associated with these problems, but prospects are excellent for the development of a unified tool. Presently, assessments are under way at NASA Goddard Space Flight Center to examine the suitability of tools such as Geant4, MCNPX, NOVICE, and other expert-level Monte Carlo-based tools for various detector-related SEE concerns.[14,15,16] In parallel, assessments are being made to evaluate the role of charge diffusion and the requirements for simulation tools to assess its role in bipolar microelectronic devices and in detectors. Until the related studies are completed, it is difficult to define the next step required for a solution to rate prediction for optocouplers. Unfortunately, at least in the interim, we expect that these solutions will require comparatively sophisticated computer codes with expert interactions before more user-friendly tools can be made available, at least when better than order-of-magnitude estimates are desired.

4.0 Physical Mechanisms for Permanent Degradation

4.1 Introduction

There are two classes of long-term radiation-induced damage: total ionizing dose (TID) and displacement damage (DD). Electrical performance degradation of an optocoupler occurs when the material properties of one of its components are altered during radiation exposure by one or both of these damage mechanisms.

Damage from TID is caused when electron-hole pairs generated by ionizing radiation are trapped in oxides near the active regions. This will gradually change the performance of an optocoupler component, with the level of change depending on the total ionizing energy absorbed—that is, on the TID.[17,18] Generally, TID changes the characteristics of the materials that make up a component, resulting in gradual parametric degradation and changes in functionality. In most cases, the basic cause of TID degradation is the trapping of charge in a medium (e.g., a field oxide or in an optical medium.)

DD, on the other hand, has a different damage mechanism. The level of performance of active optocoupler components depends on the number of defects in the crystalline lattice—a more pristine lattice will have better performance. The lattice can be damaged when an energetic particle, such as a neutron, electron, proton or heavy ion displaces one or more nuclei within the crystalline lattice, possibly creating electrically active defects. As this damage to the crystalline lattice increases, the device can degrade parametrically, and eventually stop functioning all together. For the most part, because protons are penetrating, strongly interacting and abundant, the proton environment is predominant in considerations of DD effects in shielded space applications. The 1999 Nuclear and Space Radiation Effects Conference (NSREC) Short Course explores DD effects in further detail.[19]

What follows in this section is a brief description of DD effects in LEDs, effects of TID on the passive transmission media, and an examination of the effects of TID and DD on the photodetectors and amplification circuit.

High-bandwidth optocouplers simply stop transferring data once one or more of their components have experienced sufficient performance degradation. So, in this section, we present radiation effects on low band-width devices (current transfer optocouplers), which better demonstrates the basic mechanisms for optocoupler degradation.

4.2 Permanent Degradation in LEDs

The role of the LED in an optocoupler is to convert current signals into light. For these devices, we are generally more concerned with permanent degradation due to DD. TID is often a second order effect.[3,20,21]

LEDs produce light by means of radiative recombination of injected minority carriers with majority carriers in the depletion region. As such, any damage that decreases the efficiency of radiative recombination or introduces competing processes will degrade device performance. Irradiation by heavy particles such as protons and neutrons can displace nuclei from the crystalline lattice, causing DD. These defects can serve as sites for nonradiative recombination, decreasing the efficiency of the light source.[3] See the 2000 NSREC Short Course for a detailed discussion on DD effects in LEDs.[22]

Displacement damage decreases the LED output power, which in turn can reduce the output current of the photodetector. For current transfer applications, this results in a decrease in the device output current, eventually leading to device failure. For data transfer applications, this will result in inability of the device to transmit data.

Displacement damage in the LED is believed to be the dominant mechanism for most optocoupler failures when exposed to proton-rich environments.

4.3 TID Degradation of Transmission Media

Although charged particles traversing an optical medium can generate photons, these processes are too weak and the path lengths traversed too short for these events to interfere with signals. For this reason, with respect to transmission media, we are generally more concerned with permanent, cumulative degradation resulting from TID.

Ionizing radiation causes charge to become trapped by defects in the optical medium, creating color centers. Once the color centers form, light transmission efficiency is degraded as the color centers absorb signal photons. The degradation, however, need not be permanent. As is the case for trapped charge in microelectronics, color centers can heal by annealing. The processes of formation and annealing of color centers take place in competition, and optical medium degradation depends on the relative rates of these mechanisms.

Light absorption in the medium will result in a decrease in the photodetector output current for current transfer applications. For data transfer applications, the device ability to transmit data will be compromised. We note that for most, if not all, optocouplers the signal path lengths through the coupling medium are relatively short (<0.5 cm), so these loses due to TID are likely negligible for most optocouplers.

4.4 Permanent Damage in Optical Detectors

Optical detectors in optoelectronic devices function by detecting the photocurrent generated by a photon with energy greater than the semiconductor bandgap. For this reason, any damage to the semiconductor that changes its global properties, or that degrades the semiconductor's ability to carry a current will potentially compromise the detector's efficiency.

As TID accumulates, photodetectors degrade, with leakage currents increasing up to several orders of magnitude for doses up to 1 Mrad(Si). DD can degrade carrier lifetime, which reduces responsivity and gain, thereby changing the output current. Both TID and DD increase background noise.

While there are several degradation mechanisms for photodetectors when exposed to a radiation environment, many detectors show remarkable robustness to surviving space radiation environments.

4.5 Permanent Degradation of CTR

TID and DD can severely degrade the performance of an optocoupler. The most studied effect is how radiation degrades Current Transfer Ratio (CTR), defined as the ratio of the output current (I_C) to the input current (I_F) (See **Figure 8**).

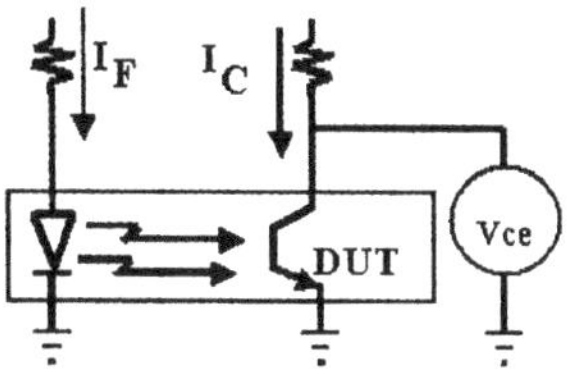

Fig. 8 Schematic showing experimental setup for measure CTR degradation.

Assessing radiation degradation of CTR is complicated by several factors that include anneal and temperature effects, lack of understanding of how physical degradation mechanisms manifest themselves as device failures, part to part variability, and others. Some of these concerns need to be addressed when estimating radiation exposure levels, others during the testing phase, and others must be considered in combination with radiation-induced degradation estimates to give overall performance. We will give a brief review of several of these complications.

4.5.1 Gamma Rays versus Protons

The observed radiation-induced degradation results from a combination of ionizing dose and displacement damage mechanisms, each affecting the individual optocoupler components differently.[4,7] **Figure 9** compares CTR degradation for gamma-ray exposures of a Texas Instruments (TI) 4N49 to that for 195 MeV proton exposures. Proton irradiations cause significantly larger degradation than gamma-ray exposures at equivalent doses. This type of optocoupler response is due to the greater amount of displacement damage occurring for protons.[4]

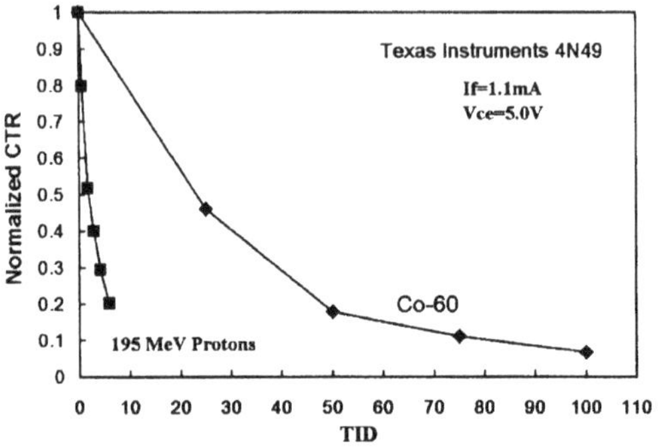

Fig. 9 Gamma and 195 MeV proton irradiations of Texas Instruments 4N49.[7]

4.5.2 *Proton Energy*

The rate of degradation not only depends on the radiation type, but also on the energy of the protons used during testing. This energy dependence is not well understood, making it very difficult to estimate the amount of degradation that would occur when the device is exposed to a spectrum of proton energies—the energy of the protons in space radiation environments can vary by over an order of magnitude.

In this section we will review radiation degradation results on LEDs—for most optocouplers, the LED is the most sensitive component to displacement damage—to qualitatively show how the damage factor could depend on proton energy and to demonstrate the lack of consistency between the damage factor and Non-Ionizing Energy Loss (NIEL)

Figure 10 plots the normalized lifetime-damage factors (K) for single heterojunction LEDs to NIEL as calculated by Burke et al. [24,25] (solid curve), as well as experimental results on amphoterically doped[26], double-heterojunction[27], and single-heterojunction[27] LEDs. For this plot, the NIEL values are normalized to the values at 10 MeV. There is an inconsistency between the experimentally-determined damage factors for LEDs and theoretical calculations for NIEL.

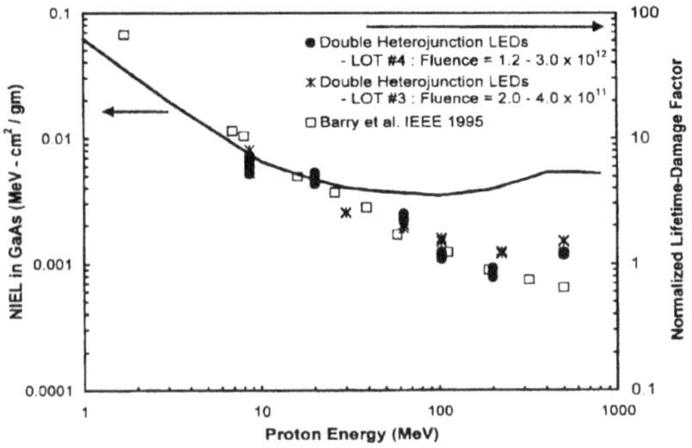

Fig. 10 Lifetime-damage-factor measurements for double-heterojunction devices (scaled to the right abscissa), normalized to agree with calculated NIEL values (solid line—scaled to the left abscissa) for 10 MeV protons.[27]

Significant error could occur in a calculation that uses NIEL to determine an equivalent fluence for a single test energy, where this equivalent fluence is expected to produce an equivalent amount of damage to that produced by a spectrum of particles with varying energy.

For example, **Figure 11** compares equivalent test fluences for 50 MeV or 200 MeV protons for a five-year satellite mission in a 600 km x 90 degree circular orbit about the earth. The equivalent fluence is given for NIEL and for the trend given by experimental data in **Figure 10**. The figure clearly shows how disastrous it could be if NIEL is used to compute the equivalent test fluence for this mission. Normalizing using NIEL correlation at 200 MeV would lead to underestimation of the damage by a factor of 7 for 50 mils of Al shielding. However, normalizing using NIEL at 50 MeV shows a factor of 3 underestimation, which is less dramatic, but significant nonetheless.

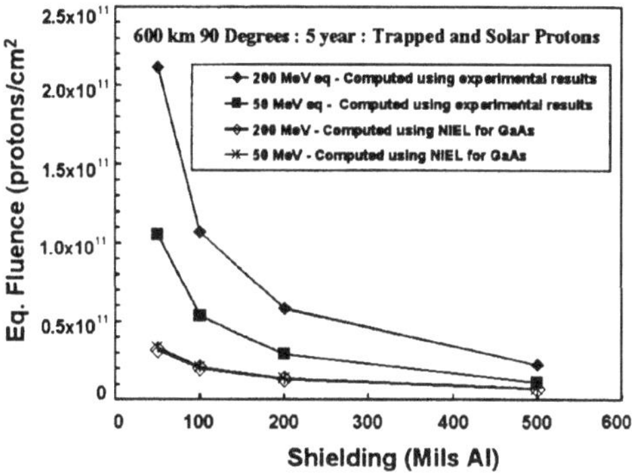

Fig. 11 On-orbit degradation for a mission can be predicted with a given fluence of protons, regardless of energy—provided displacement damage is proportional to NIEL.[27]

4.5.3 *Application specific testing*

A testing approach that is targeted towards a specific application is necessary to accurately determine the degradation expected for that application.[5,7] For example, the data represented by solid filled symbols connected by solid lines in **Figure 12** show a comparison of the CTR degradation at two different loads (RL) for the same device (V_{CE}=0.3V for both loads). The CTR for the case where the load was 2.7 ohm degraded by almost 50% of its original value after an exposure of $1x10^{11}$ protons/cm^2, while the degradation for a 1.0 kohm load was only 8%.

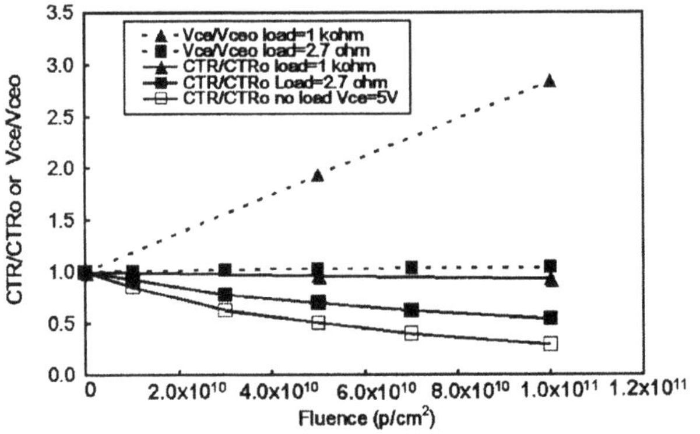

Fig. 12 Plot of either normalized CTR (solid lines) or normalized V_{CE} (dashed lines) for a 4mA forward current. This shows the load dependence on the radiation response. It also demonstrates that testing at a fixed V_{CE} would significantly over predict the CTR degradation.[7]

The dashed lines in **Figure 12** show the radiation-induced change of photodetector collector-emitter voltage (V_{CE}) with fluence for each load. For the 2.7 ohm load case V_{CE} remained constant at 0.3 V. In the case where the load is 1.0 kohm, V_{CE} increased with increasing fluence. Radiation degraded the output of the LED, as evident in the 2.7 ohm loading, but for the 1.0 kohm loaded case, the output degradation was compensated by an increase of the emitter-collector voltage. These data clearly show that an erroneous estimation of CTR degradation could occur if application specific testing was not carefully performed.

Finally, the data represented by the unfilled squares in **Figure 12** show the CTR degradation for a fixed V_{CE} of 5 V. The last data point on this curve shows that the CTR has degraded to 29% of the original value. Compare this to 92% of the original value when V_{CE} was allowed to change (1 kohm load). Data collected at V_{CE} = 5 V over predicts the degradation in the 1 kohm load case by a factor of 4.6. Several publicly accessible databases and journals give data collected at fixed V_{CE}. If these data are used for certain applications, they could significantly over predict the degradation.

4.5.4 Part-to-Part Variability in Radiation Response

Since optocouplers are typically COTS hybrids, it is not surprising that they often exhibit large part-to-part variability. Some of this variation is due to the complication of having several different components that can be sensitive to various effects. **Figure 13** shows data collected on three Mitel 3C91C optocouplers. At certain fluences there is over a factor of two variation in the response to radiation. This optocoupler is manufactured as a radiation-hardened product, and the internal components were all selected from the same device lot. These hybrids do not contain a light pipe to guide the light between the LED and the detector, so the variation must be due to the differences in the response of one of the active components.

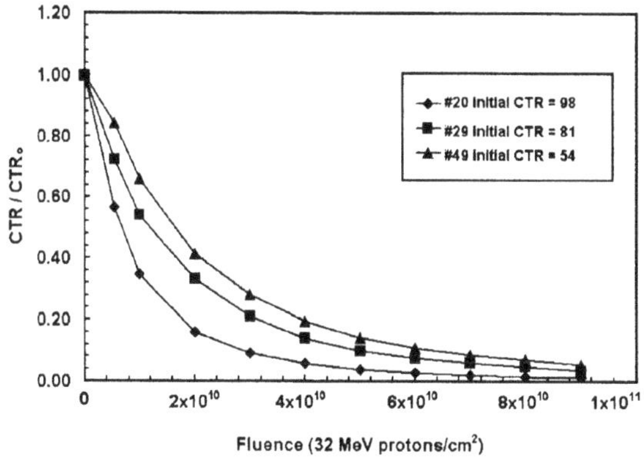

Fig. 13 Demonstration of part-to-part variability of CTR degradation within a single lot.[10]

Another reason for part-to-part variations (and the most difficult to characterize) is the flexibility the vendor has in selecting the components internal to the optocoupler. **Figure 14** gives CTR degradation data from two lots of the Micropac 4N49. These data show as much as an order of magnitude variation in the sensitivity between the two "lots". As with all COTS devices, knowledge of the manufacturing process is very limited, adding a significant risk to using these types of devices.

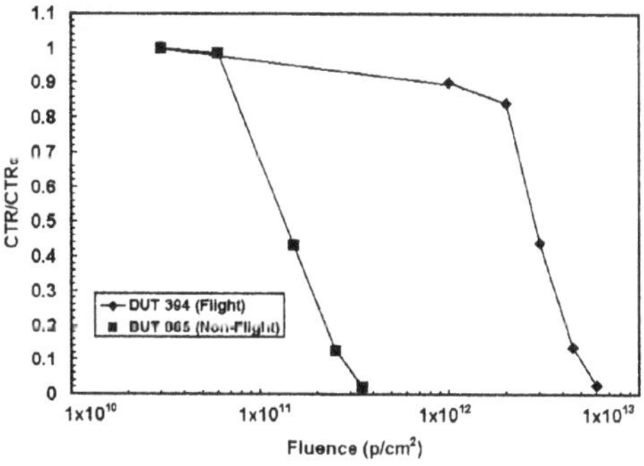

Fig. 14 Demonstration of lot-to-lot variability of CTR degradation.[10]

4.5.5 Anneal of Displacement Damage

Injection current annealing of displacement damage-induced degradation of LED light output was first reported by Barnes et al. and then further discussed by Johnston et al.[3,12] Annealing has also been observed in optocouplers.[10] **Figure 15** shows the recovery of the collector current under two annealing currents (1 mA and 10 mA) for three measurement forward currents (1, 5, and 10 mA) for an Isolink 4N49. The data connected with the solid lines were obtained after the device was annealed at 10 mA (unfilled symbols), while the data represented by the filled symbols were taken at 1 mA annealing forward current. The collector current is normalized to the pre-irradiation values. The abscissa is the amount of charge in coulombs that has passed through the LED during the annealing period. For annealing at 10 mA, the time to reach 1800 coulombs is 50 hours, while the time to reach this charge at 1 mA is 20.8 days. The recovery factor shown in the legend gives the ratio of the collector current after 1800 coulombs anneal to post-irradiation (prior to the anneal step) values for the collector current.

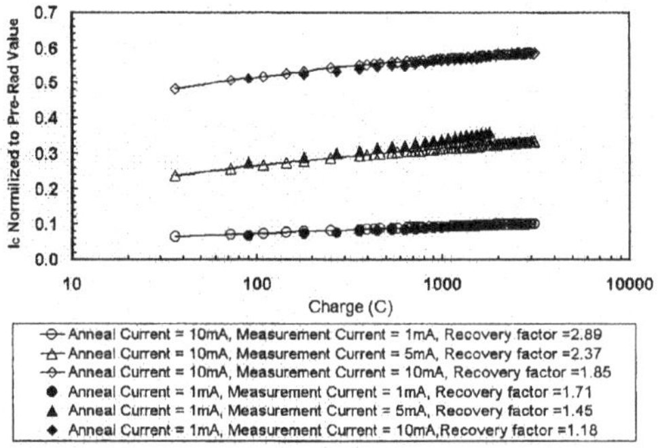

Fig. 15 Injection current annealing effect on collector current at two forward currents for three application currents. The data is normalized to the pre-irradiation collector currents.[10]

First note that for the annealing conditions between 100 and 1800 coulombs, the amount of recovery is approximately the same for the 1 and 10 mA annealing conditions for all three test conditions. This implies that an accelerated annealing test could be used for these annealing periods, reducing the time to determine the amount of recovery by a factor of 10. If this accelerated annealing approach could be generalized, then one could use it to estimate the annealing occurring over long mission times. However, in order to generalize this accelerated annealing test, more data on other part types are needed. Also, for missions with long annealing times, studies out to greater integral charge values will be required. This accelerated annealing approach should be studied in more detail before being implemented in a risk assessment approach.

The second interesting point that these data show is the collector-current recovery-factor (the ratio of the annealed collector current to the post-irradiation value) is dependent on the forward current used for the measurement and the annealing current. The recovery factor is also dependent on the anneal time. All of these are application-dependent issues.

Annealing should be minimized while collecting CTR degradation data over the test fluence. This can be best achieved if the optocoupler "on-time" (when the LED is illuminating the photodetector) is minimized and the forward current is minimized.

4.5.6 *Temperature Effects Impact CTR*

Temperature effects must also be considered when determining the likelihood of optocoupler survival when exposed to a radiation environment.[10] The data in **Figure 16** shows that for an Isolink 4N49 there is a 30 to 40% decrease in the CTR when the device is warmed from room temperature (~23°C) to 100°C.

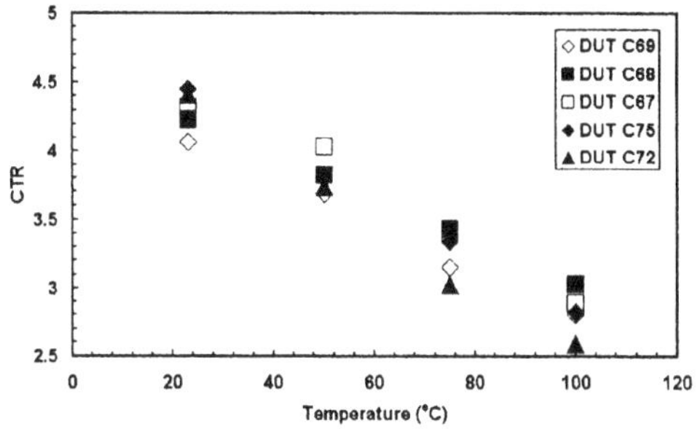

Fig. 16 Temperature effects on CTR for an Isolink 4N49 optocoupler.[10]

4.6 Final Comments on Permanent Degradation

Despite the ominous list of nightmares above it is possible to take a conservative approach for estimating optocoupler CTR degradation when exposed to space radiation. This approach is outlined by Reed et al.[10] The risk assessment techniques described by Reed require significant margin, as much as a factor of 50 for certain scenarios.[10]

5.0 Conclusions

Space radiation-induced performance degradation of optocouplers was first reported in 1996 and 1997. In 1996, Rax et. al reported on the first inflight anomaly that was reportedly caused by CTR degradation of a 4N49 optocoupler.[4] The following year LaBel et. al reported on the first space flight anomaly that could be traced to transient effects in a 6N134 high bandwidth optocoupler.[6] Since that time a significant amount of effort has been put into solving some of the mysteries that exist when assessing the space worthiness of optocouplers. The overriding result that has come out of this research is that there is no "quick-fix", "one-size fits all" solution. Every optocoupler must be assessed against the environment that it will be exposed to and the function that it will be performing. We hope that, by using the information provided in this paper and good engineering judgment, radiation effects experts can work their way through the difficult problem of space-qualifying an optocoupler.

References

1. A.S. Epstein and P.A. Trimmer. "Radiation Damage and Annealing Effects in Photon Coupled Isolators," *IEEE Trans. Nucl. Sci.*, Vol. 19, pp. 391-399, 1972.
2. K.J. Soda, C.E. Barnes, and R.A. Kiehl "The Effect of Gamma Irradiation on Optical Isolators," *IEEE Trans. Nucl. Sci.*, Vol. 22, pp. 2475-2481, 1975.
3. C.E. Barnes and J.J. Wiczer, "Radiation effects in optoelectronic devices," Sandia Report, SAND84-0771, 1984.
4. B.G, Rax, C.I. Lee, A.H. Johnston, and C.E. Barnes, "Total Dose and Proton Damage in Optocouplers." *IEEE Trans. Nucl. Sci.*, Vol. 43, pp. 3167-3173, 1996.

5. M. D'Ordine, "Proton Displacement Damage in Optocouplers." *IEEE Radiation Effects Data Workshop*, July 1997: pp. 122-124.
6. K.A. LaBel, P.W. Marshall, C.J. Marshall, M. D'Ordine, M. Carts, G. Lum, H.S. Kim, C.M. Seidleck, T. Powell, R. Abbott, J. Barth, and E.G. Stassinopoulos, "Proton-induced transients in optocouplers: in-flight anomalies, ground irradiation test, mitigation and implications," *IEEE Trans. Nucl. Sci.*, Vol. 44, pp. 1885-1892, 1997.
7. R.A Reed, P.W. Marshall, A.H. Johnston, J.L. Barth, C.J. Marshall, K.A. LaBel, M. D'Ordine, H.S. Kim, and M.A. Carts, "Emerging Optocoupler Issues With Energetic Particle-Induced Transients And Permanent Radiation Degradation," *IEEE Trans. Nucl. Sci.*, vol. 45, pp. 2833–2841, 1998.
8. A.H. Johnston, T. Miyahira, G.M. Swift, S.M. Guertin, and L.D. Edmonds, "Angular and energy dependence of proton upset in optocouplers," *IEEE Trans. Nucl. Sci.*, Vol. 46, pp. 1335-1341, 1999.
9. K.A. LaBel, S.D. Kniffin, R.A. Reed, H.S. Kim, J.L. Wert, D.L. Oberg, E. Normand, A. H. Johnston, G.K. Lum, R. Koga, S. Crain, J.R. Scwank, G.L. Hash, S. Buchner, J. Mann, L. Simpkins, M. D'Ordine, C.J. Marshall, M.V. O'Bryan, and C. Seidleck, "A Compendium of Recent Optocoupler Radiation Test Data," *IEEE Radiation Effects Data Workshop*, July 2000: 123-146.
10. R. Reed, C. Poivey, P.W. Marshall, K.A. LaBel, C.J. Marshall, S. Kniffin, J.L. Barth, and C. Seidleck, "Assessing the Impact of the Space Radiation Environment on Parametric Degradation and Single Event Transients in Optocouplers," *IEEE Trans. Nucl. Sci.*, Vol. 48, in press.
11. A.H. Johnston, G.M. Swift, T. Miyahira, S. Guertin, and L.D. Edmonds., "Single-Event Upset Effects in Optocouplers." *IEEE Trans. Nucl. Sci.*, Vol. 45, pp. 2867-2875, 1998.
12. A.H. Johnston, B.G Rax, L.E. Selva, and C.E. Barnes, "Proton degradation of light-emitting diodes," *IEEE Trans. Nucl. Sci.*, Vol. 46, pp. 1781-1789, 1999.
13. R.A. Reed, Figure 4.6 in "Performance Characterization of Digital Optical Data Transfer Systems for Use in the Space Radiation Environment," IEEE NSREC Short Course, Reno NV, 2000.
14. The Geant4 Collaboration, http://wwwinfo.cern.ch/asd/geant4/geant4.html.
15. L.S. Waters (Ed.), "MCNPX User's Manual: Version 2.4.0", LA-CP-02-408, Los Alamos National Lab, September 2002.
16. T.M. Jordan, "NOVICE: A Radiation Transport and Shielding Code", Experimental and Mathematical Physics Consultants, Report EMPC 98.01.02.01
17. P. V. Dressendorfer, "Basic mechanisms for the new millennium," Notes from 1998 IEEE Nuclear and Space Radiation Effects Conference Short Course.
18. J.L. Leray, "Total dose effects: modeling for the present and future," Notes from 1999 IEEE Nuclear and Space Radiation Effects Conference Short Course.
19. P.W. Marshall and C.J. Marshall, "Proton Effects and Test Issues for Satellite Applications," IEEE NSREC99 Short Course, IV1-110, July 1999.
20. C.E. Barnes, "Radiation hardened optoelectronic components: sources," *Proc. SPIE*, Vol. 616, pp. 248-252, 1986.
21. P.W. Marshall, C.J. Dale, and E.A. Burke, "Space radiation effects on optoelectronic materials and components for a 1300 nm fiber optic data bus," *IEEE Trans. Nucl. Sci.*, Vol. 39, pp. 1982-1989, 1992.
22. A.H. Johnston, "Optoelectronic Devices with Complex Failure Modes," IEEE NSREC Short Course, Reno NV, 2000.
23. J.J. Wiczer, "Radiation hardened optoelectronic components: Detectors," *Proc. SPIE*, vol. 616, pp. 254-266, 1986.
24. E.A. Burke, C.J. Dale, A.B. Campbell, G.P. Summers, T. Palmer, and R. Zuleeg, "Energy Dependence of Proton-Induced Displacement Damage in GaAs, *IEEE Trans. Nucl. Sci.*, Vol. 34, p. 1220, 1987.

25. G.P. Summers, E.A. Burke, M.A. Xapsos, C.J. Dale, P.W. Marshall, and E. Petersen, "Displacement damage in GaAs structures," *IEEE Trans. Nucl. Sci.*, Vol. 35, 1988.
26. A.L. Barry, A.J. Houdayer, P.F. Hinrichsen, W.G. Letourneau, and J. Vincent, "The Energy Dependence of Lifetime Damage Constants in GaAs LEDs for 1-500MeV Protons," *IEEE Trans. Nuc. Sci.*, Vol. 42, pp. 2104-2107, 1995.
27. R.A. Reed, P.W. Marshall, C.J. Marshall, R.L. Ladbury, H.S. Kim, L.X. Nguyen, J.L. Barth, and K.A. LaBel, "Energy Dependence of Proton Damage in AlGaAs Light Emitting Diodes," *IEEE Trans. Nucl. Sci.*, Vol. 47, pp. 2492-2499, 2000.

International Journal of High Speed Electronics and Systems
Vol. 14, No. 2 (2004) 419–443

World Scientific
www.worldscientific.com

RADIATION EFFECTS IN CHARGE-COUPLED DEVICE (CCD) IMAGERS AND CMOS ACTIVE PIXEL SENSORS

G. R. HOPKINSON AND A. MOHAMMADZADEH*

Sira Electro-Optics Ltd
South Hill, Chislehurst, Kent, BR7 5EH, UK
gordon.hopkinson@siraeo.co.uk

**ESA, ESTEC*
Keplerlaan 1, 2200 AG Noordwijk, The Netherlands
ali.mohammadzadeh@esa.int

This review concerns radiation effects in silicon Charge-Coupled Devices (CCDs) and CMOS active pixel sensors (APSs), both of which are used as imagers in the visible region. Permanent effects, due to total ionizing dose and displacement damage, are discussed in detail, with a particular emphasis on the space environment. In addition, transient effects are briefly summarized. Implications for ground testing, effects mitigation and device hardening are also considered. The review is illustrated with results from recent ground testing.

Keywords: CCD; APS; radiation; damage

1. Introduction

This review looks at radiation effects in silicon charge-coupled device (CCD) imagers. It also makes a comparison with effects in CMOS active pixel sensors (APSs) - which are emerging as a rival technology for many applications, particularly where low cost or high radiation tolerance are needed.

Both the CCD and the APS consist of an array of imaging detectors (pixels), either photodiodes or MOS capacitors (photogates). The main difference is in the way the collected photo-generated charge is readout. In the CCD the charge is transferred from one pixel to the next (and eventually to an output amplifier) by splitting the gate electrodes into several phases (usually 2, 3 or 4) and clocking these phases in an appropriate pattern (figure 1 shows a schematic structure and figure 2 a typical clocking pattern). In contrast, the active pixel sensor uses CMOS shift registers and transistor switches to connect each pixel in turn to the output. In order to buffer the detector from the high capacitance of the address lines it is necessary to have an amplification stage in each pixel (hence the term 'active'); usually there will also be an amplifier associated with each column. Figure 3 shows a simplified schematic. With the flexibility offered by CMOS fabrication it is often possible to include additional features, such as an analog-to-digital convertor (ADC) on-chip.

CCD formats (number and size of pixels) vary widely, from small (e.g. 14 x 14 pixels) detectors for high speed tracking or wavefront sensing to large (e.g. 9216 x 9216 pixels [1]) for scientific imaging. Large format APS arrays (e.g. 13.85 million pixel) are also starting to become available.[2] CCDs and APSs are extremely sensitive detectors for visible photons either in 2-D formats for use in 'cameras', or as 1-D (linear) devices for spectroscopy or scanning applications. Many of these applications will involve exposure to particle or gamma radiation. For example, military, scientific (e.g. high energy physics) and space (star trackers, acquisition and tracking sensors, astronomical and Earth observation instruments).

The sensitivity, which makes these solid state array detectors so useful, also makes them vulnerable to radiation-induced performance changes. Both types of imager suffer from the permanent effects of:

- total ionizing dose (due to charged particles, x-rays and gamma rays).
  This affects the MOS structures (producing a charging of the gate insulator and build-up of interface traps)
- displacement damage (usually due to heavy particles such as protons and neutrons).
  This creates defects in the crystal lattice, and hence dark current generation and charge trapping centers.

In addition, devices will show transient 'events' due to ionizing radiation which deposits charge in the photodetectors. Although these transient events will cause contamination in an image, they are short-lived and only occur in the exposure in which they were generated.

CCD design and operation have recently been reviewed in books by Janesick[3] and Theuwissen.[4] Radiation effects in CCDs have been discussed in [5], [6], [7] and [8]. In the following sections the topics of total ionizing dose damage, displacement damage and transient effects will be discussed in turn. In each section suggestions will be made concerning implications for ground testing, effects mitigation and device hardening.

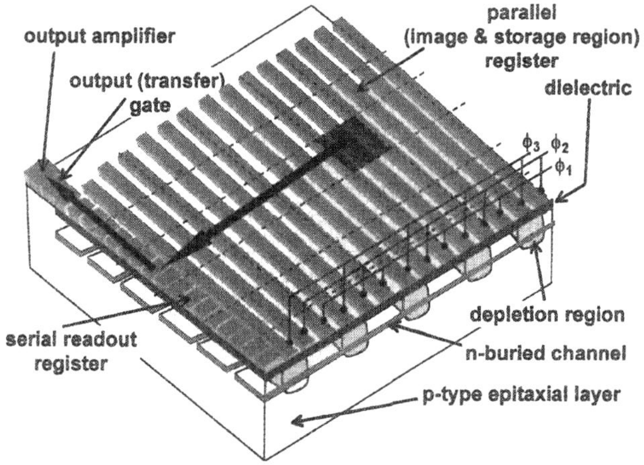

Fig. 1. Schematic of a typical 3-phase n-buried channel CCD.

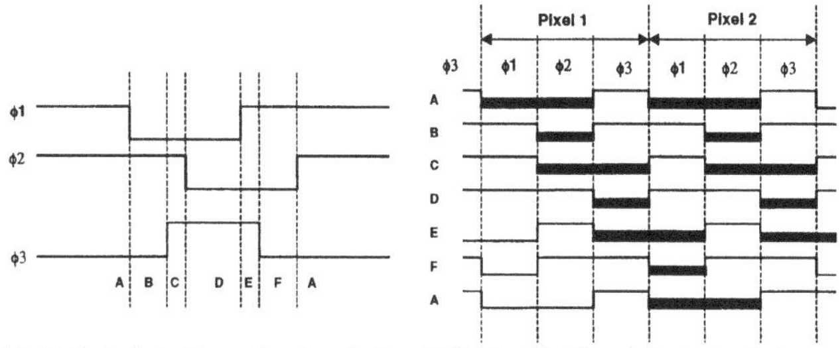

Fig. 2. Typical clocking pattern for a 3-phase CCD. The thick lines indicate stored charge.

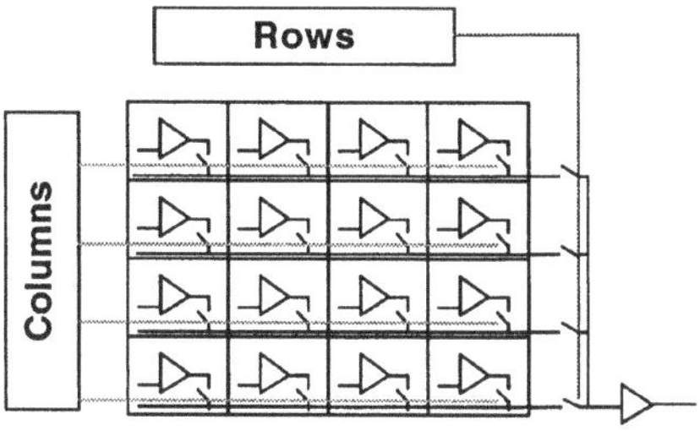

Fig. 3. Simplified schematic of a CMOS active pixel sensor.

2. Damage due to Total Ionizing Dose (TID)

Both CCDs and APSs use dielectric layers, such as silicon dioxide, to isolate gate electrodes from each other and from the silicon. In the case of the CCD the gate isolation is typically a dual layer of SiO_2 and Si_3N_4 (which acts as an etch-stop during processing of the CCD electrodes). The total thickness is of order 150 nm. For an APS the gate oxides in typical CMOS processes are thinner (on the order of a few nm) but field oxide isolations can be similarly thick. Ionizing radiation causes the generation of electron-hole pairs in these layers, the number of pairs increasing linearly with the thickness. A fraction of the charge will become trapped in the dielectric and another fraction will release hydrogen and induce interface traps. The basic mechanisms of total dose damage have been reviewed by, for example, Braunig and Wulf[9] and Schwank.[10] Damage prediction is straightforward since it is usually proportional to total ionizing dose, measured in rd(Si) or Gy(Si) - independent of the type of radiation - though saturation effects can sometimes be seen at high doses (of order 1 Mrd). In general, the response of MOS circuits is sensitive to the dose rate of the irradiation (e.g. in rd/s) and there are time dependent effects (such as annealing of trapped holes and slow build up of interface traps). In CCD imagers and APSs these dose rate effects are not usually pronounced, though they can sometimes be observed.

The gate dielectrics have a capacitance (decreasing with thickness) and so the trapped charge produces a shift in the 'effective' voltage applied to the gate (known as the flatband voltage shift). The other effect of permanent ionization damage, as mentioned above, is the build-up of interface traps and this increases dark current generation at those regions of the surface (that is, the Si/SiO_2 boundary) that are depleted. The surface has to be depleted because generation processes (which take place by carrier emission) only dominate over recombination (capture) processes when the free carrier concentrations are significantly less than their thermal equilibrium values.[11]

2.1. *Radiation-induced surface dark current due to TID*

Interface traps have energy levels within the silicon bandgap and act as centers for the generation of surface dark current in both CCDs and APSs. In a full frame or frame transfer CCD the generation is at the depleted surface under the active clock phase (held high in an n-channel CCD so as to collect charge) and in an APS (or an interline transfer or a linear CCD) at the depleted surface of the photodiode (or, with some types of APS, under the photogate). The build-up of interface traps can be a slow process, for example involving the diffusion of hydrogen through the oxide. After a cobalt-60 irradiation CCD dark current is often found to increase (by a factor of 2 or 3) over time periods of several weeks or months, though it can be accelerated by baking at 100°C for 168 hours (as is done for routine MOS device testing for 'rebound', as in ECSS Basic Specification 22900 and MIL-STD-883 method 1019.4[12]). In fact 168 hours may not always be needed and a 24 hour bake can be sufficient.[13] The final value for the increase, for the image region of a frame transfer CCD, is typically ~ 1 $nA/cm^2/krd(Si)$ at 20°C, assuming 20 μm x 20 μm pixels (corresponding to ~ 2.5 10^4 electrons/pixel/s). Some early CCDs, with thick oxide channel stop isolations, showed much larger reverse annealing[14,15] but this is rarely a problem today.

The dark current increase is often more for biased devices than for unbiased, but bias dependence depends on the device details. The dark current is normally fairly uniform across the CCD (though there will be statistical pixel-to-pixel fluctuations).

For regions of the CCD which are masked so as to form a storage region (in a frame transfer CCD) or to form dark reference pixels, the radiation-induced surface dark current can be increased (by a factor ~ 2 or 3) because of hydrogen incorporated into surface layers during processing. Sometimes a CCD is fitted with an external (non-deposited) store shield for this reason. In addition, hydrogen (or a similar interface trap enhancing species) can be generated at the periphery of the device, outside the active region and can sometimes diffuse in to give an increased dark current in the first and last columns. Hence it is advisable to avoid using these regions in any image processing.

Fortunately, for many applications, the above discussion will not be applicable because the CCD can be operated in inverted mode where surface dark current is negligible. In this mode (sometimes called multi-phase pinned or MPP), all the clock low levels are lowered with respect to the substrate voltage to the point where the surface becomes inverted and holes from the substrate and channel stop regions migrate to the surface and keep the interface traps filled. Usually the surface dark current will then be negligible, although if the surface is not completely inverted (e.g. if regions at the edge of a pixel are left non-inverted) then the dark current will be slightly increased. As will be discussed in the next section, if the total dose is large, usually several tens of krd(Si), then the flatband voltage shift can be high enough for the surface to become non-inverted. In this case the full surface dark current will appear.

A drawback of using MPP mode is that an implant is needed to create a potential well for storing the charge when all the clocks are low. This results in a reduced full well capacity (unless special 'advanced MPP' structures are used). A compromise solution is to not have an MPP implant but to operate all the clock phases except one in inverted mode and to periodically swap the phase which is used as a barrier. If the holes under this phase are trapped for longer than the time between swaps then the dark current is significantly reduced. This is known as dynamic dark charge suppression or dither clocking and has been described by Burke and Gajar.[16] As the hole trapping time increases as the CCD temperature is decreased, the reduction in surface dark current tends to be larger. This dynamic suppression happens naturally during continuous readout of the storage region of a frame transfer CCD.

In linear and interline transfer CCDs, which use photodiode detector elements, the inverted mode can still be used and the detectors are usually referred to as 'pinned photodiodes'.

In CMOS active pixel sensors it is not possible to invert the surface but the radiation induced dark current is still very small. With photogate structures the surface dark current is reduced because the gate oxide is thin. A value of 1-2 $pA/cm^2/krd(Si)$ has been reported[17] for one such device at room temperature. If a photodiode is used then this can have a small area (provided that it can be arranged to collect charge from the whole pixel area) and the surface dark current is again reduced.[18] In one case[19] it has been observed that the dark current, though small, is non-uniform across the array (see figure 4). Again, values of 1-2 $pA/cm^2/krd(Si)$ are typical for biased devices: a value of 310,000 electrons/pixel/s (3 nA/cm^2) was found after 5.3 Mrd[19] at 20°C and slightly lower values were found by Bogaerts et al.[20]

Surface dark current varies strongly with temperature, hence cooling the detector is an effective way of improving performance. An empirical relation for the change is

$$\text{Dark current} = \text{constant} \times \exp(-E_{act}/kT) \quad (1)$$

Where the activation energy, E_{act}, is in the range 0.63 - 0.65eV (though for dither clocked devices the effective activation energy will be increased).

Fig. 4. Dark current for a CMOS APS (Fillfactory STAR-250, 25 μm x 25 μm pixels) after 100 krd(Si) biased cobalt-60 irradiation, giving a peak increase (center of chip) of 0.42 nA/cm^2 at 25°C

2.2. *Flatband voltage shift*

The result of an ionizing radiation-induced flatband shift is to change the effective voltages applied to the device. In a typical 'commercial' CCD the flatband shift is roughly 0.1 V/krd(Si) when the device is biased during irradiation and a half to one quarter that when not biased (the difference being due to the effect of electric field on charge yield and transport in the dielectric). This shift will tend to increase the clock voltage differential (relative to substrate) at which the surface becomes inverted and will also tend to make the output node less attractive to clocked charge and to change the operating point of the output amplifier (which may alter the gain).

Figure 5 shows the shift in inversion point for a typical CCD. If the clock low voltage is not selected appropriately then it is possible for the device to come out of inversion mode after a high total dose and then to exhibit increased dark current.

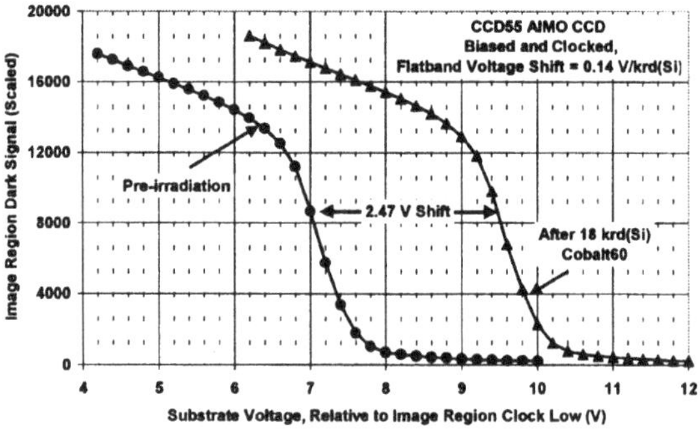

Fig. 5. Shift in inversion point for an advanced inverted mode (AIMO) CCD55 from E2V Technologies, UK, after 18 krd(Si) cobalt-60. The dark current has been scaled so that the pre-and post radiation dark currents are comparable.

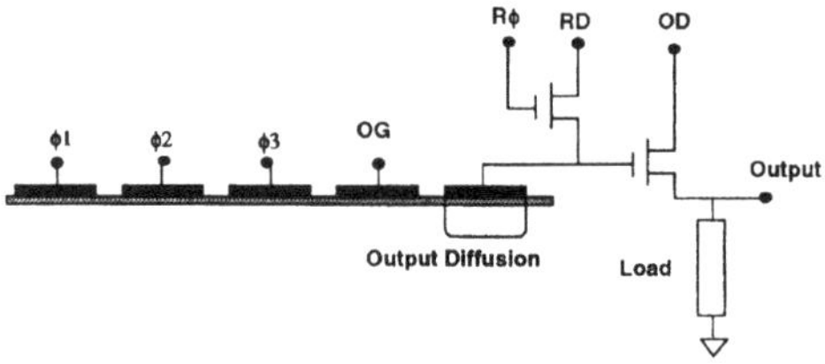

Fig. 6. Output circuit for a typical CCD

Figure 6 shows a typical CCD output circuit. The effect of total dose will depend on the CCD design and the allowed voltage margin. For TID higher than ~ 10 krd(Si) it is common for the output bias voltages (V_{OD} and V_{RD}) to need increasing and sometimes for the readout clocks to be lowered (or else charge cannot pass over the output gate (OG) into the amplifier - resulting in a complete loss of imaging capability). As a 'rule of thumb', 1-2 V voltage shift can be accommodated by choosing optimal biases prior to irradiation (which may be different from the datasheet values) and 2-4 V shift requires bias adjustment during irradiation.

CCD hardening is possible by thinning the dielectric (though this may have an adverse affect on device yield) or by using a hardened oxide process. Use of a dual SiO_2/Si_3N_4 dielectric can be advantageous since electrons trapped in the nitride can balance the charge from holes trapped in the oxide.

For a CMOS APS the gate oxide flatband voltage shifts will be much smaller because of the thinness of the oxide and are generally not an issue. For example, a 5.3 Mrd biased cobalt-60 irradiation of one device[19] showed only 80 mV shift. A change in APS responsivity was found in one case[20], probably due to TID effects.

2.3. *Parasitic leakage currents in CMOS APS devices*

For APS devices a much more important effect than gate oxide flatband voltage shift is the possibility that thick field oxide isolation regions can become inverted and so provide a parasitic leakage path leading to increased power consumption. The area of depleted silicon at the surface may also increase leading to greater thermal dark current. As for all oxide effects, there is the possibility of post-irradiation annealing.

Fortunately, even in commercial devices, it is possible to choose radiation tolerant CMOS libraries for device layout and so achieve 'hardening by design'.[21] By incorporating enclosed 'gate all around' transistors it is possible to eliminate field oxide leakage effects. Examples of such devices have been referred to above[17,19]. The drawback is that special device structures will reduce the sensitive area (the fill factor) of the device and hence the signal-to-noise performance. A fill factor of 45% can still be achieved in some cases (e.g. the APS in [19,20]).

At the same time, it is possible to harden against single event latch-up (SEL) by using a thin epitaxial layer, guard rings and frequent body ties. SEL has not been observed in the hardened device studied by Marshall et al.[22] (at a maximum LET of 106 MeV/mg/cm^2), nor in the STAR-250 APS (maximum LET of 68 MeV/mg/cm^2). The drawback of using a thin epitaxial layer is a reduction in red response (for a front illuminated device).

2.4. *Testing for total ionizing dose effects*

Total dose testing of detector arrays is almost always carried out using cobalt-60 sources. Since characterizing the detector can be time consuming, it is useful to use a low dose rate (e.g. in the range 1-10 krd/hour) so that any short-term annealing does not affect the measurements. Fortunately, post-irradiation annealing tends to take place over weeks and months after irradiation and is, in any case, small for flatband voltage shift. Any reverse annealing of surface dark current can be accelerated by a bake at 100°C, as mentioned above.

Total dose damage is dependent on the irradiation bias and it is usual for both biased and unbiased devices to be tested (with biased usually being worst case).

3. Displacement Damage

Displacement damage is caused by particles which collide with the silicon atoms within the crystal lattice of the detector array and create vacancy-interstitial pairs. Most of these will recombine after the collision but some will migrate through the lattice and form stable defects such as the divacancy, V_2 (two vacancy complex) and the E-center (phosphorous-vacancy complex).[23] These defects have energy levels within the silicon bandgap and can act as dark current generation centers (in CCDs and active pixel sensors) and minority carrier traps (which cause degradation in charge transfer efficiency in CCDs). Most of a particle's energy loss occurs by ionization, only about 0.1% goes into producing displacements.

The collision process takes place by the initial particle (e.g. a proton or a neutron) displacing a silicon atom and creating what is known as a primary knock-on atom or PKA. The recoil energy given to the PKA depends on the particle type and energy and can be worked out from the kinematics of the collision. For protons and neutrons the maximum energy, T_{max} that can be transferred to the PKA is given by[23,24]

$$T_{\max} = \frac{4Mm}{(m+M)^2}E = 0.133E \qquad (2)$$

where E is the kinetic energy of the incident proton, m the proton mass and M the mass of a silicon atom. For low energy protons (less than a few tens of MeV) most of the interactions are via elastic Coulombic (Rutherford) scattering and most of the protons carry on in the forward direction, imparting very little energy to the PKA. In fact the probability of creating a PKA with energy T is proportional to $1/T^2$ [24] so that the energy spectrum of the proton-induced PKAs falls sharply with increasing energy and the average PKA energy is low (in fact it stays ~ 200 eV for a wide range of proton energies[23]). This means that the PKAs generally do not have enough energy to displace other silicon atoms, a process which requires a PKA energy of several keV (Wood et al.[25] give a value 1-2 keV and Lindstroem[26] has recently quoted 5 keV). Hence proton elastic collisions tend to produce isolated defects.

As the incident proton energy increases above ~ 10 MeV there is an increasing chance of an inelastic collision (nuclear spallation reaction). These produce higher energy PKAs which can in turn displace other silicon atoms to produce a cascade of defects. If the PKA energy exceeds about 12 keV then multiple cascades can be formed. In these cascades and multiple cascades the lattice defects tend to be grouped together in clusters. Displacements will carry on taking place (with the production of vacancy/interstitial pairs) until the energy of the silicon atom(s) falls below the threshold energy for displacements (which, for silicon is ~ 20-25eV[9,23]).

Electrons always produce isolated point defects since they can only impart a very small energy to the PKA and the same is true for cobalt-60 gamma rays where the displacement damage is primarily due to the production of Compton electrons having a maximum energy of only 1 MeV.

Neutron irradiation, on the other hand, gives a PKA spectrum which is much flatter. In fact, at low neutron energies the collisions approximate to hard sphere interactions, leading to a perfectly flat spectrum up to the maximum transmitted energy given in equation (2). This means that neutrons tend to give much more cluster damage than do protons. It may seem, therefore, that the nature of the displacement damage will depend on the particle type and energy. However, as Wood et al.[25] first showed, all cascade damage is very similar, independent of how many cascades or sub-cascades there are. It may still be, though, that the defects created in clusters are different from isolated point defects. Nevertheless, the usual next step is to assume that the nature of ***all*** defects is independent of whether they are isolated or in clusters and this leads to the important hypothesis that damage is proportional to the total energy (per particle per cm^2/g) that goes into displacements - known as the non-ionizing energy loss, or NIEL.

The vacancies and interstitials produced are mobile and will migrate through the lattice. Roughly 90% will recombine, leaving no permanent effect but, as mentioned above, the remainder can form stable defects. Although the number of these stable defects is difficult to predict from first principles, we can expect that the final damage will always be proportional to the amount of initial damage and also to the electrical effect on the device. Hence NIEL scaling implies that we have a universal relation:

$$\text{device damage} = k_{damage} \times \text{displacement damage dose} \qquad (3)$$

Where k_{damag} is a damage constant dependant on the device and the parameter affected and the displacement damage dose (DD) is the product of the NIEL and the particle fluence. DD plays the same role for displacement damage as total dose does for ionization damage[27]. In the case where there is a spectrum of particle energies the DD has to calculated by integrating over the particle differential energy spectrum, $d\phi(E)/dE$ (and summing the contributions from the different particle types if there are more than one):

$$DD = \int NIEL(E)\frac{d\phi(E)}{dE}dE \qquad (4)$$

Since the device effects will normally depend on the characteristics of the defects involved, NIEL scaling will only be valid if the concentration of the dominant defect (for the parameter concerned) is similar for both point and cluster damage. Huhtinen[28] has recently shown that this is valid for the divacancy but it might be expected (for example) that higher order vacancy defects (V_3, V_4 or V_5) may be more common in cluster regions (and relatively more numerous after neutron damage). Nevertheless, NIEL scaling is always a good first step in allowing for the incident particle spectrum.

NIEL in silicon has been computed as a function of energy by several authors. Figure 7 shows a compilation of the available data. The proton data of Dale et al.[29] is a slightly revised version of earlier work by Burke[30] and has been extrapolated to higher energies by Huhtinen and Arnio.[31] This data is in good agreement with recent Monte Carlo simulations by Jun.[32] Other work by Akkerman et al.[33] gives somewhat higher values below 30 MeV but lower values above 60 MeV. The neutron data is from ASTM E 722-94.[34] Recent work by the authors suggests that bulk damage in CCDs more closely follows the Dale et al. curve.

Akkerman et al.[33] have also calculated the NIEL due to cobalt-60 gamma rays. Behind a few mm of Aluminum shielding, when the NIEL has reached an equilibrium value, they give a NIEL value of 1.07 10^{-7} MeV-cm^2/g.

The NIEL concept is particularly useful if ground testing is performed at one energy (e.g. 10 MeV) and results need to be extrapolated to the use environment (e.g. space). Rather than being quoted in keV-cm^2/g (or MeV-cm^2/g), the displacement damage dose is often given as an equivalent fluence of 10 MeV protons or as the displacement damage cross section, in MeVmb (100 MeVmb = 2.144 keV-cm^2/g).

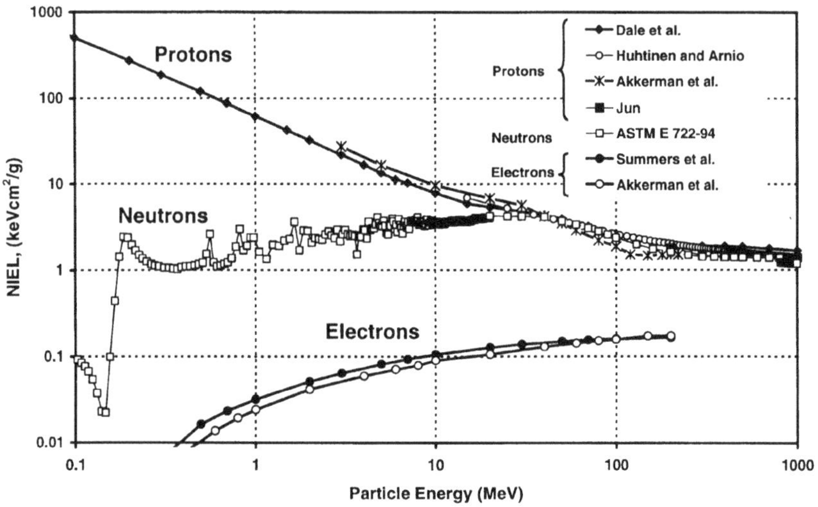

Figure 7 NIEL data for protons, neutrons and electrons in silicon

3.1. *Bulk Dark Current*

As mentioned above, displacement damage causes the formation of dark current generation centers in the silicon bandgap. These lie in the bulk of the depletion region (rather than at the surface) and give rise to what is known as bulk dark current. The most effective defects (giving the highest dark current) have an energy close to mid gap. The bandgap in silicon is 1.1 eV and so common radiation-induced defects such as the divacancy and the E center, which have energies of ~ 0.44 eV below the conduction band can be important (though any defect at ~ 0.55 eV would be more effective). The main feature of displacement damage induced bulk dark current is that it is very nonuniform as illustrated in figure 8. To characterize this dark current it is necessary to measure the distribution of pixel values.

Two typical examples from recent CCD measurements are shown in figure 9. The histograms have a broad tail extending to large signals. Although the mean dark signal increase due to displacement damage is ~ 8,000 electrons/pixel/s, the largest dark current 'spikes' are an order of magnitude greater. These large amplitude signals are due to inelastic proton collisions which deposit a large amount of energy and also to those elastic collisions which happen to occur in a region which has a high electric field. The second mechanism tends to be more important in CCDs and is known as field enhancement.[35,36] The presence of a high electric field (~ 10^5 V/cm) lowers the potential barrier for a defect to emit charge (either directly or via phonon-assisted tunneling) and so increases the generation rate. A characteristic of this mechanism is that the activation energy for dark current generation is lowered. Normally bulk generation will have roughly the same activation energy as surface generation (i.e. 0.63-0.65 eV, as mentioned above) but the brightest (field enhanced) pixels can have activation energies as low as 0.4 eV. This means that they change more slowly with temperature and are relatively more pronounced at low temperatures (that is, the dark images get more nonuniform).

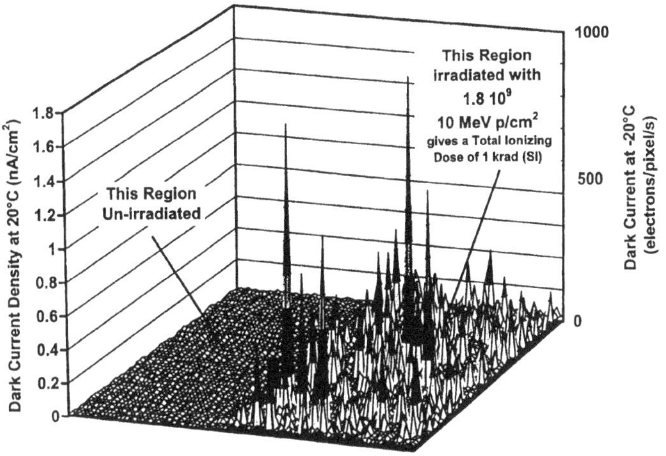

Figure 8 Dark current spikes in a proton-irradiated Atmel TH7895 CCD

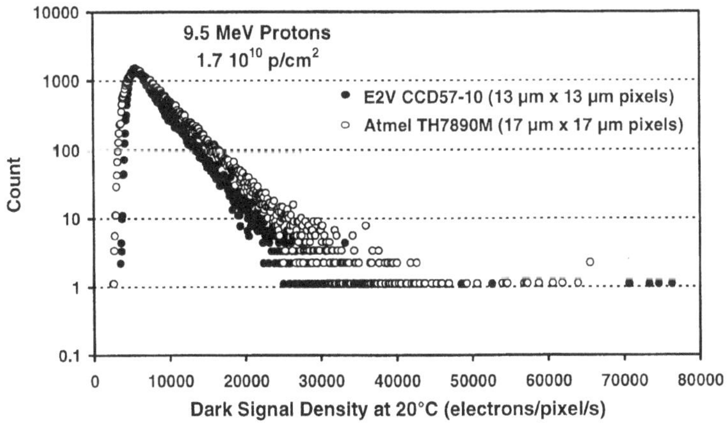

Figure 9 Histograms of dark current density for two types of CCD

Although the mean bulk dark current is determined by the displacement damage dose, the shape of the distribution of pixel-to-pixel variations (dark signal nonuniformity or DSNU) cannot, since it depends on the variations in the distribution of PKA energies (and hence on the incident particle type and energy) and on the variations in the electric field within a pixel (since this affects the proportion of defects that are influenced by field enhancement). Dale et al.[29] have used the CUPID Monte Carlo code to model the damage cascades in detail and found good agreement with dark current histograms for low field devices. Such simulations are needed in cases where the incident particle energy is high enough that the range of a typical PKA can take it outside its original sensitive volume (e.g. for protons > 60MeV and pixel dimension ~ 10 μm). Otherwise, an analytical formalism developed by Marshall et al.[37] can be used (see also Robbins[38]). In this, the elastic and inelastic collisions are treated separately. With the cross sections for these reactions (σ_e and σ_i) and an estimate of the sensitive volume (V), the number of interactions (N) in the sensitive volume can be calculated from

$$N = \sigma\rho\phi V\, N_A/A \tag{5}$$

Where ρ and A are the density and atomic mass of silicon, N_A is Avagadro's number and N can be calculated for elastic or ineleastic collisions (depending on the cross section used). For example, for 63 MeV protons σ_e and σ_i, are 318 and 0.52 barns,[37] so that the number of elastic and inelastic collisions in a 900 μm^3 volume (e.g. 15 μm x15 μm x 4 μm) for a fluence of 2 10^{10} p/cm^2 are 300 and 0.5 respectively. Thus there are far more elastic collisions than inelastic. On the other hand the inelastic collisions deposit roughly three orders of magnitude more energy. The probability density function (pdf) for the energy deposited in a single inelastic collision is highly skewed towards the higher energies and Marshall et al. modeled this with a two parameter gamma function with mean and variance determined from their calculations of PKA spectra (updated values are tabulated in Dale et al.[29]). The n-fold convolution of this distribution then represents the pdf for n collisions within the sensitive volume. Elastic collisions are modeled by a Gaussian distribution, again with a mean and variance determined from the PKA spectra (Robbins[38] found that he could obtain a better fit to experimental CCD data using a gamma distribution). The final distribution is then a weighted sum of the elastic pdf and the pdfs for 1,2, 3 etc. inelastics. Comparing this with the dark current histogram allows an estimate of the conversion factor between damage energy and dark current: k_{dark}.

In many cases the analytical model gives a good fit to experimental data but this will not be so when field enhancement is significant. Hence it is not usually possible to predict the distribution of the brightest dark current spikes – and it is these which will usually determine the cosmetic quality and scientific usefulness of the images. In practice it is found for CCDs that the maximum dark current of these dark current spikes is in the range 4-10 nA/cm^2 at 20ºC. If the electric field distribution can be estimated (and hence the distribution of field enhancement factors) then this can be used to form a weighted sum of the pdfs for each field value. This has been done by Bogaerts et al.[39] for a CMOS active pixel sensor where, as in CCDs, field enhancement can be an important factor.

With an APS there is some scope either to change the pixel geometry to reduce the electric fields or simply to reduce the applied voltage. Figure 10 shows a case where dark current spikes can be reduced to a low level (comparable with the fixed pattern noise) by reducing the pixel bias in a photodiode APS. Note that the data were obtained at a very high fluence of 1.7 10^{11} 9.5 MeV protons/cm^2. For low Earth orbits the fluence will be an order of magnitude lower and dark current nonuniformity is negligible.

The same damage factor (k_{dark}) discussed above can also be used to relate the average bulk dark charge with the displacement damage dose. Srour et al.[40] have recently compiled data for different devices, radiation sources and silicon dopings, and suggested that either the divacancy or a multivacancy complex is responsible for bulk dark current generation and defined a universal damage constant of (1.9 ±0.6) x 10^5 carriers/cm^3sec per MeV/g at 300K and 1 week after irradiation (after a large fraction of any annealing has taken place). This value is in good agreement with measurements on CCDs, but to use it requires a knowledge of the depletion volume. and, The evidence suggests that the primary defect in CCDs is not the divacancy since the dark current anneals at temperatures of 150°C or below,[41] but a vacancy or an interstitial complex is still a possibility.

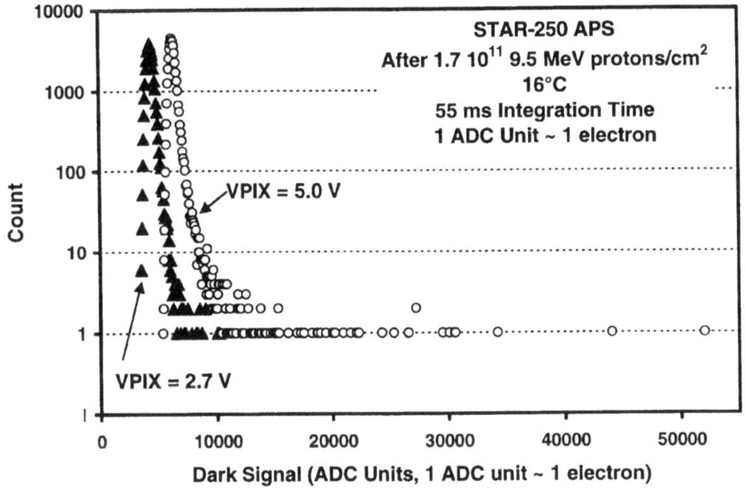

Figure 10 Histograms of dark current density for a STAR-250 APS at nominal and reduced pixel bias.

3.2 *Random Telegraph Signals*

An effective method to reduce dark signal nonuniformity is to reduce the device temperature (using a thermoelectric cooler or by passively radiating to space). However, for space applications, there is usually a significant cost and mass incentive to operate the sensor close to room temperature and to obtain calibration images to subtract out the dark current. Unfortunately, a feature of displacement damage is that the dark current generating defects often tend to have a bistable or metastable nature so that the dark current switches level, giving the appearance of a random telegraph signal (or RTS) as in figure 11. This makes the usefulness of dark calibration uncertain. RTS behavior has been studied in CCDs,[42,43] APSs[44] and IR imagers.[45,46] A significant effort is required to fully characterize a complete device and, so far, only small numbers of pixels have studied and the defect(s) responsible have not yet been identified. Based on the limited experimental work done to date one can make the following tentative conclusions:

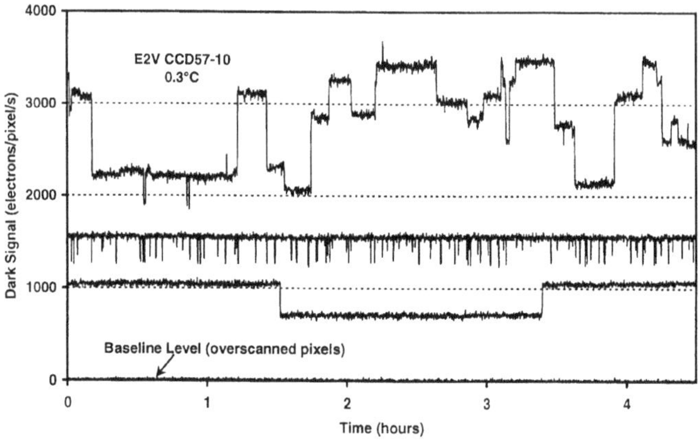

Figure 11 Examples of random telegraph behavior in CCD dark current

- RTS effects are seen in all proton irradiated CCDs and APSs over a range of proton energies (at least 1.5 to 60 MeV).
- In CCDs, the probability for an RTS defect in a pixel is ~ 0.000016 per incident 10 MeV proton and seems to scale with NIEL (though it is not clear whether the elastic NIEL or the total NIEL gives the best fit). Thus, with 20 μm x 20 μm pixels and a fluence of 10^{10} 10 MeV p/cm^2 , ~40% of pixels will show RTS effects. In the APS that was studied[39] the occurrence probability was much lower because of the small area of the photodiodes.
- There is a wide range of switching amplitudes, but in a CCD they are typically 500 to 20,000 electrons/pixel/s at 5ºC. Most large fluctuations occur for pixels with a high pedestal level, but not always – some large dark current spikes do not show RTS effects. The amplitudes decrease with cooling as expected (activation energy ~ 0.6 eV)
- There is also a wide range in time constants – near room temperature they range from seconds to hours (i.e. 4 orders of magnitude), but are mostly of order several minutes.
- The time constants for existing in the high and low dark current states are usually roughly comparable but not always – some pixels show 'spiking' - that is, short excursions to either the low or high state.
- The time constants increase as the device is cooled. In CCDs the activation energy is typically 0.9 to 1.6 eV but Bogaerts et al. measured ~0.6 eV for their APS. The fact that some defects switch very fast at room temperature means that RTS effects can still be seen (with time constants of several hours) even at -40ºC.

3.3 *Charge Transfer Degradation*

The other main effect of displacement damage in CCDs is that defects are created in the buried channel and these will trap charge as it is moved through the CCD during readout and release it some time later, usually into a following charge packet. This causes a smearing of the image and is characterized by the charge transfer inefficiency (CTI), which is the fraction of charge that is trapped during each pixel transfer. (1-CTI) is the charge transfer efficiency or CTE. In active pixel sensors the signal is read directly from the pixel and so this effect does not occur. This is one of the main advantages of the APS in a proton or neutron environment.

For a trap located at an energy, E_t, below the conduction band and with a capture cross section, σ_n, the capture and emission times, τ_c and τ_e, are given by:[47]

$$\tau_c = 1/\sigma_n v_{th} n_s \quad (6)$$

$$\tau_e = \exp(E_t/kT)/\sigma_n X_n v_{th} N_c \chi \quad (7)$$

where n_s is the signal density, v_{th} is the average thermal velocity for electrons, N_c is the effective density of states in the conduction band, T is the absolute temperature, and k is Boltzmann's constant. X_n is the 'entropy factor' associated with the entropy change for electron emission from the trap, and χ is a factor to allow for any field enhanced emission (which affects trap emission time as well as dark current generation).

The capture time depends inversely on the signal density and will be shortest in the center of a charge packet where the density is highest. If charge is transferred through a pixel faster than it can be captured, then CTE will not be degraded and we have a CTI 'capture factor' given by

$$\text{CTI capture factor} = (1 - \exp[-t_t/\tau_c]) \quad (8)$$

where t_t is a characteristic transfer time. If the charge packets of are moved through a pixel on average timescales, t_s, shorter than these emission times, then traps will be filled by the first ('sacrificial') packet readout, but will still be filled when subsequent packets arrive. The CTI 'emission factor' is then given by:

$$\text{CTI emission factor} = (1 - \exp[-t_s/\tau_e]) \quad (9)$$

Equation (6) indicates that the capture time is only weakly dependent on the temperature whereas τ_e, in equation (7), varies exponentially.

The effect of displacement damage on CTI depends on the complex interplay between the movement of charge and the trapping and release of carriers by the defect centers. Hence CTI does not have a single value but depends on the dimensions of the buried channel, clocking rate, temperature, signal level (including background charge) and, of course, on the nature and concentration of the traps themselves. Hence it is found that the CTI damage factor, k_{CTI}, (expressed as the CTI change per equivalent MeV proton/cm^2 or, alternatively, per unit displacement damage dose) can vary by several orders of magnitude though, other things being equal, it is always largest for small signals and small backgrounds (since a given trap concentration will then have a proportionately larger effect). Hence space astronomy missions such as the Hubble Space Telescope[48] and the Chandra[49] X-ray observatory have been significantly affected even though the proton fluence is low (in the latter case the damage was unexpectedly high due to low energy protons which reached some CCDs by scattering off the X-ray mirrors[50]).

Usually the CTI in the vertical (column) direction is the most important because the pixels are moved more slowly and there is more time for charge trapping. Figure 12 shows some typical results. Over recent years, several investigations[51,52,53,54,55,56,57,58,59] have produced CTI data for a variety of CCDs, particularly for the low signal/low background case where a typical k_{CTI} is ~ 4 10^{-13}/10 MeV proton/cm^2 for a 20μm x 20μm pixel at 10 ms line move rate. This data can sometimes give a confusing variation in CTI values. However these variations are reduced if some simple scaling is applied to allow for the different experimental conditions. Because the CTI (for a given trap concentration) will increase in proportion to the volume of silicon, it will tend to increase with pixel size – and the pixel area can be used as an approximate scaling factor (though the actual volume of the buried channel will give more accurate scaling if it is known). Figure 13 (from[60]) shows how the CTI can change with the dwell time within a pixel (which is determined by the line move rate). For a frame transfer CCD it has to be borne in mind that both fast (frame transfer) and slow (readout) line moves are involved and typically there will be a factor ~3 difference between the CTI at the two rates (though this factor is slightly signal and background dependent). The implications for star tracker applications have been previously discussed[61] along with the improvement in performance that can be gained by increasing the background level ('fat zero') – though at the expense of increasing the Poisson (shot) noise. CTI has also been seen to vary with temperature, decreasing by a factor 2 when the temperature decreases from -20 to -50°C.[62]

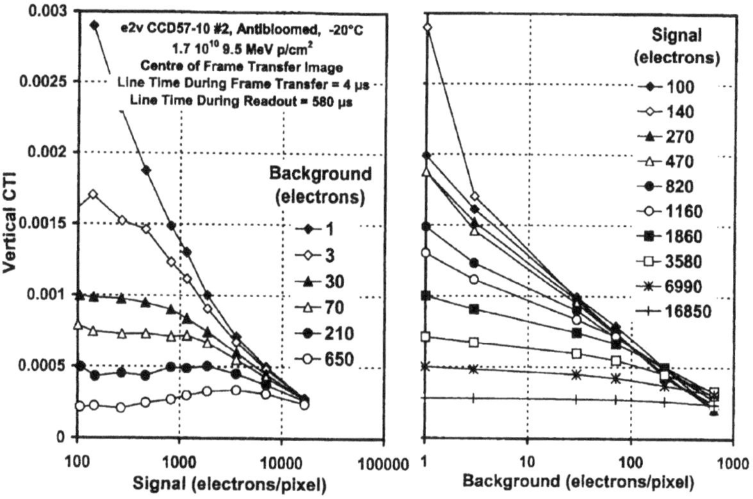

Figure 12 Vertical CTI for a typical buried channel CCD (pixel size 13 μm x 13μm, with antiblooming structures).

From the above discussion we see that CTI values can be scaled to allow predictions for different applications but the accuracy is probably a factor of 2 at best. Hence it is recommended to perform proton testing for the actual device operating conditions whenever possible. Even if the CTI is known, the effect on imaging performance still has to be calculated. Several authors[63,64] have developed models for predicting CCD performance for particular operating conditions. Care has to be taken, however, when extrapolating results.

CTI damage starts to have a major impact at equivalent 10 MeV proton fluences of around 10^{10} p/cm^2 or above. For space astronomy missions where the CCD can be cooled to low temperatures, there is the possibility to keep the trapping centers permanently filled either by special clocking sequences[65] or by using injection of an artificial charge packet ahead of the signal.[66,67] These techniques rely on knowing the emission time of the traps. Figure 14 shows experimental data on the variation of emission time with temperature for the dominant trap at moderately high temperatures for a variety of CCDs. Since the trap has an energy level ~0.44 eV it is believed to be the E center and this is supported by annealing data.[41,67]

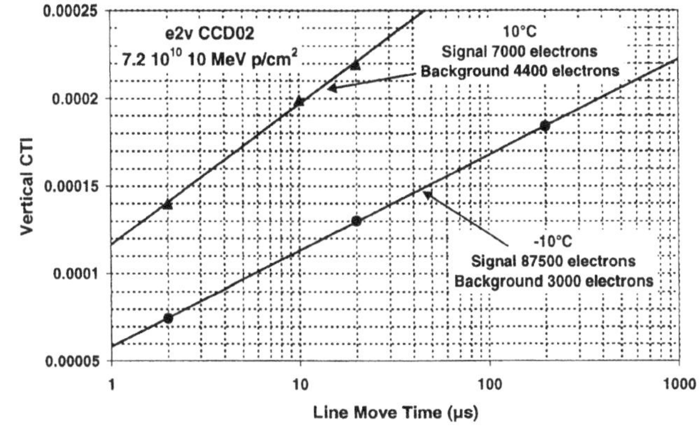

Figure 13 Illustration of the variation of vertical CTI with line move time

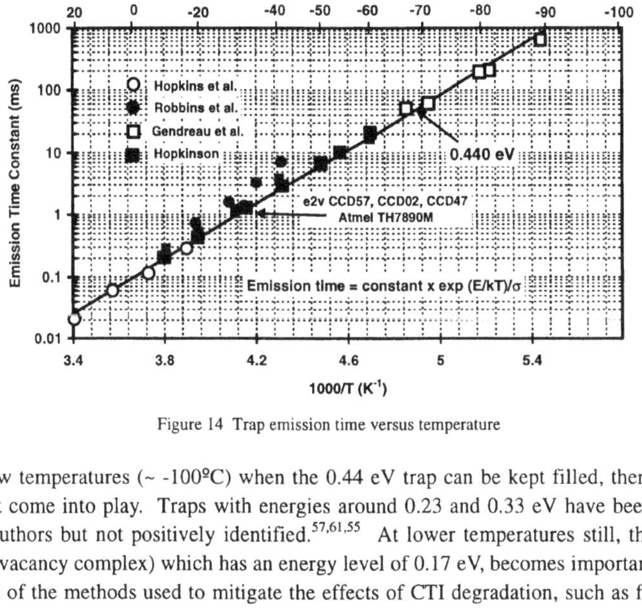

Figure 14 Trap emission time versus temperature

At low temperatures (~ -100ºC) when the 0.44 eV trap can be kept filled, there are other traps that come into play. Traps with energies around 0.23 and 0.33 eV have been found by several authors but not positively identified.[57,61,55] At lower temperatures still, the A center (oxygen-vacancy complex) which has an energy level of 0.17 eV, becomes important.

Some of the methods used to mitigate the effects of CTI degradation, such as fat zero and charge injection, have been discussed above. A further possibility is to heat the CCD to a temperature where the traps anneal out. Unfortunately, temperatures above 100ºC are needed and this is not normally practicable. Another solution is to use a p-type buried channel rather than n-type. The dominant dopant is then boron rather than phosphorous and the E center should not be present. This has been discussed by Spratt[68] et al. and Hopkinson,[69] though the improvement in CTI was inconclusive. Recent results[70] indicate that p-channel CCDs do indeed have a definite advantage for proton environments, but such devices are not routinely fabricated by commercial CCD manufacturers (though for special applications, such as space astronomy, it can be possible for devices to be custom-built).

Care in measuring CTI is important if consistent results are to be obtained. Methods include use of X-ray sources, optical spots and optical flat fields (extended pixel edge response, EPER and first pixel response, FPR). These have been discussed (for example) in [3]. Some devices are now made with an injection gate and this can be used to inject test signals for CTI measurement. The injected signal appears to have low (sub-Poisson) noise[71] and so can be used to produce a background charge (fat zero) level to reduce CTI.

3.4 *Testing for displacement damage effects*

Most CCD displacement damage data have been obtained using proton accelerators with the devices irradiated at room temperature. This can be done with the CCDs biased or unbiased and the available evidence suggests that the bias condition does not make a significant difference. The particle flux is typically ~ 10^8 p/cm^2/s – much higher than in space – but this does not appear to affect the displacement damage (significant effects have only been seen when the flux becomes an order of magnitude higher[72]). The protons will also produce ionization (total dose) damage and this is known to be dose rate (i.e. flux) dependent. Hence proton beams are not good simulators for total dose damage. When testing dark current (total dose does not affect CTE) the two damage components need to be carefully separated and allowed for separately (with the ionization component tested using cobalt-60 gamma rays).

At room temperature it is usually observed that some annealing takes place immediately after irradiation[73] and in the first few weeks. After that the damage is permanent (unless the device is deliberately heated). Unless otherwise stated, it is usually safe to assume that test results are for devices that have been stored for several weeks at room temperature so that the short term annealing has been completed. There is some evidence from space astronomy missions that cooled CCDs show an annealing of the dark current spikes when heated to near room temperature.[48,74] This implies that irradiation when cold gives greater damage than when at room temperature. There is also evidence of CTI changes after cold irradiation.[75] However, very little work on low temperature irradiations has so far been performed.

4. Transient Events

Transient events are produced by the ionization of silicon atoms along the tracks of charged particles. The energy deposited depends on the product of the ionizing energy loss, dE/dx, of the particle and the track length. Approximately one electron-hole pair is created for every 3.65 eV ionizing energy deposited. In a space environment the main concern is usually transients caused by primary protons and by secondary particles produced by proton interactions with shielding material.[76,77] Heavy ion transients from cosmic rays will also occur, though in greatly reduced numbers compared with proton events. To take an example, Hopkinson et al.[6] showed that, for a polar Earth orbit at 800 km altitude and the 10 mm aluminum shielding, the average flux during transit through the South Atlantic Anomaly (SAA) is ~ 1500 protons/cm^2/s. In some applications, for example missions to Jupiter or its moons, electron events may also be important.[78]

Minimum ionizing particles create ~ 80 electron hole pairs /μm so that, for a typical CCD charge collection depth of 20 μm, a proton transient will generate at least 1,600 electrons and each particle strike results in a significant signal. There will be some variation in the amount of charge deposited in each pixel.[6] This is primarily because, in a space environment, there are variations in the proton energy and path length. There is also a statistical spread in the charge deposited, even for mono-energetic beams.

Lomheim et al.[79] have carried out an experimental investigation of single particle transient signals from mono-energetic protons and heavy ions incident at various incident angles to a Kodak KAF-1400 CCD. The results were in agreement (to within 20%) with the signal charge expected for a particular ion track length including contributions from the pixel depletion and diffusion volumes. McCarthy et al.[80] have performed a similar investigation (using protons in the range 50 to 300 MeV) for a deep depletion CCD, specially designed for X-ray astronomy. They found that in many cases the proton events could be distinguished from their X-ray signals. However, in general, the large spread in the amount of charge deposited in a CCD makes it very difficult to discriminate against proton transients. Another way of mitigating the effects is to use a thinner active volume so that the track lengths are shorter, but this will reduce the detector responsivity for wavelengths absorbed deep in the device. Instruments (e.g. star trackers) which continuously stare at a scene will often have software which discriminates against signals that are only present for a single image frame. If the observing situation allows, this is the best way of ensuring a tolerance to transient effects.

5. Future Work

The high performance of CCDs makes them vulnerable to even low doses of radiation and, since the radiation effects vary with operating conditions, a continuing effort is needed to characterize devices for particular applications. In addition there are some fundamental aspects, particularly relating to displacement damage, where further work is needed. For example, the mechanisms of dark current fluctuations (random telegraph signals) have not been investigated enough to make accurate predictions of the effects on images or the calibration strategies that are needed. Also, the full inventory of defects, that affect dark current and charge transfer inefficiency, has not been identified. Modeling and prediction of proton-induced CTI remains a major concern for many missions.

The CMOS active pixel sensor is emerging as a good alternative to the CCD. Although these do not suffer from CTI degradation and gate oxide effects can usually be neglected, a substantial test effort is still needed to fully characterize these devices for radiation effects since they can still be prone to parasitic leakage effects (total ionizing dose) and to increases in dark signal. In addition there are the possibilities of single event latch-up or functional interrupt and for degradation in the performance of on-chip circuitry (e.g. ADCs).

Acknowledgements

Much of the data presented in this review has arisen from recent ground test programs funded by the European Space Agency. Wojtek Hajdas (PSI), Keith Jones (Ebis Iotron Ltd) and Bob Nickson (ESA) are thanked for their help in performing the irradiations.

References

1. CCD595 datasheet, available from http://www.fairchildimaging.com/main/area_595.htm

2. See, for example, http://www.fillfactory.com

3 J. R. Janesick, Scientific Charge-Coupled Devices, SPIE Press, Bellingham, 2001.

4 A. J. P. Theuwissen, Solid-State Imaging with Charge-Coupled Devices, Klewer, Dordrecht, 1995.

5. C. J. Marshall and P. W. Marshall, "Proton effects and test issues for satellite designers, part B: displacement damage effects", 1999 IEEE NSREC Short Course Notes, pp. IV-50-IV-110.

6. G. R. Hopkinson, C. J. Dale, and P. W. Marshall, "Proton effects in CCDs", IEEE Trans. Nucl. Sci., vol. 43, no. 2, pp. 614-627, Apr 1996.

7 J. C. Pickel, A. H. Kalma, G. R. Hopkinson, and C. J. Marshall, "Radiation effects on photonic imagers – a historical perspective", IEEE Trans. Nucl. Sci., vol. 50, no. 3, pp. 671-688, Jun 2003.

8. T. D. Hardy, M. J. Deen and R. Murowinski, "Effects of radiation damage on scientific charge coupled devices", Adv. Imaging and Electron Phys. Vol. 106, pp. 1-96, 1999.

9. D. Braunig and F. Wulf, "Atomic displacement and total ionizing dose damage in semiconductors", Radiat. Phys. Chem. Vol. 43, pp. 105-107, 1994.

10. J. R. Schwank, "Total-Dose Effects in MOS Devices", 2002 IEEE NSREC Short Course Notes, pp. III-1-III-123.

11. S. M. Sze, Physics of Semiconductor Devices, 2nd Ed., Wiley-Interscience, New York, 1981, p 90.

12. P. S. Winokur, M. R. Shaneyfelt, T. L. Meisenheimer and D. M. Fleetwood, "Advanced qualification techniques", IEEE Trans. on Nucl. Sci., vol. 41, no. 3, pp 538-548, Jun 1994.

13. G. R. Hopkinson, "Radiation-induced dark current increases in CCDs", Proc. RADECS 93, St Malo, France, Sept 1993, IEEE 93TH0616-3, pp. 401-408.

14. J. C. Boudenot and P. Augier "Total dose effects on charge coupled devices (CCD) reverse annealing phenomena", Proc ESA Electronic Components Conference, ESTEC, The Netherlands, November 1990, ESA SP-313 (1991), pp. 319-324.

15. G. R. Hopkinson, "Radiation effects on CCDs for spaceborne acquisition and tracking applications", Proc. First European Conference on Radiation and its Effects on Devices and Systems (RADECS 91), September 1991, IEEE 91TH04000-2, (1992), pp. 368-372

16. B. Burke and S. A. Gajar, "Dynamic suppression of interface-state dark current in buried channel CCDs", IEEE Trans on Electron Devices, vol. 38, no. 2, pp. 285-290, 1991.

17. E.-S. Eid et al. "Design and characterization of ionizing radiation-tolerant CMOS APS image sensors up to 30 Mrd(Si) total dose", IEEE Trans. Nucl. Sci., 48, no. 6, pp. 1796-1806, Dec 2001.

18. J. Bogaerts and B. Dierickx, "Total dose effects on CMOS active pixel sensors", Proc SPIE, vol. 3965, pp. 157-167, 2000.

19. G. R. Hopkinson, M. D Skipper and B. Taylor "A radiation tolerant video camera for high total dose environments", Workshop Record, 2002 IEEE Radiation Effects Data Workshop, IEEE 02TH8631, (2002), pp. 18-23.

20 J. Bogaerts, B. Dierickx, G. Meynants and D. Uwaerts, "Total dose and displacement damage effects in a radiation hardened CMOS APS", IEEE Trans. Elec. Dev., vol. 50, no. 1, pp. 84-90, Jan 2003.

21. M. Hollreiser, "Strategy for low power radiation tolerant commercial technologies", Proc ESCCON2000, 21-23 March, 2000, ESA SP-439, pp. 101-107, 2000.

22. C. J. Marshall et al., "Heavy ion transient characterization of a hardened-by-design active pixel sensor array", Workshop Record, 2002 IEEE Radiation Effects Data Workshop, IEEE 02TH8631 (2002), pp. 187-193.

23. J. Bourgoin and M. Lanoo, "Point Defects in Semiconductors", Springer-Verlag, Berlin, 1983.

24. Chr. Lehmann, "Interaction of Radiation with Solids and Elementary Defect Production", North Holland, Amsterdam, 1977.

25. S. Wood et al. "Simulation of radiation damage in solids", IEEE Trans. on Nucl. Sci., vol 28, no. 6, pp. 4107-4112, Dec 1981.

26. G. Lindstroem, "Radiation damage in silicon detectors, "Nucl. phys. and Meth. in Phys. Res. A, in press.

27. C. J. Dale, P. W. Marshall, G. P. Summers and E. A. Wolicki, "Displacement damage equivalent of dose in silicon devices", Appl. Phys. Lett., vol. 54, no. 5, pp 451-453, 1989.

28. M. Huhtinen, "Simulation of non-ionising energy loss and defect formation in silicon", Nucl. Inst. Meth. In Phys. Res. A, vol. 491, no. 1-2, pp 194-215, 2002.

29. C. J. Dale, L. Chen, P. J. McNulty, P. W. Marshall and E. A. Burke, "A comparison of Monte Carlo and analytical treatments of displacement damage in Si microvolumes", IEEE Trans Nucl. Sci., vol. 41, no. 6, pp. 1974-1983, Dec. 1994.

30. E. A. Burke, "Energy dependence of proton-induced displacement damage in silicon", ", IEEE Trans Nucl. Sci., vol. 33, no. 6, pp. 1276-1281, Dec. 1986.

31. M. Huhtinen and P. A. Arnio, "Pion induced displacement damage in silicon devices", Nucl. Inst. Meth. In Phys. Res. A, vol. 335, pp 581-582, 1993.

32. I. Jun, "Effects of secondary particles on the total dose and the displacement damage in space proton environments", IEEE Trans. Nucl. Sci., vol. 48, no. 1, pp. 162-175, Feb.2001.

33. A. Akkerman, J. Barak, M. B. Chadwick, J. Levinson, M. Murat and Y. Lifshitz, "updated NIEL calculations for estimating the damage induced by particles and g rays in Si and GaAs", Radiation Physics and Chemistry, vol. 62, pp. 301-310, 2001.

34 ASTM E-722-94 (Reapproved 2002), "Standard practice for characterizing neutron energy fluence spectra in terms of an equivalent monoenergetic neutron fluence for radiation-hardness testing of electronics".

35. C. J. Dale, P. W. Marshall and E. A. Burke, "Particle-induced spatial dark current variations in focal plane arrays", IEEE Trans. Nucl. Sci., vol. 37, pp. 1784-1791, 1990.

36. J. R. Srour and R. A. Hartmann, "Enhanced displacement damage effectiveness in irradiated silicon devices", IEEE Trans. Nucl. Sci., vol. 36 no.6, pp. 1825-1830, 1989.

37. P. W. Marshall, C. J. Dale and E. A. Burke, "Proton-Induced Displacement Damage Distributions in Silicon Microvolumes", IEEE Trans Nucl. Sci., vol. 37, no. 6 , pp. 1776-1783, Dec. 1990.

38. M. S. Robbins, "High-energy proton-induced dark signal in silicon charge coupled devices", IEEE Trans Nucl. Sci., vol. 47, no. 6 , pp. 2473-2479, Dec. 2000.

39. J. Bogaerts, B. Dierickx and R. Mertens, "Enhanced dark current generation in proton-irradiated CMOS active pixel sensors", IEEE Trans. on Nucl. Sci., vol. 49, no. 3, pp 1513-1521, Jun. 2000.

40. J. R. Srour and D. H. Lo, "Universal damage factor for radiation-induced dark current in silicon devices", IEEE Trans. on Nucl. Sci., vol. 47, no. 6, pp 2451-2459, Dec. 2000.

41. A. D. Holland, "Annealing of proton-induced displacement damage in CCDs for space use", Proc 10th Symposium on Photoelectronic Image Devices, Institute of Physics Conference Series 121, pp. 33-40, 1992.

42. I. H. Hopkins and G. R. Hopkinson, "Random telegraph signals from proton-irradiated CCDs", IEEE Trans. Nucl. Sci., vol. 40, no. 6, pp. 1567 -1574, Dec. 1993.

43. I. H. Hopkins and G. R. Hopkinson, "Further measurements of random telegraph signals in proton irradiated CCDs" IEEE Trans. Nucl. Sci., vol. 42, no. 6, pp. 2074 -2081, Dec. 1995.

44. J. Bogaerts. B. Dierickx, and R. Mertens, "Random telegraph signals in a radiation-hardened CMOS active pixel sensor", IEEE Trans. on Nucl. Sci., vol. 49, no. 1, pp. 249-257, Feb. 2002.

45 D. Pogany and G. Guillot, "Random telegraph signal noise instabilities in lattice-mismatched InGaAs/InP photodiodes", Microel. Reliab., vol. 39, pp. 341-345, 1999.

46. S. Barde, R. Ecoffet, J. Costeraste, A. Meygret, and X. Hugon, "Displacement damage effects in InGaAs detectors: experimental results and semi-empirical model prediction", IEEE Trans. Nucl. Sci., vol. 47, no. 6, pp. 2466-2472, Dec. 2000.

47. A. Mohsen and M. F. Tompsett, "The effects of bulk traps on the performance of bulk channel charge-coupled devices", IEEE Trans. Elec. Dev., vol. 21, pp. 701-712, 1974.

48. J. Holtzmam et al., "The performance and calibration of WFPC2 on the Hubble Space Telescope", Publ. Astron. Soc. Pac., vol. 107, pp.156-178, 1995.

49. L.K. Townsley, P. S. Boos, J. A. Nousek and G. P. Garmire, "Modeling charge transfer inefficiency in the Chandra advanced CCD imaging spectrometer", Nucl. Inst. Meth. In Phys. Res. A, vol. 486, pp. 751-784, 2002.

50. R. Nartallo, E. Daly, H. Evans, P. Nieminen, Fan Lei; and P. Truscott, "Low-angle scattering of protons on the XMM-Newton optics and effects on the on-board CCD detectors" IEEE Trans. on Nucl. Sci., vol. 48, no. 6, pp 1815-1821, Dec. 2001.

51. A. Holmes-Seidle, A. Holland and S. Watts, "The impact of space protons in x-ray sensing with charge coupled devices (CCDs)", IEEE Trans. Nucl. Sci., vol. 43, no. 6, pp. 2998-3003, Dec. 1996.

52. J. Janesick, T. Elliott and F. Pool, "Radiation damage in scientific charge-coupled devices", IEEE Trans. on Nucl. Sci. vol. 36, no. 2 pp. 572-578, Jun. 1988.

53 C. Dale, P. Marshall, B. Cummings, L. Shamey and A. Holland, "Displacement damage effects in mixed particle environments for shielded spacecraft CCDs", IEEE Trans. Nucl. Sci., vol. 40, no. 6, pp. 1628-1637, Dec. 1993.

54 M. S. Robbins, T. Roy and S. J. Watts, "Degradation of the charge transfer efficiency of a buried channel charge coupled device due to radiation damage by a beta source", RADECS 91, IEEE Proceedings, pp. 327-332 (1992).

55 T. Hardy, R. Murowinsky and M. J. Deen, "Charge transfer efficiency in proton damaged CCD's", IEEE Trans. on Nucl. Sci., vol. 45, no. 2, pp. 154-163, Apr. 1998.

56. K. C. Gendreau, G. Y. Prigozhin, R. K. Huang and M. W. Bautz, "A technique to measure trap characteristics in CCD's using X-rays", IEEE Trans. Elec. Dev. vol. 42, no. 11, pp. 1912-1917, Nov. 1995.

57. A. Holland, "The effect of bulk traps in proton-irradiated EEV CCDs" Nucl. Inst. Meth. vol. A326, pp. 335-343, 1993.

58. N. Meidinger, B. Schmalhofer and L. Struder, "Particle and x-ray damage in pn-CCDs", Nucl Inst. Meth. vol. A439, pp. 319-336, 2000.

59. K. D. Stefanov, T. Tsukamoto, A. Miyamoto, Y. Sugimoto, N. Tamura, K. Abe, T. Nagamine and T. Aso, "Electron and neutron radiation damage effects on a two-phase CCD", IEEE Trans. Nucl. Sci., vol. 47, no. 3, pp. 1280-1291, Jun. 1997.

60. I. H Hopkins, G. R. Hopkinson and B. Johlander, "Proton-Induced Charge Transfer Degradation in CCDs for Near-Room Temperature Applications", IEEE Trans. Nucl. Sci., vol. 41, no. 6, pp. 1984-1990, Dec. 1994.

61. G. R. Hopkinson, "Proton-induced changes in CTE for n-channel CCDs and the effect on star tracker performance", IEEE Trans. Nucl. Sci., vol. 47, no. 6, pp. 2460-2465, 2000.

62 "Radiation testing of CCD and APS imaging devices", European Space Agency Contract Report, https://escies.org/public/radiation/esa/database/239DO53FinalReport_iss2.pdf

63. D. Gallagher, R. Demara, G. Emerson, W. Frame and A. Delamere, "Monte Carlo model for describing charge transfer in irradiated CCDs", Proc SPIE, vol. 3301, pp. 80-88, 1998.

64. R. H. Philbrick, "Modeling the impact of preflushing on CTE in proton irradiated CCD-based detectors", IEEE Trans. on Nucl. Sci., vol. 49, no. 2, pp. 559 -567, April 2002.

65. G. Prigozhin et al. "Characterization of the radiation damage in the Chandra X-ray CCDs". Proc. SPIE, vol. 4140, pp. 123-134, Aug. 2000.

66. G R Hopkinson, "Proton-induced charge transfer degradation at low operating temperatures", IEEE Trans. on Nucl. Sci., vol. 48, no. 6, pp. 1790-1795, Dec. 2001.

67. J. E. Brau and N. Sinev, "Operation of a CCD particle detector in the presence of bulk neutron damage", IEEE Trans. Nucl. Sci., vol. 47, no. 6, pp 1898-1901, Dec. 2000.

68. J. Spratt, B. C. Passenheim and R. E. Leadon, "The effects of nuclear radiation on P-channel CCD imagers", 1997 IEEE Radiation Effects Data Workshop Record, pp. 116-121 1998.

69. G. R. Hopkinson, " Proton damage effects on p-channel CCDs", IEEE Trans. on Nucl. Sci., vol. 46, no. 6, pp. 1790 -1796, Dec. 1999.

70. C. Bebek et al., " Proton radiation damage in P-channel CCDs fabricated on high-resistivity silicon", IEEE Trans. Nucl. Sci., vol. 49, no. 3, pp. 1221 -1225, June 2002.

71 M. Giavalisco, "Minimizing CTE losses in the WFC3 CCDs: post flash vs. charge injection", Instrument Science Report WFC3 2003-01, Space Telescope Science Institute. Available on-line at http://www.stsci.edu/instruments/wfc3/ISRs/WFC3-2003-01.pdf.

72 A. Hallen, D. Fenyo and B. U. R. Sundqvist, "The influence of ion flux on defect production in MeV proton-irradiated silicon", J. Appl. Phys., vol. 70, pp. 3025-3030, 1991.

73. J. R. Srour, R. A. Hartmann and S. Othmer, "Transient and permanent effects of neutron bombardment on a commercially available n-buried channel CCD", ", IEEE Trans. Nucl. Sci., vol. 27, no. 6, pp. 1402 -1410, Dec. 1980.

74. R. A. Kimble, P. Goudfrooij and R. L. Gilliland, "Radiation damage effects on the CCD detector of the Space Telescope Imaging Spectrograph", Proc. SPIE, vol. 4013, pp. 532-545, July 2000.

75 G. Prigozhin, M. Bautz, S. Kissel, B. LaMarr and C. E. Grant, "Anomalous annealing behavior of the high resistivity CCD irradiated at cold temperature", presented at the 2003 IEEE Workshop on Charge-Coupled Devices and Advanced Image Sensors, Elmau, Germany, May 15-17, 2003.

76. C. Pickel, "Radiation-induced charge collection in infrared detector arrays", IEEE Trans. Nucl. Sci., vol. 49, no. 6, pp. 2822-2829, Dec. 2002.

77. R. Ladbury et al., "Characteristics of the Hubble Space Telescope's radiation environment inferred from charge-collection modeling of near-infrared camera and multi-object spectrometer darkframes", IEEE Trans. Nucl. Sci., vol. 49, no. 6, pp. 2765-2770, 2002.

78. C.C. Liebe,"Charged particle-induced noise in camera systems", IEEE Trans. Nucl. Sci., vol. 48, no. 4, pp. 1541 -1549, Aug. 2001.

79. T. S. Lomheim et al. "Imaging Charge-Coupled Device (CCD) Transient Response to 17 and 50 MeV Proton and Heavy-Ion Irradiation", IEEE Trans. Nucl. Sci., vol. 37, no. 6 pp. 1876-1885, 1990.

80. K. J. McCarthy, A. Owens, A. Wells, W. Hajdas, F. Mattenberger and A. Zehnder, "Measured and Modelled CCD Response to Protons in the Energy Range 50 to 300 MeV", Nucl. Inst. Meth. in Phys. Res. A, vol. 361, pp. 586-601, 1995.

International Journal of High Speed Electronics and Systems
Vol. 14, No. 2 (2004) 445–463

THE EFFECTS OF SPACE RADIATION EXPOSURE ON POWER MOSFETS: A REVIEW

K. SHENAI[†]
Electrical and Computer Engineering, University of Illinois at Chicago
Chicago, IL 60607, U.S.A.
[†]shenai@ece.uic.edu

K.F. GALLOWAY[§] and R.D. SCHRIMPF[‡]
Electrical Engineering and Computer Science, Vanderbilt University
Nashville, TN 37235, U.S.A.
[§]ken.galloway@vanderbilt.edu [‡] ron.schrimpf@vanderbilt.edu

Power MOSFETs are a commonly used device for many switching and power control applications. Their upper frequency limit spans a fairly broad range, from 1 MHz to 10 MHz. These devices are frequently used in spaceborne electronic systems where they encounter radiation exposure during operation. This paper reviews the current technology, its high frequency capability, the future trends for power MOSFET technology, and the degradation that the power VDMOS technology experiences in the space radiation environment.

Keywords: MOSFETs, power MOSFETs, radiation effects, space radiation, VDMOS.

1. Introduction

The electrical characteristics of power metal-oxide-semiconductor (MOS) field-effect-transistors (FETs) make them of interest for use in space systems where they may be exposed to radiation. Power MOSFETs, in general, have higher switching speeds than their power bipolar transistor counterparts.[1] Power MOSFETs also have a negative temperature coefficient of carrier mobility, which reduces the potential for thermal runaway, second breakdown, and power hogging in parallel devices when compared to bipolar transistors.[1] In addition, power MOSFETs have simpler input drive requirements than bipolar transistors.[1]

The space radiation environment can be very detrimental to electronic components. For example, the natural space radiation environment of the earth consists of (1) trapped particles in the magnetic field of the earth - primarily protons and electrons; and (2) cosmic rays - heavy ions or protons of solar or galactic origin.[2] There are high-energy photons, protons, heavy ions, and electrons that can cause total-dose ionizing radiation damage and high-energy protons, neutrons, and heavy ions that can cause displacement damage. The high-energy protons and heavy ions can introduce dense charge filaments along the path of the particle through an electronic device that can interact with sensitive regions of the device to trigger catastrophic failure mechanisms. The concentrations and types of ions vary significantly with altitude and inclination of angle of an orbit, recent solar activity, and the amount of protective shielding. These variations make it difficult to specify a typical space environment.[3] Also, the precise environment (total ionizing dose and high-energy protons and heavy ions) encountered by a specific electronic system will be highly dependent on the mission.

This paper will briefly discuss the characteristics of power MOSFETs with an emphasis on the vertical double-diffused (VDMOS) technology that is widely used commercially. Switching speed constraints will also be discussed. With this background, the effects of radiation on power MOSFETs will be introduced and reviewed based on a large body of work appearing in the literature.

2. Power MOSFET Technologies

2.a. *Current commercial technology*

The VDMOS transistor is a special type of MOS transistor well suited to power applications. Figure 1 shows the basic structure of a VDMOS power transistor.[1] The device is termed double-diffused because, typically, both the source and the body are defined using the polysilicon gate as a mask. The gate and source are self-aligned processes in commercial products. However, in some radiation hardened power MOSFETs, these processing steps may not be self-aligned. Usually then, the channel length is determined by the difference in lateral diffusion of the dopants, rather than by photolithography limitations, and thus, the channel is not uniformly doped. Contact to the drain is made on the bottom surface of the silicon chip, rather than on the top surface as in a lateral MOSFET. The thick epitaxial drain region is required to drop the large drain to source voltages that the power transistor must support while operating in the *OFF* state. When turned *ON* with an appropriate gate bias, an inversion layer forms along the p^- surface forming a channel between the n^+ source region and the n^- epi drain region allowing current to flow. In practice, a large number of devices (or cells) are connected in parallel to create a wide effective channel while retaining the short channel lengths of each individual cell. The large width allows large currents to flow when placed in the *ON* state.

The on-resistance of a VDMOS transistor is one of its most important parameters because it limits the amount of current flow that can be conducted by the device before the power dissipation within the device induces junction temperatures that lead to device failure. Therefore, the on-resistance should be as low as possible. Another important parameter is the maximum blocking voltage. This is the maximum voltage applied between the source and drain contacts with zero gate-source voltage that does not lead to breakdown.

Due to its many attractive features, there has been a concerted effort to optimize the design, structure, and process technology of VDMOSFETs in terms of the on-state resistance. The on-resistance of a VDMOSFET is primarily defined by the resistance of the low-doped epitaxial layer. By increasing the doping in the epi-layer, the on-resistance can be reduced. However, the breakdown voltage of the device is inversely proportional to the epi-layer doping, so a compromise between on-resistance and breakdown voltage must be found. In addition, the thickness of the epi-layer must increase with the desired breakdown voltage in order to accommodate the depletion region, further increasing the on resistance. For low breakdown voltage devices ($\leq$ 100V), the parasitic resistances (drain-substrate resistance, channel, source, contact, electrode, etc.) must be considered.

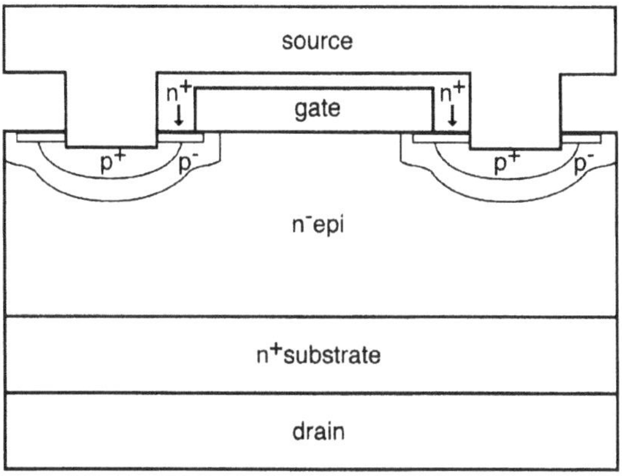

Fig.1. The basic structure of a VDMOS transistor.

In one approach to reduce the on-resistance, the n-doping concentration in the epi-layer is increased uniformly. To retain the voltage blocking capability, the additional n-doping must be compensated by appropriate p-doping. Several different configurations are possible. Figure 2 shows a configuration where the doping concentration of the n⁻ epi-layer has been increased. To compensate, the higher n-doping concentration a vertical p⁻ doped area is placed below the source contact.[4,5] Because of the higher doping concentration, the resistance of the n epi-layer is lower than in the original device (compare to Figure 1). Width and concentration of the vertical p⁻ doped area have to be chosen in such a way that the space charge region in the *OFF* state can be accommodated.

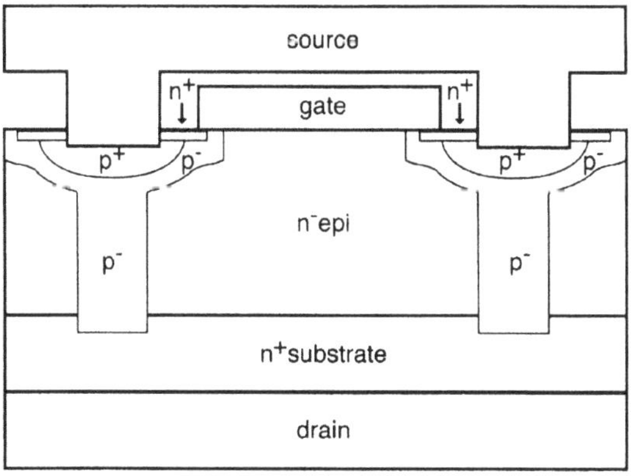

Fig. 2. Modified device structure to reduce on-resistance with increased doping concentration in the epi-layer and an additional vertical p⁻ doped area under the source.

Most VDMOS processes use a poly-silicon gate structure rather than the metal-gate type. If the resistance of the gate structure is high, the switching time of the device is increased, thereby reducing its upper operating frequency. Compared to a metal gate, a poly-silicon gate has a higher gate resistance. This property accounts for the frequent use of metal-gate MOSFETs in very high-frequency (typically more than 20 MHz) applications, and poly-silicon gate MOSFETs in higher-power, but lower-frequency systems. Power MOSFETs are usually faster than power bipolar devices because they do not rely on injection of excess minority carriers that must be supplied by the forward biased junction to turn the device on and then removed to turn the device off. An equivalent circuit model shown in Figure 3 indicates the components that have the greatest effect on the switching characteristics of a MOSFET.[6] R_G is the distributed resistance of the gate and is approximately inversely proportional to active area. L_S and L_D are source and drain lead inductances and are around a few tens of nano-Henry (nH).[6]

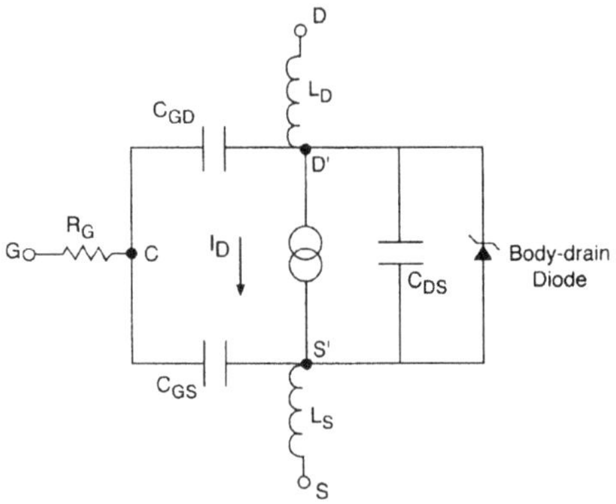

Fig. 3. Equivalent circuit showing components that have the greatest effect on MOSFET switching.

Since the effective resistance and capacitance of the gate terminal control the frequency response of a MOSFET, a rough estimate can be made of the maximum operating frequency of the MOSFET. The resistance (R_G) depends on the sheet resistance of the poly-silicon gate overlay structure. One of the dynamic characteristics of a power MOSFET that can therefore affect its switching performance is its parasitic capacitance. There are three main capacitance parameters of a power MOSFET, namely, gate-to-drain capacitance C_{GD}, gate-to-source capacitance C_{GS}, and drain-to-source capacitance C_{DS}. C_{GS} and C_{GD} are voltage-dependent capacitors while C_{DS} is nearly constant. Gate-to-drain capacitance, C_{GD}, is a nonlinear function of voltage and is the most important parameter because it provides a feedback loop between the output and the input of the circuit. C_{GD} is also called the Miller capacitance because it causes the total dynamic input capacitance to become greater than the sum of the static capacitances. A combination of these parameters determines the typical input and output capacitance parameters: C_{ISS} (C_{GD} + C_{GS}), C_{OSS} (C_{GD} + C_{DS}) and C_{RSS} (C_{GD}). The value of C_{ISS} is closely related to chip size; the larger the die area, the greater the capacitance value. The switching performance of a device is determined by the time required to establish voltage changes across capacitances. Since the RC combination of the input circuit must be charged and discharged by the driving circuit, and since the capacitance value dominates, chips with a larger die area will have slower switching times than smaller chips. In general, the upper frequency limit of most power MOSFETs spans a fairly broad range, from 1 MHz to 10 MHz.

2.b. *Future trends in power MOSFET technology*

The focus in power MOSFET design has been to minimize the on-resistance of the MOSFETs, improve the reverse recovery characteristics and increase the reverse voltage blocking capability of the MOSFETs.[5] New device structures have been proposed for the reduction in on-state resistance. Prominent technologies today are the MDMESH™ from ST Microelectronics, CoolMOS™ from Infineon Technologies, and the SuperJunction FET from International Rectifier, each of which is based on the charge balance concept.[7,8] These devices have a smaller die area than their equivalent conventional MOSFET counterparts and substantially less output capacitance. Results indicate an improvement in the body-drain diode switching recovery behavior. This tends to minimize the losses in switching converter applications.

Silicon carbide (SiC) power MOSFETs are also the subject of active research. SiC offers significant advantages for power-switching devices because the critical field for avalanche breakdown is about ten times higher than in silicon. SiC power devices have made remarkable progress in the past five years, demonstrating currents in excess of 100 A and blocking voltages in excess of 19,000 V.

Several SiC devices have demonstrated performance figures that far exceed comparable silicon devices and, unlike silicon, the theoretical performance limits for SiC do not impose any near-term constraints on continued progress.[9] SiC wafers are commercially available in 75-mm diameters with micropipe densities below 10 cm. This permits high-yield fabrication of transistors and diodes with 20 A capability as single devices and 100 A or higher capability in modules. SiC wafers are expected to continue to increase in size and improve in quality in coming years.[10] SiC power devices will first appear in specialized commercial applications where silicon devices cannot effectively compete. One such application is freewheeling rectifiers in motor control circuits, where SiC Schottky or p-i-n diodes will displace conventional silicon p-i-n diodes. Substitution of SiC diodes in these systems provides significant systems advantages due to greatly reduced switching losses. Further into the future, SiC transistors and thyristors may begin to displace silicon insulated gate bipolar transistors (IGBTs). It is not clear at this point whether these SiC transistors will be MOSFETs, junction field effect transistors (JFETs), or bipolar junction transistors (BJTs). JFETs and BJTs have fewer materials issues and are easier to fabricate than SiC MOSFETs, but MOSFETs ultimately offer the greatest advantages over silicon IGBTs because of their high input impedance and low dynamic power dissipation. However, considerable development work is needed to bring SiC MOSFETs to their potential.

3. Total Dose Ionizing Radiation Effects on Power MOSFETs

3.a. *Ionizing radiation effects basics*

The topic of total ionizing dose radiation effects in MOS systems has been actively studied for several decades.[11,12] When the dielectric films (assumed to be SiO_2) in a power MOS device are exposed to ionizing radiation, electron-hole pairs are created. These electron-hole pairs either recombine or begin a transport process through the dielectric under the influence of the applied fields. In a gate oxide with positive bias on the gate, the electrons are quickly swept to the gate and collected. The holes experience a stochastic transport process to the oxide - silicon (SiO_2-Si) interface where a fraction is trapped in deep hole-traps in close proximity to the interface. These trapped holes give rise to the fixed oxide charge N_{ot}.[13]

In addition to the fixed oxide charge, interface traps are produced at the SiO_2-Si interface. Interface traps are formed when radiation-generated holes liberate protons in the oxide that transport to the SiO_2-Si interface, where they depassivate dangling bonds. Good reviews of the current state of understanding of radiation-induced interface traps in MOS devices are available.[11,12,14]

Interface traps are manifested as energy levels within the silicon band-gap. The occupancy of these levels depends on the potential at the SiO_2-Si interface. It is usually assumed that states lying above mid-gap are acceptor-like, while those below mid-gap are donor-like. When charged, the interface traps give rise to the interface trapped charge N_{it}. Both N_{it} and N_{ot} are considered as areal concentrations at the SiO_2-Si interface measured in number of charges per square centimeter. The technique introduced by McWhorter and Winokur is the most frequently used to separate the relative contributions of oxide trapped charge N_{ot} and interface trapped charge N_{it} in device electrical behavior.[15]

The interface and oxide trapped charge introduced into device dielectrics by ionizing radiation leads directly to degradation of MOS device characteristics. The most commonly observed and reported effect is a shift in threshold voltage; however, ionizing radiation will also lead to mobility degradation, increase in leakage currents, and reduction of breakdown voltage.

3.b. *Effects on threshold voltage*

For a MOSFET exposed to ionizing radiation, the gate oxide will contain trapped charge, which is normally positive, and charge in interface traps. In n-channel enhancement-mode MOSFETs with the gate bias above threshold, interface traps contribute negative charge. The threshold voltage shift for an n -channel MOSFET can be expressed as:

$$\Delta V_T = \frac{-q}{C_{ox}}\Delta N_{ot} + \frac{q}{C_{ox}}\Delta N_{it} = \Delta V_{ot} + \Delta V_{it}$$

where q is the electron charge and C_{ox} is the gate oxide capacitance per unit area. The threshold voltages of n-channel MOSFETs normally exhibit a negative shift following exposure to ionizing radiation, due to a net build-up of positive trapped charge in the gate oxide. This positive oxide trapped charge can be partially compensated by the build-up of negative interface trapped charge. For p-channel devices, the charge in interface traps is predominantly positive, as is the oxide trapped charge. Figure 4 is a schematic of the shift in threshold voltage for both n-channel and p-channel MOSFETs.[16] For a device that has been radiation hardened, the threshold voltage may not change significantly to levels of several Mrad(Si); however, a non-hardened n-channel device could become depletion mode at a few tens of krad(Si). Reducing the gate oxide thickness usually improves the total dose tolerance, but also decreases the maximum voltage that can be applied to the gate and may increase susceptibility to single event gate rupture as described below. The rate at which the threshold voltage changes with dose is very technology-dependent and in some n-channel devices the threshold voltage may recover, depending on the relative magnitudes of oxide charge and interface charge as a function of dose. This is schematically illustrated in Figure 5 where the threshold shift for an n-channel device is decomposed into its oxide trapped charge component and interface trapped charge component.[15]

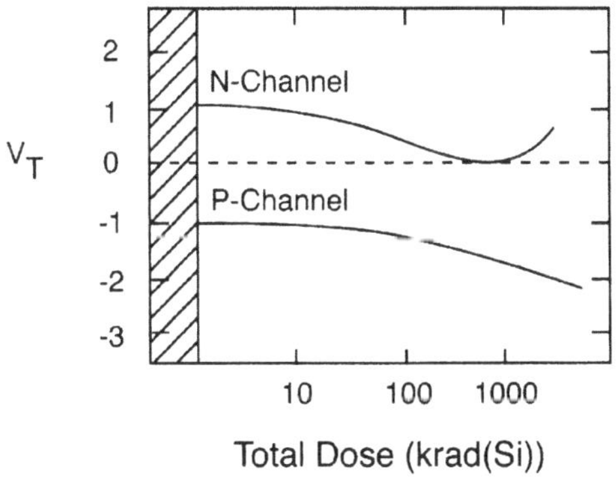

Fig. 4. Schematic representation of the threshold voltage for n- and p- channel MOSFETs as a function of total ionizing radiation dose. After ref. 16.

The effects of N_{ot} and N_{it} on device behavior can be somewhat complicated in the natural space environment. Both interface and oxide trapped charge may exhibit complex temporal behavior. Oxide trapped charge anneals with time and temperature, while interface traps often require long periods of time following irradiation in which to form. This behavior is particularly important for space applications. Because of the long time over which the total ionizing dose is accumulated, significant in-situ annealing of oxide trapped charge and buildup of interface traps may occur. As the positive oxide trapped charge anneals and the negatively charged interface traps form in an n-channel device, the threshold voltage may begin to increase. In some devices exposed to ionizing radiation at dose rates approaching those in natural space, the threshold voltages may actually shift to levels more positive than their pre-irradiation values.[17-20] This phenomenon is most commonly encountered in radiation-hardened devices because they have been engineered to balance the densities of interface and oxide trapped charge.

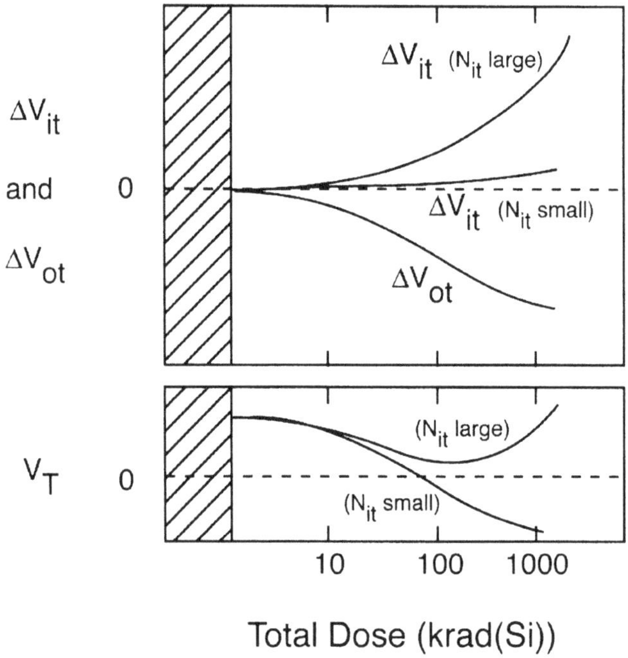

Fig. 5. Schematic representation of the decomposition of threshold voltage shift into components due to oxide trapped charge and interface trapped charge. After ref. 16.

Figure 6 illustrates this effect for power VDMOS transistors exposed to radiation from a Co-60 radiation source at dose rates approaching those encountered in natural space.[20] The IRH-254 is a radiation hardened device while the IRF-440 is an "off-the-shelf" commercial device. Both device types were biased at +9 V on the gate during exposure. The interesting point is that at "space-like" dose rates the threshold voltage of the hardened devices increased with total dose. The same effect has been observed for n-channel devices designed for rad-hard CMOS circuits.[19] In contrast, the threshold voltage of the unhardened devices decreased monotonically and did not exhibit significant dose rate dependence.

Thus far, the discussion has considered power MOSFETs exposed to ionizing radiation with constant bias. Experiments have shown that the threshold voltage shift for n-channel MOSFETs alternately biased on and off does not correspond to the average of the threshold shifts observed for devices biased continuously on or off.[21,22] For simulated space radiation conditions, the threshold shift of both hardened and unhardened power

MOSFETs operated under switching conditions to simulate a switched mode power supply did not exhibit recovery to pre-rad levels.[22]

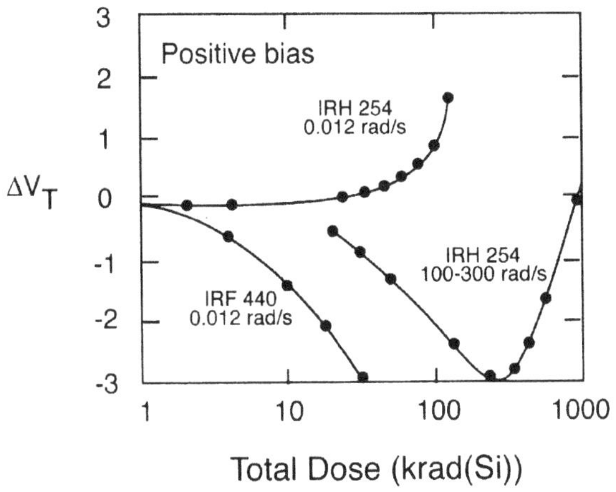

Fig. 6. Threshold voltage shift versus total ionizing dose for positively biased power MOSFETs. After ref. 20.

3.c. *Effects on channel mobility*

The threshold voltage is not the only parameter affected by ionizing radiation. Radiation-induced charge significantly degrades the channel mobility in MOSFETs.[23,24] Figure 7 illustrates the mobility degradation for a power VDMOS device. The following expression has been used to describe the effect on mobility of adding an incremental number of interface traps:[23]

$$\mu = \mu_o (1 + \alpha \Delta N_{it})^{-1}$$

where ΔN_{it} is the increase in interface-trapped charge density resulting from radiation, μ_o is the pre-irradiation mobility, and α is an empirical parameter. A typical value for α might be 5×10^{-12} cm^2. Oxide trapped charge has a measurable but smaller effect on mobility degradation than interface trapped charge.[24]

Mobility degradation in MOSFETs can lead to significant reduction in transconductance and in current-drive capability.[22] The problems with drive and saturated transconductance are compounded in devices that exhibit an increase in threshold voltage when irradiated at low dose-rates. The maximum current that a device can provide at fixed gate voltage depends on the carrier mobility and the difference between the gate voltage and the threshold voltage. In unhardened devices, the threshold voltage shift is usually negative and failure occurs when the device becomes depletion mode. In hardened devices, the threshold shift may be positive and lack of current drive may be the failure mode.

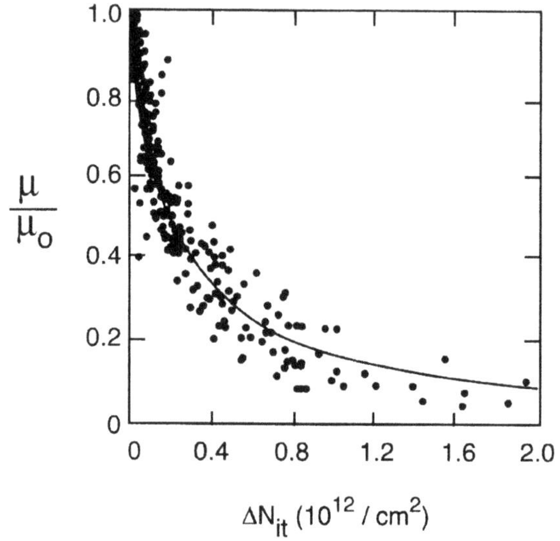

Fig.7: Reduction in mobility versus radiation induced interface trapped charge for a typical VDMOS transistor. After ref. 16.

3.d. *Effects on breakdown voltage*

The maximum breakdown voltage obtainable for a p–n junction of specified doping concentration occurs for a planar structure.[25] In actual devices utilizing planar diffusion technology, however, the high-voltage junction must intersect the surface at some finite position. The resulting junction curvature compresses the equipotential lines where the junction bends to the surface and increases the peak electric field. Junction-termination structures are used in power devices to reduce the peak electric field and to allow the breakdown voltage to approach its ideal one-dimensional value. The most commonly used termination methods are field rings and field plates.

Avalanche multiplication is the physical mechanism that causes the primary breakdown of high-voltage reverse-biased p–n junctions. The most important factor that influences the magnitude and location of the peak electric field in planar semiconductor technology is the method used to terminate the junction.[25-27] The objective of the different termination methods is to reduce the peak electric field either at the surface or at the curved junction regions in the bulk.

The drain-source breakdown voltage of power MOSFETs is strongly affected by ionizing radiation.[28-31] This is a result of the introduction of trapped oxide and interface charge in the field oxide. These trapped charges alter the potential at the surface of the junction and, in turn, become a part of the junction termination. The effect of ionizing radiation on the functioning of junction termination structures and thus power MOSFET breakdown voltage has been successfully modeled.[30,31] In general, the breakdown voltage of n-channel devices decreases with total dose, while that of p-channel devices may increase.

4. Single Event Radiation Effects

4.a. *Single event radiation effects basics*

The natural space radiation environment contains contains high-energy protons and heavy ions that can introduce transient, dense charge filaments along the path of the ion through an electronic device that can interact with sensitive regions of the device to trigger single event effects.[2,3]

The sensitivity of a semiconductor device to single-event effects is expressed as the ratio of the number of events to the total particle fluence. The resulting ratio is referred to as the cross-section since it has units of cm^2. Laboratory cross-section data are often presented as a function of linear energy transfer (LET), which is related to the energy of the incident ion. The energy loss of the incident ion is expressed in terms of LET with typical units of MeV-cm^2/mg.[32] The energy loss results in the creation of electron-hole pairs along the ion track. There are two key parameters from such data: the saturation cross-section, σ_{sat}, and the threshold LET, L_{th}. The saturation cross-section is related to the total area of the sensitive regions of the component, and the threshold LET is a measure of the critical, or minimum, charge required to trigger the event. These parameters are essential in estimating the single-event error rate for a particular environment. In Figure 8, the concepts of saturation cross-section and threshold LET are illustrated.

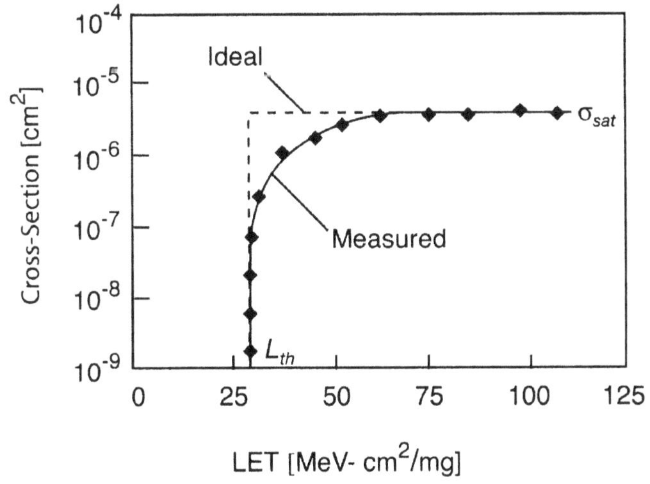

Fig. 8. A typical measured and ideal cross-section curve plotted against the LET of the incident ion.

Two single-event effects are of particular interest for power MOSFETs: single-event burnout (SEB) and single-event gate rupture (SEGR). These are catastrophic failure mechanisms that are initiated by the passage of a heavy ion through sensitive regions of the device structure. SEB of power MOSFETs was first reported by Waskiewicz, *et. al.* and SEGR of power MOSFETs was first reported by Fischer.[35,36]

4.b. *Single event burnout*

Recall that the VDMOS power transistor is capable of conducting large currents when turned *ON* and withstanding large voltages when turned *OFF*. A schematic cross-section of an n-channel VDMOS power transistor appears in Figure 1. Note the parasitic npn bipolar junction transistor (BJT) inherent to the DMOS structure. The source, body, and drain regions of the MOSFET comprise the emitter, base, and collector regions of the parasitic BJT, respectively. During normal operation of the power MOSFET, this parasitic BJT is always turned off due to the common source-body metallization that shorts out the base-emitter junction.

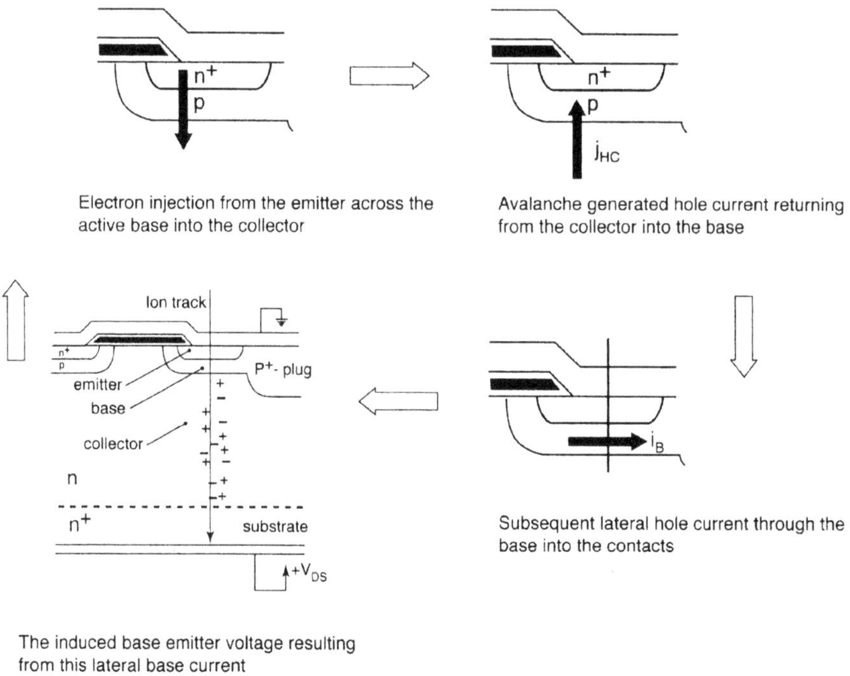

Fig. 9. SEB Triggering Mechanism

As shown in Figure 9, SEB is triggered when a heavy ion passes through a power MOSFET that is biased in the *OFF* state (blocking a high drain-source voltage). Transient currents generated by the heavy ion turn on the parasitic BJT of the power MOSFET.[37] Due to a regenerative feedback mechanism, collector currents in the parasitic BJT increase to the point where thermal failure creates a *permanent* short between the source and drain, rendering the MOSFET useless. A key component of the regenerative feedback mechanism is avalanche-generated hole current in the collector region of the BJT. SEB of p-channel power MOSFETs has not been reported in the literature. While p-channel power MOSFETs may be susceptible to SEB, it is less probable than for n-channel devices because the impact ionization rate for holes is less than that for electrons, making the magnitude of avalanche-generated currents lower than for n-channel devices.

A considerable amount of data related to single event burnout has appeared in the literature including the results on a space experiment.[38-40] A typical cross section for SEB is shown in Figure 10. In addition, numerous papers have been published to elucidate the physical effects and provide simulation and modeling of SEB.[37,41-45] The references here are not inclusive; however, these will point the interested reader to many other results.

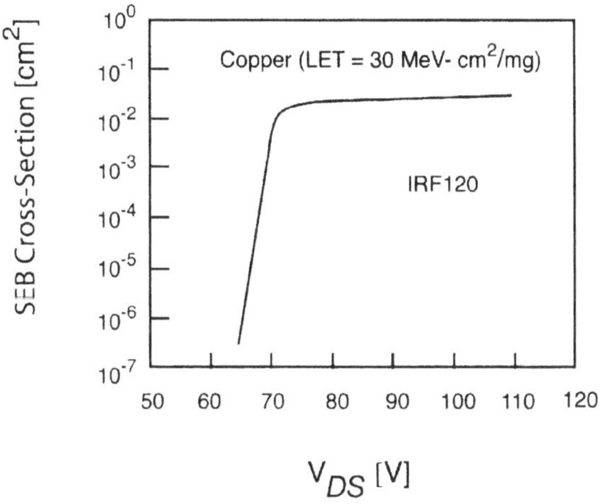

Fig. 10. Typical SEB cross-section versus VDS for a power MOSFET exposed to a mono-energetic copper ion beam giving a LET of 30 MeV-cm^2/mg. After ref 39.

4.c. ***Single event gate rupture***

Although SEB and single-event gate rupture (SEGR) are both caused by the passage of a heavy ion through the device, the failure mechanisms are altogether different. Recall, SEB occurs when the parasitic BJT is turned on and the regenerative feedback mechanism is triggered. SEGR, on the other hand, is caused by heavy-ion-induced localized dielectric breakdown of the gate oxide.[46,47] In Figure 11, a DMOS device is shown with an incident ion track in the gate-drain overlap (neck) region. Under appropriate bias conditions, accumulation of charge in the silicon (generated by the heavy ion) at the Si-SiO_2 interface in the gate-drain overlap region (*i.e.*, the neck region) can result in sufficiently high electric fields across the gate oxide to cause a localized gate rupture (*i.e.*, localized dielectric breakdown). In addition, the charge generated in the oxide briefly reduces the breakdown field of the oxide. Subsequent gate to drain current results in a thermal runaway condition, melting a *permanent* short between the gate and drain and rendering the MOSFET useless.

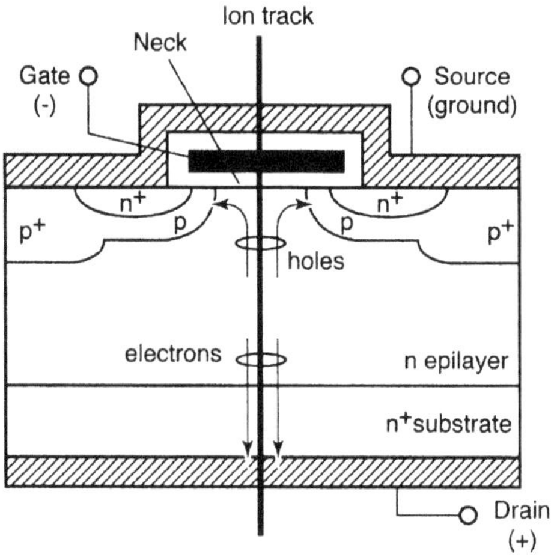

Fig. 11. SEGR is initiated when an incident ion passes through the gate-drain overlap region of the VDMOS power transistor.

Wheatley *et. al* obtained an extensive matrix of the gate to source biases, V_{GS}, and drain to source biases, V_{DS}, necessary to induce SEGR in a particular power MOSFET structure exposed to mono-energetic heavy ions at a given LET.[48] Over 400 transistors from the same wafer lot and heavy ions with LETs that ranged from 0 to 83 MeVcm^2/mg were used in the study (an LET of zero means no ion was incident and represents the control case for the experiment). Figure 12 schematically depicts their results. Later Titus *et al.* extended their matrix and included gate oxide dependence.[49] Also, Nichols *et al.* provided SEGR data for a variety of commercially available power MOSFETs.[50] Work has also been done by Titus *et al.* to predict SEGR failures in commercial space systems.[51]

Due to the destructive nature of SEGR, it is difficult to display data as a cross-section (see Figure 10). SEGR data have usually been displayed as in Figure 12. The single-event cross-section is the ratio of the total number of events to the total ion fluence. In the case of SEGR, the total number of events is one and the ion fluence is the number of ions required to trigger SEGR. An example of an SEGR cross-section is given in Figure 13.[52]

The single-event gate rupture phenomenon has been successfully modeled using two-dimensional device simulations.[45,53-55] Technologies have been developed that reduce SEGR susceptibility.[56] Again, the references given here for SEGR are not inclusive; however, they will point the interested reader to many other results.

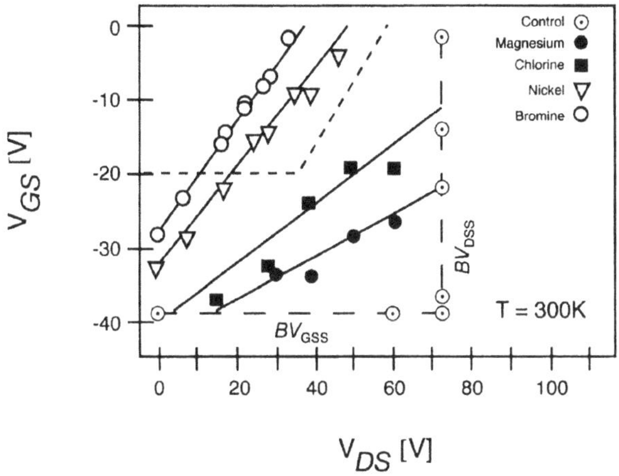

Fig. 12. SEGR experimental data after ref. 48 Bromine (LET = 37.3); Nickel (LET = 27); Chlorine (LET = 11.5), Magnesium (LET = 6).

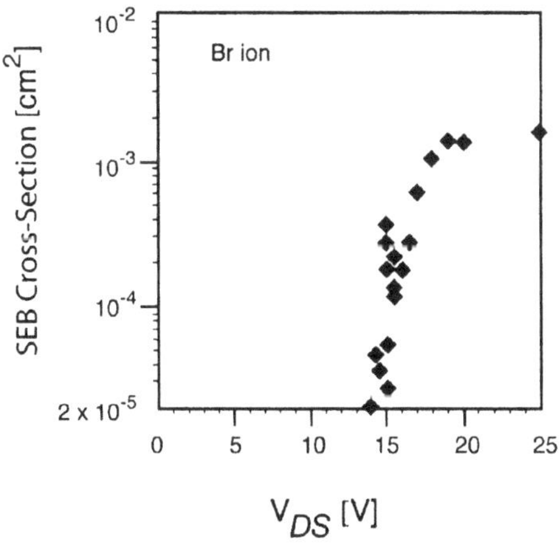

Fig. 13. Measured SEGR cross-section versus drain-source bias after ref. 52.

5. Conclusions

Power MOSFET technology is important to both military and civilian spaceborne efforts. It has a number of advantages over power bipolar technologies. This paper has briefly reviewed the current technology, its high frequency capability, and the future trends for power MOSFET technology. In addition, it has delineated a number of the susceptibilities that the power VDMOS technology has in the space environment. These susceptibilities that the power VDMOS technology has in the space environment. These susceptibilities must be considered when designing circuits for spaceborne systems utilizing these devices.

Acknowledgements

The authors would like to acknowledge the contributions of Frank Wheatley, Jeff Titus, Dave Blackburn, John Brews, Ron Pease, Mark Allenspach, Greg Johnson, Lew Cohn, Steve Kosier, Dragan Zupac, and many other colleagues to their understanding of this topic.

References

1. D. A. Grant and J. Gowar, *Power MOSFETS: Theory and Applications*. Wiley-Interscience, New York, 1989.

2. E.G. Stassinopolous and J.P Raymond, "The Space Radiation Environment for Electronics," Proc. IEEE **76** (1988) 1423-1442.

3. J.H. Adams, Jr., "The Natural Radiation Environment Inside a Spacecraft," IEEE Trans. Nucl. Sci. **29** (1983) 2095-2100.

4. G. DeBoy, M. Marz, J.P. Stengl, H. Strack, J. Tihanyi, and H. Weber, "A New Geneeration of High Voltage MOSFETs Breaks the Limit Line of Silicon," Tech. Digest Int. Electron Device Mtg. (1998) 683-685.

5. B.J. Baliga, "The Future of Power Semiconductor Device Technology," Proc. IEEE **89** (2001) 822-832.

6. N. Mohan, T.M. Undeland, and W.P. Robbins, *Power Electronics: Converters, Applications, and Design* (2nd ed.), John Wiley & Sons, Inc., New York, 1995.

7. M. Saggio, D. Fagone, and S. Musumeci; "MDmesh™: Innovative Technology for High Voltage Power MOSFETs," Proc. 12th Intl. Symp. on Power Semiconductor Devices and ICs (2000) 65-68.

8. L. Lorenz, G. Deboy, A. Knapp, and M. Marz, "COOLMOS™ - A New Milestone in High Voltage Power MOS," Proc. 11th Intl. Symp. On Power Semiconductor Devices and ICs (1999) 3-10.

9. G. Deboy, H. Hulsken, H. Mitlehner, and R. Rupp, "A Comparison of Modern Power Device Concepts for High Voltage Applications: Field Stop-IGBT, Compensation Devices and SiC Devices," Proc. Bipolar/BiCMOS Circuits and Technology Mtg. (2000) 134-141.

10. R. Schomer, P. Friedrichs, D. Peters, and D. Stephani, "Significantly Improved Performance of MOSFETs on Silicon Carbide Using the 15R-SiC Polytype," IEEE Elect. Dev. Lett. **20** (1999) 241-244.

11. T.P. Ma and P.V. Dressendorfer (eds.), *Ionizing Radiation Effects in MOS Devices and Circuits*. Wiley-Interscience, New York, 1989.

12. Timothy R. Oldham (ed.), *Ionizing Radiation Effects in MOS Oxides*. World Scientific, Singapore, 1999.

13. D.M. Fleetwood, R.A. Reber, Jr., L.C. Riewe, and P.S. Winokur, "Thermally Stimulated Current in SiO_2," Microelectronics Reliability **39** (1999) 1323-1336.

14. D.M. Fleetwood, "Effects of Hydrogen Transport and Reactions on Microelectronics Radiation Response and Reliability," Microelectronics Reliability **42** (2002) 523-541.

15. P.J. McWhorter and P.S. Winokur, "Simple Technique for Separating the Effects of Interface Traps and Trapped-Oxide Charge in Metal-Oxide-Semiconductor Transistors," Appl. Phys. Lett. **48** (1985) 133-135.

16. K.F. Galloway and R.D. Schrimpf, "MOS Device Degradation due to Total Dose Ionizing Radiation in the Natural Space Environment: A Review," Microelectronics J. **21** (1990) 67-81.

17. A.H. Johnston, "Super Recovery of Total Dose Damage in MOS Devices," IEEE Trans. Nucl. Sci. **31** (1984) 1427-1433.

18. A.H. Johnston and S.B. Roeske, "Total Dose Effects at Low Dose Rates," IEEE Trans. Nucl. Sci. **33** (1986) 1487-1492.

19. P.S. Winokur, F.W. Sexton, G.L. Hash, D.C. Turpin, "Total-Dose Failure Mechanisms of Integrated Circuits in Laboratory and Space Environments," IEEE Trans. Nucl. Sci. **34** (1987) 1448-1454.

20. R.D. Schrimpf, P.J. Wahle, R.C. Andrews, D.B. Cooper, and K.F. Galloway, "Dose-Rate Effects on the Total-Dose Threshold-Voltage Shift of Power MOSFETs," IEEE Trans. Nucl. Sci. **35** (1988) 1536-1540.

21. D.L. Blackburn, D.W. Berning, J.M. Benedetto, and K.F. Galloway, "Ionizing Radiation Effects on Power MOSFETs During High Speed Switching," IEEE Trans. Nucl. Sci. **29** (1982) 1555-1558.

22. P.J. Wahle, R.D. Schrimpf, and K.F. Galloway, "Simulated Space Radiation Effects on Power MOSFETs in Switching Power Supplies," IEEE Trans. Industry Appl. **26** (1990) 798-802.

23. K.F. Galloway, M. Gaitan, and T.J. Russell, "A Simple Model for Separating Interface and Oxide Charge Effects in MOS Device Characteristics," IEEE Trans. Nucl. Sci. **31** (1984) 1497-1501.

24. D. Zupac, P. Khosropour, S. Anderson, R.D. Schrimpf, and K.F. Galloway, "Separation of Effects of Oxide-Trapped Charge and Interface-Trapped Charge on Mobility in Irradiated Power MOSFETs," IEEE Trans. Nucl. Sci. **40** (1993) 1307-1315.

25. B. J. Baliga, "Silicon Power Field Controlled Devices," in: Silicon Integrated Circuits, Part B, (D. Kahng, ed.), Academic Press, New York, pp. 158-174, 1981.

26. B. J. Baliga, "High-Voltage Device Termination Techniques – A Comparative Review," IEE Proc. **129** (1982) 173-179.

27. B. J. Baliga, *Modern Power Devices.* John Wiley, New York, 1987.

28. D. L. Blackburn, J. M. Benedetto, and K. F. Galloway, "The Effects of Ionizing Radiation on the Breakdown Voltage of Power MOSFETs," IEEE Trans. Nucl. Sci. **30** (1983) 4116-4121.

29. R. D. Pugh, A. H. Johnson, and K. F. Galloway, "Characteristics of the Breakdown Voltage of Power MOSFETs After Total Dose Irradiation," IEEE Trans. Nucl. Sci. **33** (1986) 1460-1464.

30. K.R. Davis, R.D. Schrimpf, F.E. Cellier, K.F. Galloway, D.I. Burton, and C.F. Wheatley, Jr., "The Effects of Ionizing Radiation on Power-MOSFET Termination Structures," IEEE Trans. Nucl. Sci. **36** (1989) 2104-2109.

31. S.L. Kosier, R.D. Schrimpf, F.E. Cellier, and K.F. Galloway, "The Effects of Ionizing Radiation on the Breakdown Voltage of P-Channel Power MOSFETs," IEEE Trans. Nucl. Sci. **37** (1990) 2076-2082.

32. J.F. Zeigler, J.P. Biersack, and U. Littmark, *The Stopping and Range of Ions in Solids.* Pergamon Press Inc., Elmsford, NY, 1985.

33. W.J. Stapor and P.T. McDonald, "Practical Approach to Ion Track Energy Distribution," J. Appl. Phys **64** (1988) 4430-4434.

34. H. Dussault, J.W. Howard, Jr., R.C. Block, M.R. Pinto, W.J. Stapor, and A.R. Knudson, "High Energy Heavy-Ion-Induced Single Event Transients in Epitaxial Structures," IEEE Trans. Nucl. Sci. **41** (1994) 2018-2025.

35. A.E. Waskiewicz, J.W. Groninger, V.H. Strahan, and D.M. Long, "Burnout of Power MOS Transistors with Heavy Ions of Californium-252," IEEE Trans. Nucl. Sci. **33** (1986) 1710-1713.

36. T.A. Fischer, "Heavy-Ion Induced Gate Rupture in Power MOSFETs," IEEE Trans. Nucl. Sci. **34** (1987) 1786-1791.

37. G.H. Johnson, J.H. Hohl, R.D. Schrimpf, and K.F. Galloway, "Simulating Single-Event-Burnout of N-Channel Power MOSFETs," IEEE Trans. Electron Dev. **40** (1993) 1001-1008.

38. D.K. Nichols, K.P. McCarty, J.R. Cross, A. Waskiewicz, J. Groninger, D. Oberg, J. Wert, P. Majewski, and R. Koga, "Observations of Single-Event Failure in Power MOSFETs," IEEE Radiation Effects Data Workshop Record (1994) 41-54.

39. J. L. Titus and C. F. Wheatley, "Experimental Studies of Single-Event Gate Rupture and Burnout in Vertical Power MOSFETs," IEEE Trans. Nucl. Sci. **43** (1996) 533-545.

40. J.W. Adolphsen, J.L. Barth, and G.B. Gee, "First Observation of Proton Induced Power MOSFET Burnout in Space: The CRUX Experiment on APEX," IEEE Trans. Nucl. Sci. **43** (1996) 2921-2926.

41. T.F. Wrobel, F.N. Coppage, G.L. Hash, and A.J. Smith, "Current Induced Avalanche in Epitaxial Structures." IEEE Trans. Nucl. Sci. **32** (1985) 3991-3995.

42. T.F. Wrobel and D.E. Beutler, "Solutions to Heavy Ion Induced Avalanche Burnout in Power Devices," IEEE Trans. Nucl. Sci. **39** (1996) 1636-1641.

43. C. Dachs, F. Roubaud, J. M. Palau, G. Bruguier, J. Gasiot, P. Tastet, "Evidence of the Ion's Impact Position Effect on SEB in N-channel Power MOSFETs," IEEE Trans. Nucl. Sci. **41** (1994) 2167-2171.

44. C. Dachs, F. Roubaud, J. M. Palau, G. Bruguier, J. Gasiot, P. Tastet, M-C. Calvet, and P. Calvel, "Simulation Aided Hardening of N-channel Power MOSFETs to Prevent Single Event Burnout," IEEE Trans. Nucl. Sci. **42** (1995) 1935-1939.

45. G. H. Johnson, J. M. Palau, C. Dachs, K. F. Galloway, and R. D. Schrimpf, "A Review of the Techniques Used for Modeling Single-Event Effects in Power MOSFETs," IEEE Trans. Nucl. Sci. **43** (1996) 546-560.

46. T.F. Wrobel, "On Heavy Ion Induced Hard-Errors in Dielectric Structures," IEEE Trans. Nucl. Sci. **34** (1987) 1262-1268.

47. J.R. Brews, M.Allenspach, R.D. Schrimpf, K.F. Galloway, J.L. Titus, and C.F.Wheatley, "A Conceptual Model of Singel-Event Gate Rupture in Power MOSFETs," IEEE Trans. Nucl. Sci. **40** (1993) 1959-1966.

48. C. F. Wheatley, J. L. Titus, and D. I. Burton, "Single-Event Gate Rupture in Vertical Power MOSFETs: An Original Empirical Expression," IEEE Trans. Nucl. Sci. **41** (1994) 2152-2159.

49. J. L. Titus, C. F. Wheatley, D. I. Burton, M. Allenspach, J. Brews, R.D. Schrimpf, K.F. Galloway, I. Mouret, and R.L. Pease,"Impact of Oxide Thickness on SEGR Failure in Vertical Power MOSFETs: Development of a Semi-Empirical Expression," IEEE Trans. Nucl. Sci. **42** (1995) 1928-1934.

50. D.K. Nichols, J.R. Cross, and K.P. McCarty, "Single-Event Gate Rupture Commercial Power MOSFETs," Conf. Record, RADECS 93 (1993) 462-467.

51. J.L. Titus, C.F. Wheatley, T.H. Wheatley, W.A. Levinson, D.I. Burton, J.L. Barth, R.A. Reed, K.A. LaBel, J.W. Howard, and K.M. VanTyne, "Prediction of Early Lethal Failures of VDMOSFETs for Commercial Space Systems," IEEE Trans. Nucl. Sci. **46** (1999) 1640-1651.

52. I. Mouret, P. Calvel, M. Allenspach, J. L. Titus, C. F. Wheatley, Jr., K. A. Label, M. C. Calvet, R. D. Schrimpf, and K. F. Galloway, "Measurement of a Cross-Section for Single-Event Gate-Rupture in Power MOSFETs," IEEE Electron Dev. Lett. **17** (1996) 163-165.

53.M. Allenspach, J.R. Brews, I. Mouret, R.D. Schrimpf, and K.F. Galloway, "Evaluation of SEGR Threshold in Power MOSFETs," IEEE Trans. Nucl. Sci. **41** (1994) 2160-2166.

54. M. Allenspach, I. Mouret, J. L. Titus, C. F. Wheatley, Jr., R. L. Pease, J. R. Brews, R. D. Schrimpf, and K. F. Galloway, "Single-Event Gate-Rupture in Power MOSFETs: Prediction of Breakdown Biases and Evaluation of Oxide Thickness Dependence," IEEE Trans. Nucl. Sci. **42** (1995) 1922-1927.

55. M. Allenspach, C. Dachs, G.H. Johnson, R.D. Schrimpf, E. Lorfevre, J.M. Palau, J.R. Brews, K.F. Galloway, J.L. Titus, and C.F. Wheatley, "SEGR and SEB in N-Channel Power MOSFETs," IEEE Trans. Nucl. Sci. **43** (1996) 2927-2931.

56. M.W. Savage, D.I. Burton, C.F. Wheatley, J.L. Titus, and J.E. Gillberg, "An Improved Stripe-Cell SEGR Hardened Power MOSFET Technology," IEEE Trans. Nucl. Sci. **48** (2001) 1872-1878.

International Journal of High Speed Electronics and Systems
Vol. 14, No. 2 (2004) 465–487

World Scientific
www.worldscientific.com

INTRODUCTION TO SOI MOSFETs: CONTEXT, RADIATION EFFECTS, AND FUTURE TRENDS

SORIN CRISTOLOVEANU
Institute of Microelectronics, Electromagnetism and Photonics (UMR CNRS, INPG & UJF)
ENSERG, B.P. 257, 38016 Grenoble Cedex 1, France
sorin@enserg.fr

VÉRONIQUE FERLET-CAVROIS
Commissariat à l'Energie Atomique (CEA-DIF)
B.P. 12, 91680 Bruyères-le Châtel, France
veronique.ferlet@cea.fr

The context of SOI technology is briefly presented in terms of wafer fabrication, configuration/performance of SOI devices, and operation mechanisms in partially and fully depleted MOSFETs. Typical radiation effects, induced by single particles and cumulated dose, are evoked: BOX degradation, parasitic bipolar action, coupling effects, transistor latch, and back-channel conduction. The future of SOI is tentatively explored, by discussing the further scalability of SOI-MOSFETs as well as the innovating architectures proposed for the ultimate generations of SOI transistors.

Keywords: SOI; radiation; transistor.

1. Introduction

Silicon-On-Sapphire (SOS) was the initial Silicon On Insulator (SOI) technology, developed for the fabrication of radiation-hard circuits. Four decades of research in material science and device physics resulted in new types of SOI structures and integrated circuits with high performance.[1,2] The basic SOI concept is to dielectrically separate, using a buried oxide (BOX), the active device volume from the Si substrate. The adjustability of the film thickness is actually the main asset of the SOI technology in terms of enhanced performance and superior scalability. It is hard to imagine the CMOS scaling beyond the 25–35 nm generation without making use of thin-film SOI MOSFETs.[3,4,5,6]

Although interest in SOI expanded from the niche of radiation-hard devices to the mainstream microelectronics, the integrated circuits continue to be exposed to radiations.[7,8] The natural environment, from high altitudes to ground level, involves a significant flux of heavy and light particles, which contribute to degrade the component reliability. This implies that the relevance of radiation effects is accentuated as the SOI centroid was shifted from defense and space electronics towards commercial components.

The aim of this paper is to provide an introduction to SOI, with particular focus on radiation effects, which can be used to decode more specialized publications. In sections 2, we briefly present the current status of the SOI technology. The typical mechanisms of operation of fully and partially depleted MOSFETs are addressed in section 3. Section 4 contains an overview of the main radiation effects in SOI MOSFETs. More detailed analysis and systematic data is available in other papers of this special journal issue. In section 5, we attempt to give a flavor of the future trends in SOI.

2. Current Status of SOI Technology

2.1. *Prime merits of SOI devices*

SOI circuits consist of single-device silicon islands, dielectrically isolated from each other and from the underlying substrate (Fig. 1). The lateral isolation offers more compact design and simplified technology than in bulk silicon: there is no need of wells or interdevice trenches. In addition, the vertical isolation prevents the *latch-up* mechanism which is detrimental in bulk-Si.

The source/drain regions normally extend down to the buried oxide (BOX) so that the junction surface is minimized. The leakage current and the off-state power dissipation are therefore reduced. The junction capacitance, which governs the switching speed as well as the dynamic power dissipation, is also dramatically improved. The limited cross-section of the junctions enables SOI circuits, especially fully depleted CMOS, to operate in a wider temperature range and to be less affected by short-channel effects (section 5).

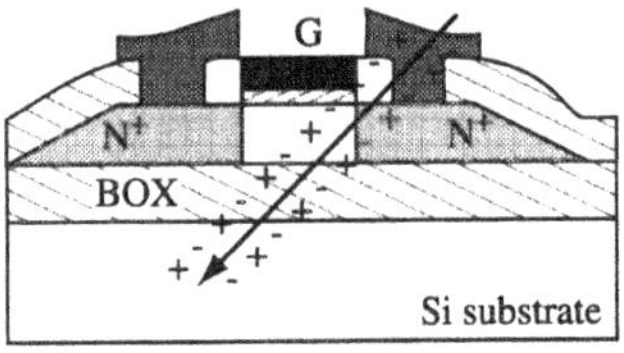

Fig. 1. Schematic architecture of an SOI MOSFET under radiation.

It is in the highly competitive domain of low-power & low-voltage circuits that SOI is most attractive. SOI offers a quasi-ideal subthreshold slope (60 mV/decade at room temperature), hence the opportunity to shrink the threshold voltage below 0.3 V. Frequencies beyond 150 GHz and delay times below 10 ps/stage[9,10] have been reached and operation at 0.5–0.9 V has been demonstrated. It has been repeatedly proven that SOI circuits operate 20–30 % faster, in other words the SOI generation (n) and the *next* bulk-Si generation ($n+1$) perform comparably. We do not include specific examples because new records are frequently set, their ephemeral existence reflecting the health and progress of SOI technology.

As far as the reliability is concerned, SOI MOSFETs are extremely robust to *transient* radiation effects because the charge collection volume is, in principle, proportional to the film thickness. We will see in the following sections that the *permanent* radiation-induced damage in the BOX is the main issue. The integrity of the gate oxide is another problem which can be solved by reducing the defect density in the film. SOI MOSFETs are rather resistant to hot carrier damage because the electric-field peak is lower than in bulk Si. Another difference with bulk-Si MOSFETs is the impact of electrostatic discharge (ESD) and the design solutions to eliminate the charge from the isolated body.

Besides the fully and partially depleted MOSFETs, the family of SOI devices also includes bipolar and high-voltage DMOS transistors (with lateral configuration), smart power devices, 3-D circuits, optical switches, waveguides and modulators, microwave transistors on high resistivity SOI wafers, etc.[1,2] Most innovative devices make use of the possibility to (i) combine bulk-Si power devices with low-power control SOI MOSFETs on a single chip containing localized SOI regions, (ii) adjust the thickness of the Si overlay and buried oxide, and (iii) implement metal layers or additional gates underneath the BOX. SOI is also an ideal material for microsensors and MEMS, where very thin membranes are fabricated by using the interface between the BOX and substrate as a perfect etch-stop mark.

2.2. *SOI materials*

The market of device grade SOI wafers is currently shared by Unibond and SIMOX Unibond process uses the deep implantation of hydrogen to generate a plane of microcavities in an oxidized wafer.[11] After bonding to a second wafer and annealing, this plane allows the wafers to separate naturally (Smart–Cut), leaving a thin SOI film (Fig. 2c). Eltran wafers are also prepared by bonding, except that the splitting is defined by a sacrificial layer of porous silicon, on which the Si film has been epitaxied.[12] In these bonding processes, only the wafer which provides the Si film needs to be premium quality. The multiple slicing and recyclability of these wafers is a key economic argument.

Unibond and Eltran are very flexible and can both provide, in 300 mm wafers, unlimited combinations of BOX and film thickness. The BOX and film–BOX interface have thermal-oxide quality. The bonding interface is more defective and more receptive to radiation effects. In particular, electron-paramagnetic resonance studies in Unibond have shown the presence of hydrogen-related defects near the bonding region.[13] However, the bonding interface has a minor impact on the device performance because it is located underneath the BOX. As far as the *active* interface is concerned, the density of interface states that may be induced by the released hydrogen is not significant up to very high total doses (about 1 Mrad). The electrical properties being excellent, it is not clear whether the film quality can reflect the impact of either implanted hydrogen (Unibond) or epitaxy on porous Si (Eltran). Stress effects in ultra-thin structures also need evaluation.

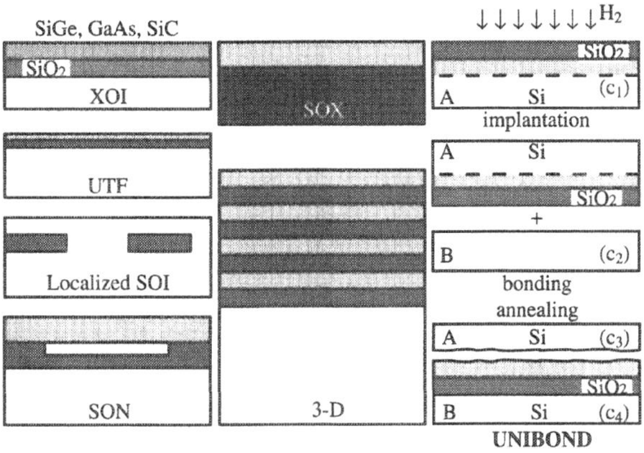

Fig. 2. Unibond process (c_1–c_4) and innovative SOI–like structures: *other* semiconductor on insulator (XOI), ultra-thin film and buried oxide (UTF), interrupted BOX, silicon on nothing (SON), silicon on *other* dielectric (SOX), 3–D Si–BOX 'super-lattices'.

SIMOX is synthesized by deep implantation of large doses of oxygen and high temperature annealing, which results in a BOX with special microstructure. Typical defects are interface traps, hole and electron traps in the BOX, the latters being activated by exposure to radiation.[14,15,16,17] The quality of the BOX and interfaces has dramatically been improved and the defects in the film have been erased, except the threading dislocations. The trend is to achieve thinner films (< 100 nm) and BOX (< 150 nm) by reducing the implant energy and dose, and also by adding special implantation/anneal steps (ITOX and new Ibis/IBM materials).

SOS material is still handicapped by the mediocre quality of the epitaxial Si film and interface. SOS hardly survives for the benefit of radiation-hard and RF circuits, for which the presence of an 'infinite' dielectric substrate is an ideal solution. Silicon on Nothing (SON) has recently been proposed as a solution for synthesizing *localized* SOI–like structures within a bulk-Si wafer (Fig. 2).[18] Several other SOI materials have temporarily emerged and disappeared:[1] silicon on zirconia, FIPOS (*Full Isolation by Porous Oxidized Silicon*), ZMR (*Zone Melting Recrystallization*). *Epitaxial Lateral Overgrowth* (ELO or other acronyms) is still being considered for 3–D integration and double-gate MOSFETs.

The fabrication of very advanced devices will require not only extremely thin SOI films but also a new class of SOI-like materials (Fig. 2). Using bonding techniques, the silicon film can be replaced by other semiconductors (SiGe, III–V or II–VI compounds, etc) with attractive optical and electrical properties. Alternatively, the BOX can be traded in for other types of insulators (sapphire, diamond, glass, flexible plastic, air) which may offer improved thermal dissipation, enhanced RF performance, new optical functions, and portability.

The large variety of SOI structures makes the signature of radiation-induced defects to differ according to the conditions of synthesis. The pseudo–MOS transistor (Ψ–MOSFET) is a simple and elegant method to monitor the generation of BOX, film and interface defects during *radiation.*[19,20] SOI is a natural upside-down MOS structure, where the Si substrate acts as a gate terminal and can be biased to induce a conduction channel (inversion or accumulation) at the interface. The BOX plays the role of a gate oxide and the Si film represents the transistor body. Low-pressure probes are placed on the film and form source and drain point contacts (Fig. 3).[19]

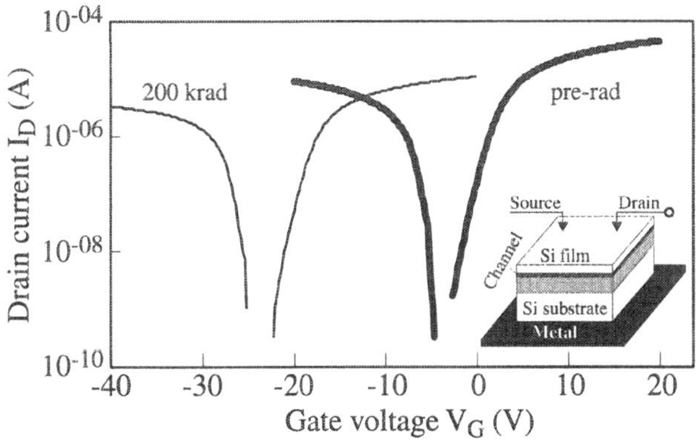

Fig. 3. Configuration of the pseudo-MOS transistor and typical $I_D(V_G)$ characteristics in a thin SOI film, before and after radiation.

Very pure MOSFET–like characteristics are produced. In strong inversion/accumulation and ohmic region, the drain current is given by

$$I_{DS} = f_g \mu C_{ox} V_D (V_G - V_{T,FB}) \tag{1}$$

where $f_g \simeq 0.75$ is the aspect ratio. The slope of $I_D/\sqrt{g_m}$ *vs.* V_G curves yields the mobility of electrons and holes, whereas the intercept with the V_G axis gives the threshold (V_T) or the flat-band (V_{FB}) voltage.[20]

The density of traps at the film–BOX interface is calculated from the subthreshold slope in weak inversion, the fixed charge density from V_{FB}, and the film doping from the difference $V_T - V_{FB}$. The carrier lifetime can be evaluated by recording the transient drain current after the gate is pulsed in strong inversion.

3. Operation Mechanisms in SOI MOSFETs

3.1. *Fully depleted MOSFETs*

In fully depleted MOSFETs, the depletion region covers the whole transistor body and does not extend with gate bias. Interface coupling causes the front-gate measurements to depend on the back gate bias V_{G_2} as well as on BOX thickness and interface defects. Totally different $I_D(V_{G_1})$ relations describe the complex characteristics of the front-channel transistor (Fig. 4).

Threshold voltage. In depletion regime, the threshold voltage $V_{T_1}^{dep}$ decreases linearly with V_{G_2} between two plateaus corresponding respectively to accumulation and inversion at the back interface (Fig. 4d):[21]

$$V_{T_1}^{dep} = V_{T_1}^{acc} - \frac{C_{si} C_{ox_2} \left(V_{G_2} - V_{G_2}^{acc}\right)}{C_{ox_1}(C_{ox_2} + C_{si} + C_{it_2})} = V_{T_1}^{acc} - k_1 \left(V_{G_2} - V_{G_2}^{acc}\right) \tag{2}$$

where C_{si}, C_{ox}, C_{it} are the capacitances of the fully-depleted film, oxide, and interface traps, respectively. The subscripts 1 and 2 hold for the front or the back channel parameters. They can be interchanged in order to account for the variation of the back-channel threshold voltage V_{T_2} with the front-gate voltage V_{G_1}. Usually, C_{ox_2} and C_{it_2} are much smaller than C_{si}, hence the slope k_1 of the $V_{T_1}(V_{G_2})$ curve can be approximated by the ratio of the front-gate oxide and buried oxide thickness $k_1 \simeq t_{ox_1}/t_{ox_2}$.

Subthreshold slope. The front-channel subthreshold swing S_1 is given by:

$$S_1 = 2.3\frac{kT}{q}\left(1 + \frac{C_{it_1}}{C_{ox_1}} + \alpha_1 \frac{C_{si}}{C_{ox_1}}\right) \tag{3}$$

For depletion at the back interface, the coupling coefficient α_1

$$\alpha_1 = \frac{C_{ox_2} + C_{it_2}}{C_{si} + C_{ox_2} + C_{it_2}} < 1 \tag{4}$$

accounts for the influence of back interface traps C_{it_2} and BOX thickness C_{ox_2}.[22]

By accumulating the back channel, the front inversion channel becomes decoupled from back interface defects but, in turn, coefficient α_1 tends to unity (as in bulk–Si or partially-depleted MOSFETs), causing an overall degradation of the swing (Fig. 4e).

The above equations are valid only when the buried oxide is thick enough so that substrate effects occurring underneath the BOX can be ignored. Otherwise, 3-interface models become necessary. The general trend is that the subthreshold slope improves for thinner silicon films and thicker BOX.

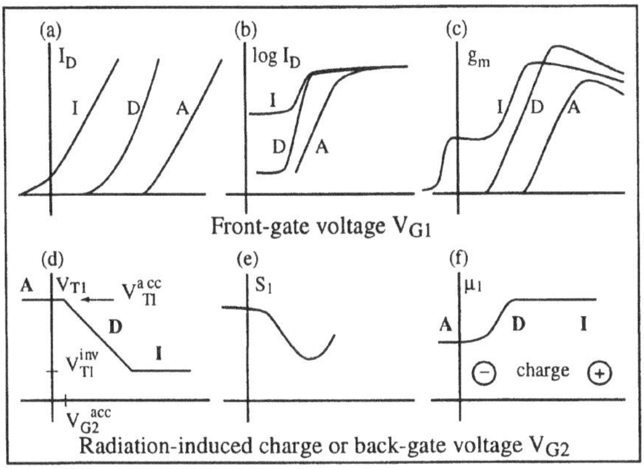

Fig. 4. Generic front-channel characteristics of a fully-depleted n-channel SOI MOSFET for different conditions at the back interface: accumulation (A), depletion (D), and inversion (I). The top characteristics are: (a) $I_D(V_{G_1})$ in strong inversion, (b) $\log I_D(V_{G_1})$ in weak inversion, and (c) transconductance $g_m(V_{G_1})$. The bottom curves indicate the variation of the corresponding parameter with radiation dose (or back gate bias): (d) threshold voltage, (e) subthreshold swing, and (f) electron mobility.

Transconductance. In strong inversion and ohmic region, the transconductance is

$$g_{m_1} = \frac{C_{ox_1} W V_D}{L} \cdot \frac{\mu_1}{[1 + \theta_1(V_{G_1} - V_{T_1}(V_{G_2}))]^2} \tag{5}$$

where μ_1 is the mobility of front channel carriers, and θ_1 is the mobility attenuation coefficient.

The complicated transconductance curves in fully depleted MOSFETs (Fig. 4c) are explained by the influence of the back gate bias on front-channel threshold voltage (Eq.(2)), and carrier mobility. The effective mobility and transconductance are maximum for depletion at the back interface (Fig. 4f), due to combined effects of reduced vertical field and series resistances.[23]

The distortion of the transconductance (curve I, Fig. 4c) reflects the activation of the back channel: increasing V_{G_1} reduces the back threshold voltage and opens the back channel, while the front interface is still depleted.[23]

3.2. *Partially depleted MOSFETs*

In partially-depleted SOI MOSFETs, a neutral region subsists and cancels the interface coupling effects. When the body is grounded, most of the standard $I_D(V_G, V_D)$ equations developed for bulk-Si MOSFETs apply. In contrast, if body contacts are not supplied, several *floating-body* effects arise.

The *kink effect* is triggered by majority carriers, which are generated by impact ionization and accumulate in the body: the body potential is gradually increased and the threshold voltage is lowered. The kink results in excess drain current and low-frequency noise.[24]

In weak inversion and for high drain bias, a similar positive feedback is responsible for negative resistance regions, *hysteresis* in $\log I_D(V_G)$ curves, and eventually transistor *latch* (see Fig. 9).

The charging of the body also leads to the forward biasing of the source-to-body junction and to a parasitic bipolar effect. The activation of the *parasitic bipolar transistor* yields more current but causes premature breakdown.

Transient currents are other typical floating body effects. A drain current *overshoot* occurs when the gate is turned on. Majority carriers are expelled from the depletion region and collect in the body, so increasing the potential. Equilibrium is reached through electron-hole recombination, which makes the drain current decrease gradually with time. A reciprocal *undershoot* is observed when the gate is switched from strong to weak inversion: the current increases with time as the generation process of majority carriers allows the depletion depth to shrink. In advanced MOSFETs, the transients are remarkably reduced, either by the proximity of source and drain junctions which efficiently remove the majority carriers, or by the gate tunneling current.[25]

During high-frequency switching of integrated circuits, the transistor body does not always succeed to reach equilibrium. The charging and discharging of the body is an iterative process which may cause *history* and *memory* effects as well as dynamic instabilities.

Partially-depleted transistors with dynamic threshold (DT–MOSFETs) are very attractive for low-voltage/low-power CMOS circuits. The gate and body are interconnected: increasing the gate voltage in weak inversion causes a simultaneous raise in body potential and a gradual decrease in threshold voltage (dynamic V_T). The gate-charge coupling is excellent and results in improved subthreshold slope, current, transconductance, and short-channel effects.

4. Radiation Effects on SOI MOSFETs

The radiation effects, critical for space applications, also appear at the ground level. The charge stored in elementary nodes decreases with technology integration, down to a few femto-coulomb, and can be easily upset by the radiation deposited charge. Charge is deposited by either direct or indirect ionizing effect.[26] The heavy charged particles encountered in the space environment produce a direct ionizing effect by interaction with the electronic cortege of atoms. On the other hand, the light particles, protons in space and neutrons in the atmospheric environment, induce an indirect ionizing effect through their nuclear interaction with silicon nucleus or other chemical elements of the chip. The charge is then deposited by the produced recoil heavy charged particles along their track in the device.

Depending on the struck region and the applied electric field, the charges are separated and induce either transient or permanent damage:

(i) **Soft Errors.** Charges are swiftly transported to elementary transistor terminals and give rise to a parasitic transient current. If this current is high enough, permanent damage (gate rupture, burnout) can occur. In most cases, the deposited charge induces a Single Event Upset (SEU), *i.e.*, a local change of the logic information contained in DRAMs and SRAMs. These SEUs are 'soft errors' as the memory cell can be re-written without permanent damage.

(ii) **Cumulated Dose.** The cumulative charge deposited in the insulators tends to build a trapped charge density, which causes a permanent shift of device parameters (threshold voltage, off and on currents). Thick oxide parts, mainly lateral isolations and buried oxides, are particularly sensitive to radiation-induced trapped charge.

4.1. *Parasitic conduction and impact of radiations*

When an ionizing particle crosses the drain region of a bulk-Si MOSFET, a typical funneling phenomenon occurs. It consists in the distortion of the drain junction electric field along the particle track, which induces a collection of the generated charges on several microns. The funneling can activate the parasitic latch-up structure (between junctions, wells, and substrate) which is inherent to bulk devices. In SOI transistors, the total dielectric isolation suppresses the latch-up structure, reduces the role of the drain junction, and limits the collection to the charges generated in the thin silicon film only. However, this advantage is mitigated by several parasitic structures, typical to SOI transistors.[27]

(i) *The parasitic bipolar transistor* can be triggered under irradiation, when parasitic charges accumulate in the body and forward bias the body-source diode. To avoid the bipolar onset, the body is tied to the source so that the generated charge is evacuated.[28,29,30] The body ties are space consuming. Other solutions are Ge implantation to reduce the majority carrier lifetime,[31] partial trenches,[32] BUSFET.[33]

(ii) *The parasitic back-gate transistor* is triggered under cumulative dose effect. Usually, a net positive charge is trapped, leading to a possible parasitic conduction in N-channel MOSFETs.

(iii) *The lateral parasitic transistor* is activated by charge trapping in the lateral isolation (LOCOS, STI), and induces a leakage current on the transistor edges.[34] A similar structure also exists in irradiated bulk devices.[35]

The front-gate oxide is so thin that the dose effect is negligible. We will see next that more important is the difference between fully or partially depleted architectures in terms of SEU and total dose sensitivity.

4.2. *Single event effects*

The parasitic bipolar action is effective when the body is left floating. After an ionizing particle strikes, the electron-hole pairs generated in the body are separated by the local electric field. Minority carriers drift towards the drain, while majority carriers tend to accumulate in the body. They contribute to forward bias the body-source junction, and to trigger the parasitic bipolar.

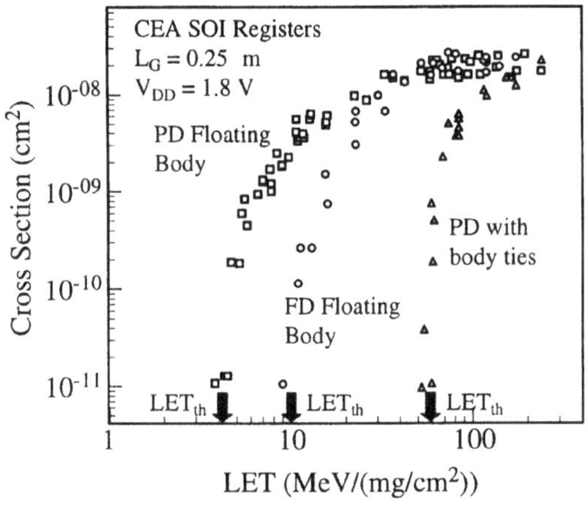

Fig. 5. Heavy ion cross-section versus effective Linear Energy Transfer (LET) in typical SOI circuits (registers). The graph shows the improvement achieved by using body ties.[27,38]

The bipolar current amplifies the charge collection.[36,37] The total collected charge, Q_{coll}, corresponds to the amplification of the irradiation deposited charge Q_{dep}. A soft error may arise if Q_{coll} is higher than the (minimum) critical charge, Q_{crit}, able to upset a memory cell:

$$Q_{coll} = (1 + \beta) \times Q_{dep} \geq Q_{crit} \tag{6}$$

where β is the bipolar gain. The higher β, the lower Q_{dep} able to induce an upset. Q_{dep} is defined with the Linear Energy Transfer (LET) of the ion, when crossing the sensitive region. In normal incidence, the deposited charge in the silicon film is given by:

$$LET[MeV \cdot mg^{-1}cm^2] \times t_{si}[\mu m] = Q_{dep}[pC] \times 96.9 \tag{7}$$

where t_{si} is the film thickness and the conversion factor 96.9 is valid in silicon.

For analysis of the device sensitivity, components are tested with different LET particles (Figure 5). The sensitive cross-section σ is deduced from the number of errors N_{err}, and the total particle fluence. The cross-section can be normalized by the number of bits for memories or registers, which enables the comparison of components with different complexity. The cross section then represents the *sensitive* surface of the memory (or register) cell.

$$\sigma[cm^2] = N_{err}/(Fluence[cm^{-2}] \times Bits) \tag{8}$$

Figure 5 shows the sensitivity of fully and partially depleted devices. The threshold LET or LET_{th} corresponds to the lowest deposited charge able to induce an upset: $Q_{dep} = Q_{crit}/(1+\beta)$. Fully depleted circuits are less sensitive than the equivalent partially depleted ones.[27,38] This is due to a thinner silicon layer and a lower bipolar gain for the fully depleted devices.

The adjunction of body ties allows to reduce the bipolar amplification and to increase the threshold LET. It is an efficient method for both fully and partially depleted devices.[27,39]

The bipolar amplification can even lead to snap-back.[40] The impact ionization in the drain region generates additional carriers, which maintain the body-source junction forward biased, so that the transistor can not return instantly to its initial off-state.

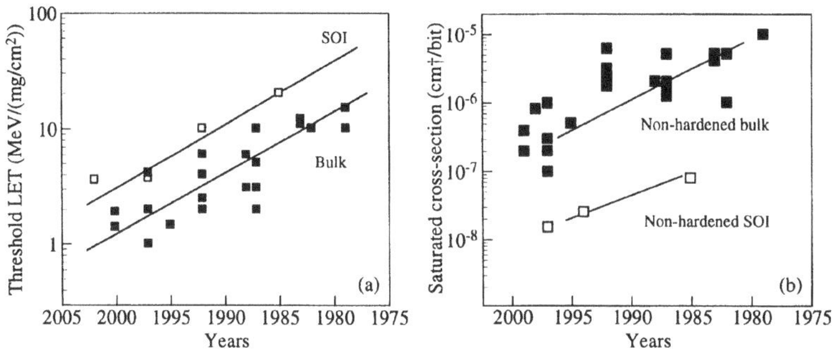

Fig. 6. (a) Threshold LET and (b) saturated cross-sections in non-hardened SOI and bulk-Si circuits as a function of the technology generation. The data is extracted from literature on SOI and bulk SRAMs.[39,41,42,43,44,45,46,47,48]

Only few references compare standard thin SOI (non-hardened and without body ties) with bulk devices of the same generation. A compilation of experimental data obtained on SOI and bulk devices shows that commercial SOI (without body ties) and bulk circuits have a rather similar threshold LET (Fig. 6a).[39,41,42,43,44,45,46,47,48] On the other hand, SOI has a significantly lower sensitive cross-section (Fig. 6b).[41] For the 2000 generation node, an SOI device has a typical cross-section of about $1\mu m^2$/bit, which is one order of magnitude lower than in bulk. Furthermore, technology down-scaling induces a decrease of both the cross-section and the threshold LET of SOI and bulk devices. This is consistent with respectively a more compact design and a reduced critical charge with integration.

In SOI, it is commonly admitted that the sensitive cross-section is reduced to the gate surface which covers the body region. However, recent micro-beam experiments on a 0.35 μm SOI technology showed that the drain region can also be sensitive to high LET particle strikes.[49] Actually the drain doping has been shown to be the key parameter of the SOI drain sensitivity.[38] When the drain is not strongly doped down to the buried oxide, the generated charge does not recombine completely. Subsisting charges are drifted towards the body region where they contribute to the bipolar amplification effect. The drain sensitivity can be avoided by carefully tailoring the doping profile.

4.3. *Total dose effects*

The typical effect in SOI is the charge accumulation in the BOX. Back-channel measurements provide a direct insight into the BOX defects responsible for radiation-induced degradation. The two main types of buried oxides, SIMOX and UNIBOND, present similar qualitative trapping properties:

(i) An uniform trap density in the volume of the BOX.[50] This special feature is presumably due to the high temperature anneal (1100°C for UNIBOND, 1350°C for SIMOX) that ends the SOI substrate process. By contrast, in conventional thermal oxides, the trap density is located near the oxide-silicon interface.
(ii) A positive charge trapping, partially compensated, especially at high radiation dose, by electron trapping. The trapping centers are due to structural defects, like constraint or dangling bonds, oxygen vacancies, silicon clusters, etc.[51]

The contributions of both species, holes and electrons, have been demonstrated by back-channel measurements.[14] Figure 7 reveals several distinct mechanisms according to the bias applied to the back gate during exposure to radiations. For positive bias, radiation-induced electron-hole pairs are separated by the field and the holes are driven towards the upper interface of the BOX. A net positive charge is build-up, yielding a very large decrease in the back-channel threshold voltage V_{T_2}. For zero bias, most of the electron-hole pairs recombine instantly, hence V_{T_2} shift is attenuated. But, for negative bias, the pairs again separate and the holes are attracted to the bottom BOX interface. The positive charge effect still dominates for low doses, whereas beyond 100 krad a rebound is observed. The large increase in V_{T_2} accounts for electron trapping near the film-BOX interface, which modifies the surface potential as well as the series resistance of the transistor.[14]

In *partially-depleted* devices, charge trapping in the BOX induces the parasitic conduction of the back-gate transistor, as illustrated in Figure 8. Under usual bias conditions (back gate grounded), the leakage current is observed in n-type MOS transistors, due to a net positive trapped charge in the buried oxide. Above a threshold dose (300 krad(SiO_2) in Fig. 8), the back-channel transistor is clearly

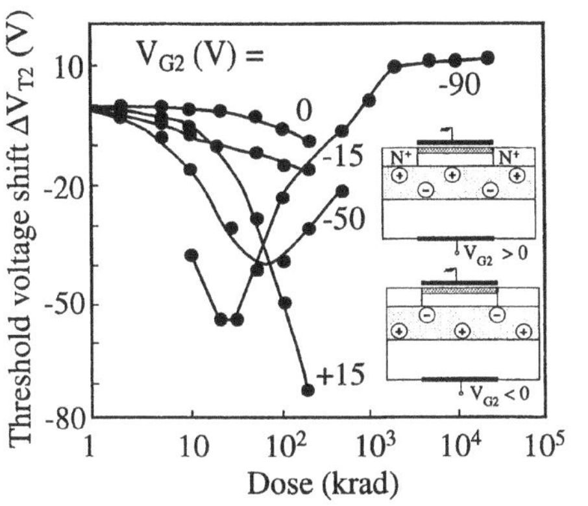

Fig. 7. Radiation-induced shift of the back-channel threshold voltage in a SIMOX MOSFET.[14]

conducting even for $V_{G_2} = 0$. The parasitic current flows at the bottom interface of the silicon film (Fig. 8a), and adds to the main transistor current. In partially depleted devices, such a 'leakage' current cannot be controlled by the front gate (no coupling action); this prevents the main transistor from switching off (Fig. 8b).

Below the threshold dose, the neutral region can in principle protect the front channel from the detrimental impact of radiation-induced damage in the BOX. This is not totally true, because several subtle mechanisms can arise. For example, a positive charge in the BOX will lower the series resistance of LDD regions, increasing the impact ionization rate and the body charging. As a consequence, hysteresis and latch can be activated during radiation. The transient and history effects are also modified as soon as the generation-recombination rate in the volume or at the BOX and sidewall interfaces is increased.

Fully-depleted devices are characterized by the strong coupling effect between the front and the back gate transistor. Carriers flowing at one interface are influenced by the presence of defects at the opposite interface. This *defect coupling* is responsible for an apparent degradation of the front-channel characteristics, which is actually due to radiation-induced damage in the BOX. For example, the generation of back interface traps alters the threshold voltage and subthreshold swing according to Eqs.(2) and (3).

When irradiating fully depleted devices, the trapped charge in the BOX acts as a build-in positive back-gate voltage. It induces a strong shift of the front-channel threshold voltage (for example as in Fig. 4a,d).[52,53] As long as the transistor remains in full depletion mode, ΔV_{T_1} is directly proportional to the back-channel threshold voltage shift induced by charge trapping into the BOX: $\Delta V_{T_1} \simeq k_1 \Delta V_{T_2}$ (see Eq.(2)). In addition, the build-up of a fixed charge in the BOX can result in

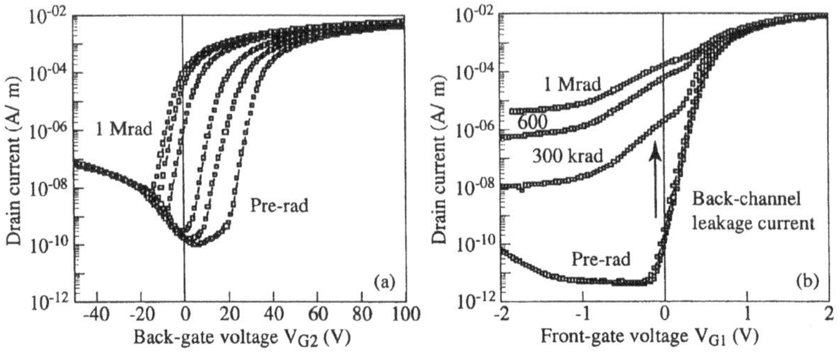

Fig. 8. Drain current versus back gate (a) and front gate (b) bias in a partially depleted n-channel SOI MOSFET, at low drain bias and after several doses of radiation. The transistor was in off-state during radiation. The trapping of positive charges in the BOX induces a negative shift of the back-channel threshold voltage. The back channel current increases the apparent leakage current of the front-channel transistor.[27]

a modification of the back surface potential (accumulation, depletion or inversion), changing the mode of operation of the front channel, as shown in Figure 4. The transconductance may also be affected by the modulation of the series resistance.

An apparent solution to improve the radiation hardness of fully-depleted transistors is to use a thinner BOX. However, the coupling coefficient is enhanced and the front transistor becomes more sensitive to the charge trapped in the BOX. This is why a thinner BOX does not significantly improve the radiation hardness of fully depleted devices.[34] A more effective solution is to use a hardened buried oxide, with a low trap density.[54,55]

Total dose deposition can also result in a transistor latch effect. Figure 9 shows the static characteristics $I_D(V_{G_1})$, at high drain bias, for irradiated SOI transistors.[56] For a 'critical' dose, a hysteresis phenomenon appears in the subthreshold region of both partially and fully depleted devices. Below the critical dose, no leakage current is visible on the $I_D(V_{G_1})$ curves at high drain voltage, whereas above the critical dose the transistor is latched. The strong 'leakage' current causes circuit failure. Again, the latch can be effectively suppressed, in partially and fully depleted devices, by (i) lowering the drain voltage, (ii) designing body ties, and (iii) using thinner buried oxides.

4.4. *Hardness assurance: worst case bias and irradiation source*

The worst-case bias conditions have been described for transistors processed on SIMOX and Unibond wafers.[57,58] Similar conclusions have been reached. Biasing has been studied by systematic measurements and by modeling the total-dose-induced charge trapping in the buried oxide (Fig. 10).[58]

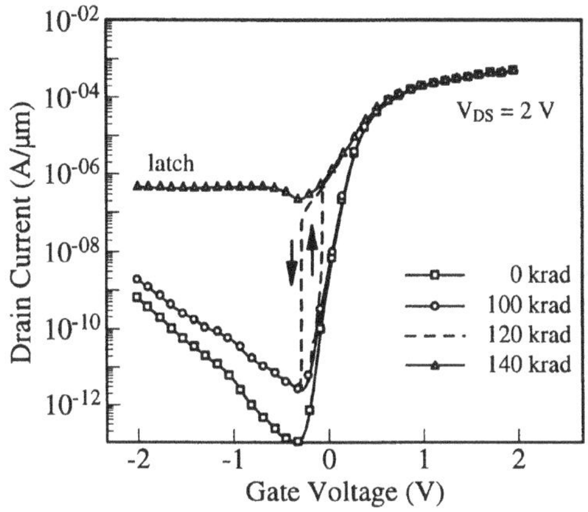

Fig. 9. Onset of a hysteresis on the current-voltage characteristics of a partially depleted SOI MOSFET, during exposure to radiation.[56]

For floating body transistors, the worst case corresponds to the off-state (drain at the supply voltage, other terminals grounded). The distribution of trapped charges in the BOX or at the back interface becomes laterally inhomogeneous. More trapping occurs on the drain side, where the field is larger. This inhomogeneity can be revealed by performing tests in reverse-mode (by exchanging the roles of the source and drain).[59]

For transistors with body ties, the worst case rather corresponds to the "transmission-gate" (TG) state (encountered for example in SRAM cell access transistors), where both the source and drain are biased, while the other terminals are grounded. Additional experiments and simulations have shown the impact of the transistor geometry (gate length, buried oxide thickness) and doping profiles. In very short-channel devices, the fringing fields from drain and source to the body (insct of Fig. 11), modify the potential distribution in the BOX, even leading to a modification of the worst-case configuration.

The radiation source impacts on the total dose response of the SOI transistors. The two main irradiation sources actually used to study the total dose hardness of integrated technologies, Co-60 or 10 keV X-rays, have different signatures. Depending on the bias and the device architecture, a difference by a factor of two can be obtained on the back-gate threshold voltage shift between X-rays and Co-60. The main conclusion drawn by Schwank *et al*[60] is that laboratory X-rays sources more closely simulate a rich proton environment (like on low earth orbits), while Co-60 gamma irradiation provides a better modeling for rich electron environment (high orbits).

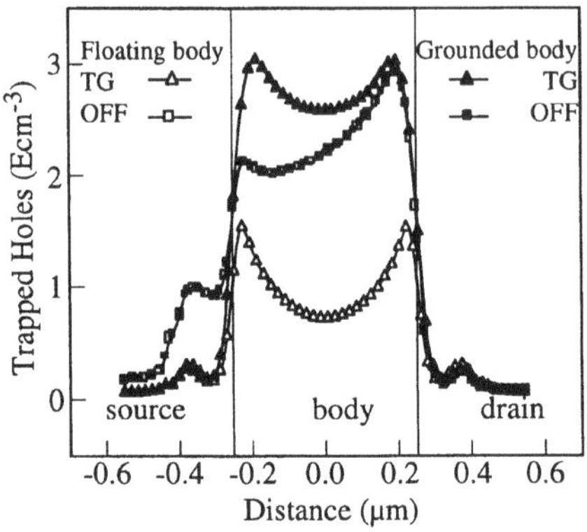

Fig. 10. Simulated distribution of trapped holes in the BOX, 6 nm under the interface. The n-channel MOSFETs are either floating or grounded body, with TG bias or no bias ($L = 0.5\,\mu m$, 1 Mrad (SiO_2) dose).[58]

5. Future Trends

Several factors impeded the earlier commercial development of SOI circuits: extra cost of SOI wafers, unavailability of dedicated libraries for design and, more importantly, the fabulous progress of bulk-Si technology. There was no real need for an alternative technology such as SOI, before now when the frontiers of Si become imminent. SOI offers two avenues to break the 10–nm gate-length barrier: more or less conventional scaling and new paradigm in device architecture.

5.1. *Scalability of SOI MOSFETs*

The scaling rules for bulk-Si MOSFETs require tremendous doping levels, which adversely affect the carrier mobility, junction capacitances, and circuit speed. Fortunately, the scaling rules and design windows are different for fully-depleted SOI transistors, where additional parameters (film and BOX thickness, substrate doping and biasing) are available for device optimization.

The minimum channel length of SOI MOSFETs, where the short-channel effects are maintained at a tolerable level, is $L \simeq 3\lambda$. The 'intrinsic' length λ, calculated by solving the Poisson equation, can be dramatically reduced by using ultra-thin Si films.[61] More importantly, below a critical film thickness (15 nm in Fig. 11), the doping becomes irrelevant and the film can be left undoped for the benefit of the carrier mobility.[62,63] However, for light doping, metal gates with selected work function, will be needed to adjust the threshold voltage.

Besides charge sharing and *drain-induced barrier lowering* (DIBL), a serious short-channel effect is the lateral penetration of the electric field into the BOX and underlying substrate (Fig. 11). This fringing field is responsible for an increase in the potential at the film–BOX interface, via the *drain-induced virtual substrate biasing* (DIVSB). Due to interface coupling, the front-channel threshold voltage is lowered and the subthreshold swing is degraded.

The fringing fields in the BOX can be remarkably attenuated by using thin oxides or low–K insulators. Counter effects of thin BOX are the increase in parasitic capacitances and interface coupling effects. A Ground–Plane (GP), located underneath a relatively thin (50 nm) BOX, has the merit to suppress the field penetration and the depletion region in the substrate.[63]

The ideal architecture of sub-50-nm-long FD MOSFETs will combine very thin films, low doping, moderately thin buried insulators, mid-gap gate, elevated source and drain terminals to reduce series resistances, and perhaps a ground-plane. We can expect that fully depleted, single-gate MOSFETs will reach the limit of 20-nm-long and 5-nm-thick bodies.[5,64]

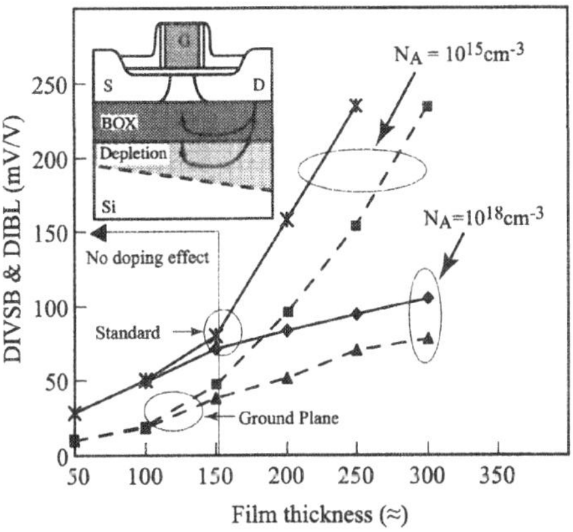

Fig. 11. Threshold voltage roll-off, including standard DIBL and fringing field (DIVSB) effects, versus film thickness for various doping levels and architectures (with and without a ground plane).

5.2. *Innovative transistors*

Double-Gate (DG) SOI MOSFETs are recognized as the final metamorphosis of the MOS transistor. The ideal DG architecture is symmetrical and the two gates are interconnected. A compact variant (GAA : Gate–All–Around) consists of a unique gate that surrounds a 'suspended' Si membrane (body of the transistor).[2] DG–MOSFETs were fabricated by Delta process, epitaxial lateral overgrowth, wafer bonding, FinFET process, tunnel epitaxy, SON, and other impressive technologies . . . still apart from the mainstream CMOS technology.[3,65,66,67]

The DG concept has initially been demonstrated on conventional fully-depleted SOI MOSFETs by simultaneously biasing the front and back gates with $V_{G_2} - V_{T_2} = (t_{ox2}/t_{ox1})(V_{G_1} - V_{T_1})$, such as to compensate for the asymmetry of the two oxides. The formation of front and back inversion channels causes, by continuity, the spreading of minority carriers in the volume of a thin SOI film.[68] This *volume inversion* results in enhanced drain current and transconductance. The total inversion charge in DG–mode is roughly twice the inversion charge in SG–mode and the subthreshold swing is ideal (60 mV/decade at room temperature). The important point is that the minority carriers flow in the middle of the film and experience less surface scattering, hence the mobility,[69,70] radiation hardness,[2,71] and $1/f$ noise are improved. Measurements in 3-nm-thick transistor reveal an outstanding increase, by more than 200%, in transconductance for DG-mode.[4]

The two gates collaborate to ensure an excellent control of the potential and inversion charge, so that short-channel effects (DIBL, DIVSB, punchthrough) are highly reduced. The channel length can be remarkably shorter in DG, as compared to single gate (SG) transistors: 15–30 nm long DG–MOSFETs have already been demonstrated.[67,72] Numerical simulations including quantum effects, band-to-band tunneling, and direct source-to-drain tunneling indicate excellent characteristics for 2–8 nm DG-MOSFETs.[6] The body thickness should be roughly 1/3 of its length.

A fascinating 4-gate accumulation-mode transistor, G^4–FET, has recently been demonstrated.[73] Apart from the standard front and back MOS gates, there are two lateral P–N junctions (side gates) which control the effective width of the body. The conductive path is therefore modulated by mixed MOS–JFET effects: from a tiny wire, surrounded by depletion regions, to a strongly accumulated body. Each gate has the capability of switching the transistor on and off which opens new perspectives for mixed-signal applications, quaternary logic schemes, and quantum wire physics. Preliminary measurements have shown that the G^4–FET can be rather immune to radiation effects.

6. Conclusions

SOI is no longer just a technology to alleviate radiation effects in defense or space circuits; SOI devices now operate in a non radiation-free world and they better survive. Knowledge and experience accumulated so far in radiation-hard devices is valuable, however this needs to be updated with fresh information collected in commercial (non rad-hard) devices fabricated with state of the art SOI technology.

Future work will gradually focus on fully depleted SOI MOSFETs—with tunneling gate oxide, ultra-thin Si film and rather thin BOX— which are more talented than partially depleted transistors in the CMOS scaling competition. Exciting opportunities for research are also offered by multiple-gate transistors with two, three or four channels. Finally, the conventional SOI structure is rapidly evolving by absorbing other technologies and materials (strained Si, SiGe or Ge films as well as various dielectrics) with still unknown radiation properties.

Acknowledgements. This work is a cooperation between CEA–DAM (Bruyères-le Châtel) and the Center for Projects in Advanced Microelectronics (CPMA), operated by CNRS, LETI and universities. Thanks are also due to our numerous colleagues.

References

1. S. Cristoloveanu and S.S. Li, *Electrical Characterization of SOI Materials and Devices*, Kluwer, Boston, 1995.
2. J–P. Colinge, *SOI Technology: Materials to VLSI* (2nd ed.), Kluwer, Boston, 1997.
3. H–S. Wong, D.J. Frank and P.M. Solomon, "Device design considerations for double-gate, ground-plane, and single-gated ultra-thin SOI MOSFETs at the 25 nm channel length generation", in *IEDM Techn. Dig.*, 1998, pp. 407–410.
4. S. Cristoloveanu, T. Ernst, D. Munteanu, and T. Ouisse, "Ultimate MOSFETs on SOI: ultra thin, single gate, double gate, or ground plane", *Int. J. High Speed Electronics and Systems* **10** (2000) 217–230.
5. D. Franck, S. Laux, and M. Fischetti, "Monte Carlo simulations of a 30 nm dual gate MOSFET: how short can Si go?", in *IEDM Techn. Dig.*, 1992, pp. 553–556.
6. Y. Naveh, V. Sverdlov, and K. Likharev, "Nanoscale MOSFETs scaling", in *Proc. Dev. Res. Conf. ISDRS'2001*, 2001, pp. 61–62.
7. J. Barth, "Modeling space radiation environments", Short Course, *IEEE Nuclear and Space Radiation Effects Conf.*, Snowmass, Colorado, July 1997.
8. J.F. Ziegler, "IBM experiments in soft fails in computer electronics", *IBM Journal Research and Development* **40(1)** (1996) p. 3.
9. F. Assaderaghi and G. Shahidi, "Advanced SOI circuits for microelectronics market", in *Silicon-On-Insulator Technology and Devices IX*, eds. P.L.F. Hemment, S. Cristoloveanu, T.W. Houston, K. Izumi, H. Hovel, Electrochemical Soc., Pennington, vol. 99–3, 1999, pp. 1–10.
10. F. Assaderaghi et al, "A 7.9/5.5 psec room/low temperature SOI CMOS", in *IEDM Techn. Dig.*, 1997, pp. 415–418.
11. M. Bruel, "Silicon on insulator material technology", *Electronics Letts.* **31** (1995) p. 1201.
12. H. Isaji, J. Nakayama, and T. Yonehara, "Volume production in ELTRAN SOI-epi wafer," in *Silicon-On-Insulator Technology and Devices X*, eds. S. Cristoloveanu, P.L.F. Hemment, K. Izumi, G.K. Celler, F. Assaderaghi, and Y–W. Kim, Electrochemical Soc., Pennington, vol. 2001–3, 2001, pp. 45–50.
13. D.M. Fleetwood, "Hydrogen-related reliability issues for advanced microelectronics," *Microelectron. Reliab.* **42** (2002) 1397–1403.
14. T. Ouisse, S. Cristoloveanu, and G. Borel, "Electron trapping in irradiated SIMOX buried oxides", *IEEE Electron Device Letts.*, **12** (1991) 312–314.
15. R.E. Stahlbush, G.J. Campisi, J.B. McKitterick, W.P. Maszara, P. Roitman, and G.A. Brown, "Electron and hole trapping in irradiated SIMOX, ZMR and BESOI buried oxides", *IEEE Trans. Nucl. Sci.* **39** (1992) 2086–2097.
16. P. Paillet, D. Hervé, J.L. Leray, and R.A.B. Devine, "Evidence of negative charge trapping in high temperature annealed thermal oxide", *IEEE Trans. Nucl. Sci.* **41** (1994) 473–478.
17. P. Paillet, J.L. Autran, J.L. Leray, B. Aspar, and A.J. Auberton-Hervé, "Trapping-detrapping properties of irradiated ultra-thin SIMOX buried oxides", *IEEE Trans. Nucl. Sci.* **42** (1995) 2108–2113.

18. T. Skotnicki, "Silicon On Nothing (SON) – fabrication, material and devices", in *Silicon-On-Insulator Technology and Devices X*, eds. S. Cristoloveanu, P.L.F. Hemment, K. Izumi, G.K. Celler, F. Assaderaghi, and Y–W. Kim, Electrochemical Soc., Pennington, vol. 2001–3, 2001, pp. 391–402.
19. S. Cristoloveanu and S. Williams, "Point-contact pseudo-MOSFET for in-situ characterization of as-grown silicon on insulator wafers", *IEEE Electron Device Letts.* **13** (1992) 102–104.
20. S. Cristoloveanu, D. Munteanu, and M. Liu, "A review of the pseudo-MOS transistor in SOI wafers: operation, parameter extraction, and applications", *IEEE Trans. Electron Devices* **47** (2000) 1018–1027.
21. H.K. Lim and J.G. Fossum, "Threshold voltage of thin-film silicon-on-insulator (SOI) MOSFETs", *IEEE Trans. Electron Devices* **30** (1983) 1244–1251.
22. B. Mazhari, S. Cristoloveanu, D.E. Ioannou, and A.L. Caviglia, "Properties of ultra-thin wafer-bonded silicon on insulator MOSFETs," *IEEE Trans. Electron Dev.* **38** (1991) 1289–1295.
23. T. Ouisse, S. Cristoloveanu, and G. Borel, "Influence of series resistances and interface coupling on the transconductance of fully depleted silicon-on-insulator MOSFETs", *Solid-State Electron.* **35** (1992) 141–149.
24. J. Jomaah, F. Balestra, and G. Ghibaudo, "Experimental investigation and numerical simulation of low frequency noise in thin film SOI MOSFETs", *Physica Status Solidi (a)* **142** (1994) 533.
25. J. Pretet, T. Matsumoto, T. Poiroux, S. Cristoloveanu, R. Gwoziecki, C. Raynaud, A. Roveda, and H. Brut, "New mechanism of body charging in partially depleted SOI–MOSFETs with ultra-thin gate oxide", in *Proc. ESSDERC 2002*, eds. G. Baccarani, E. Gnani, and M. Rudan, University of Bologna, 2002, pp. 515–518.
26. J.R. Schwank and P.E. Dodd, "Radiation effects in SOI microelectronics", Short course, *2001 IEEE International SOI Conf.*, Durango, Colorado, October 2001.
27. O. Musseau and V. Ferlet-Cavrois, "Silicon on insulator technologies, radiation effects", Short Course, *IEEE Nuclear and Space Radiation Effects Conf.*, Vancouver, July 2001.
28. C.B. Oh, J. Ahn, and Y.W. Kim, "The effect of body contact arrangement on thin SOI MOSFET characteristics", *Microelectronic Eng.* **28** (1995) 467-470.
29. M.L. Alles, "SPICE analysis of the SEU sensitivity of a fully depleted SOI CMOS SRAM cell", *IEEE Trans. Nucl. Sci.* **41** (1994) 2093-2097.
30. V. Ferlet-Cavrois, C. Marcandella, G. Giraud, G. Gasiot, T. Colladant, O. Musseau, C. Fenouillet, and J. du Port de Poncharra, "Characterization of the parasitic bipolar amplification in SOI technologies submitted to transient irradiation", *IEEE Trans. Nucl. Sci.* **49** (2002) 1456–1461.
31. H.F. Wei, J.E. Chung, N.M. Kalkhoran, F. Namavar, N.K. Annamalai, and W.M. Shedd, "Improvement of radiation hardness in fully-depleted SOI n-MOSFETs using Ge-implantation", *IEEE Trans. Nucl. Sci.* **41** (1994) 2291–2296.
32. Y. Hirano, T. Iwamatsu, K. Shiga, K. Nii, K. Sonoda, T. Matsumoto, S. Maeda, Y. Yamaguchi, T. Ipposhi, S. Maegawa, and Y. Inoue, "High soft-error tolerance body-tied SOI technology with partial trench isolation (PTI) for next generation devices", in *Proc. 2002 Symposium on VLSI Technology*, 2002.
33. J.R. Schwank, M.R. Shaneyfelt, B.L. Draper, and P.E. Dodd, "BUSFET – a radiation-hardened SOI transistor", *IEEE Trans. Nucl. Sci.* **46** (1999) 1809-1816.
34. V. Ferlet-Cavrois, O. Musseau, J.L. Leray, J.L. Pelloie, and C. Raynaud, "Total dose effects on a fully-depleted SOI NMOSFET and its lateral parasitic transistor", *IEEE Trans. Electron Dev.* **44** (1997) 965–971.

35. C. Brisset, V. Ferlet-Cavrois, O. Flament, O. Musseau, J.L. Leray, J.L. Pelloie, R. Escoffier, A. Michez, C. Cirba, and G. Bordure, "Two-dimensional simulation of total dose effect on NMOSFET with lateral parasitic transistor", *IEEE Trans. Nucl. Sci.* **43** (1996) 2651–2658.
36. V. Ferlet-Cavrois, O. Musseau, J.L. Leray, Y.M. Coïc, G. Lecarval, and E. Guichard, "Comparison of the sensitivity to heavy ions of SRAMs in different SIMOX technologies", *IEEE Electron Dev. Lett.* **15** (1994) 82–84.
37. O. Musseau, J.L. Leray, V. Ferlet-Cavrois, Y.M. Coïc, and B. Giffard, "SEU in SOI SRAMs – a static model", *IEEE Trans. Nucl. Sci.* **41** (1994) 607–612.
38. V. Ferlet-Cavrois, G. Gasiot, C. Marcandella, C. D'hose, O. Flament, O. Faynot, J. de Pontcharra, and C. Raynaud, "Insights on the transient response of full and partially depleted SOI technologies under heavy ion and dose rate irradiations", *IEEE Trans. Nucl. Sci.* **49** (2002) 2948–2956.
39. K. Hirose, H. Saito, Y. Kuroda, S. Ishii, Y. Fukuoka, and D. Takahashi, "SEU resistance in advanced SOI-SRAM fabricated by commercial technology using a rad-hard circuit design", *IEEE Trans. Nucl. Sci.* **49** (2002) 2965–2968.
40. P.E. Dodd, M.R. Shaneyfelt, J.R. Schwank, G.L. Hash, B.L. Draper, and P.S. Winokur, "Single-event upset and snapback in silicon-on-insulator devices", *IEEE Trans. Nucl. Sci.* **47** (2000) 2165–2174.
41. G. Gasiot, V. Ferlet-Cavrois, P. Roche, P. Flatresse, C. D'Hose, O. Musseau, and J. du Port de Poncharra, "Comparison of the sensitivity to heavy ions of 0.25 μm bulk and SOI technologies, *IEEE Trans. Nucl. Sci.* **49** (2002) 1450–1455.
42. C. Brothers, R. Pugh, P. Duggan, J. Chavez, D. Schepis, D. Yee, and S. Wu, "Total-dose and SEU characterization of 0.25 micron CMOS/SOI integrated circuit memory technologies", *IEEE Trans. Nucl. Sci.* **44** (1997) 2134–2139.
43. G.E. Davis, L.R. Hite, T.G.W. Blake, C.E. Chen, H.W. Lam, and R. DeMoyer Jr., "Transient radiation effects in SOI memories", *IEEE Trans. Nucl. Sci.* **32** (1985) 4432–4437.
44. L.Z. Scheick, G.M. Swift, and S.M. Guertin, "SEU evaluation of SRAM memories for space applications", in *Proc. IEEE NSREC Data Workshop*, July 2000, pp. 61–63.
45. J.R. Coss, T.F. Miyahira, L.E. Selva, and G.M. Swift, "Device SEE susceptibility update : 1996–1998", *Proc. IEEE NSREC Data Workshop*, July 1999, pp. 60–81.
46. R. Ecoffet, M. Labrunee, S. Duzellier, and D. Falguero, "Heavy ion test results on memories", *Proc. IEEE NSREC Data Workshop*, July 1992, pp. 27–33.
47. D.K. Nichols, L.S. Smith, G.A. Soli, R. Koga, and W.A. Kolasinski, "Latest Trends in parts SEP susceptibility from heavy ions", *IEEE Trans. Nucl. Sci.* **36** (1989) 2388–2397.
48. R. Koga and W.A. Kolasinski, "Heavy ion-induced single event upsets of microcircuits; a summary of the aerospace corporation test data", *IEEE Trans. Nucl. Sci.* **31** (1984) 1190–1195.
49. P.E. Dodd, M.R. Shaneyfelt, K.M. Horn, D.S. Walsh, G.L. Hash, T.A. Hill, B.L. Draper, J.R. Schwank, F.W. Sexton, and P.S. Winokur, "SEU-sensitive volumes in bulk and SOI SRAMs from first-principles calculations and experiments", *IEEE Trans. Nucl. Sci.* **48** (2001) 1893–1903.
50. D. Hervé, P. Paillet, and J.L. Leray, "Space charge effects in SIMOX buried oxides", *IEEE Trans. Nucl. Sci.* **41** (1994) 466–472.
51. W.L. Warren, M.R. Shaneyfelt, D.M. Fleetwood, J.R. Schwank, P.S. Winokur, and R.A.B. Devine, "Microscopic nature of border traps in MOS devices", *IEEE Trans. Nucl. Sci.* **41** (1994) 1817–1827.

52. F.T. Brady, H.L. Hughes, P.J. McMarr, and B. Mrstik, "Total dose hardening of SIMOX buried oxides for fully-depleted devices in rad-tolerant applications", *IEEE Trans. Nucl. Sci.* **43** (1996) 2646–2650.
53. W.C. Jenkins and S.T. Liu, "Radiation response of fully-depleted MOS transistors fabricated in SIMOX", *IEEE Trans. Nucl. Sci.* **41** (1994) 2317–2321.
54. H. Hughes and P. McMarr, "Radiation-hardening of SOI by ion implantation into the buried oxide", *U.S. Patent*, No. 5,795,813.
55. R.E. Stahlbush, H.L. Hughes, and W.A. Krull, "Reduction of charge trapping and electron tunneling in SIMOX by supplemental implantation of oxygen", *IEEE Trans. Nucl. Sci.* **40** (1993) 1740–1747.
56. V. Ferlet-Cavrois, S. Quizola, O. Musseau, O. Flament, J.L. Leray, J.L. Pelloie, C. Raynaud, and O. Faynot, "Total dose induced latch in short channel NMOS/SOI transistors", *IEEE Trans. Nucl. Sci.* **45** (1998) 2458–2466.
57. S.T. Liu, S. Balster, S. Sinha, and W.C. Jenkins, "Worst case total dose radiation response of 0.35 μm SOI CMOS FETs, *IEEE Trans. Nucl. Sci.* **46** (1999) 1817–1823.
58. V. Ferlet-Cavrois, T. Colladant, P. Paillet, J.L. Leray, O. Musseau, J.R. Schwank, M.R. Shaneyfelt, J.L. Pelloie, and J. du Port de Poncharra, "Worst-case bias during total dose irradiation of SOI transistors", *IEEE Trans. Nucl. Sci.* **47** (2000) 2183–2188.
59. S. Cristoloveanu, D.E. Ioannou, R.K. Lawrence, G.J. Campisi, and H.L. Hughes, "Asymmetrical irradiation effects in SIMOX MOSFETs", in *Proc. RADECS'93*, IEEE, 1994, p. 173.
60. J.R. Schwank, M.R. Shaneyfelt, P. Paillet, D.E. Beutler, V. Ferlet-Cavrois, B.L. Draper, R.A. Loemaker, P.E. Dodd, and F.W. Sexton, "Optimum laboratory radiation source for hardness assurance testing", *IEEE Trans. Nucl. Sci.* **48** (2001) 2152–2157.
61. S–H. Oh, D. Monroe, and J.M. Hergenrother, "Analytic description of short-channel effects in fully-depleted double-gate and cylindrical, surrounding-gate MOSFETs", *IEEE Electron Device Letts.* **21** (2000) 445–185.
62. L.T. Su, J.B. Jacobs, J.E. Chung, and D.A. Antoniadis, "Deep-submicrometer channel design in silicon-on-insulator (SOI) MOSFETs", *IEEE Electron Device Letts.*, **15** (1994) 183–185.
63. T. Ernst and S. Cristoloveanu, "The ground-plane concept for the reduction of short-channel effects in fully-depleted SOI devices", in *Silicon-On-Insulator Technology and Devices IX*, eds. P.L. Hemment, S. Cristoloveanu, T.W. Houston, K. Izumi, H. Hovel, Electrochemical Soc., Pennington, vol. 99–3, 1999, pp. 329–334.
64. S. Cristoloveanu, "Far-future trends in SOI technology: a guess", *Int. J. High Speed Electronics and Systems* **12** (2002) 343–351.
65. D. Hisamoto, T. Kaga, and E. Takeda, "Impact of the vertical SOI 'DELTA' structure on planar device technology", *IEEE Trans. Electron Dev.* **38** (1991) 1419–24.
66. H-S. Wong, K.K. Chan, and Y. Taur, "Self-aligned (top and bottom) double-gate MOSFET with a 25 nm thick silicon channel", in *IEDM Techn. Dig.*, 1997, pp. 427–430.
67. X. Huang *et al*, "Sub 50–nm FinFET: PMOS", in *IEDM Techn. Dig.*, 1999, pp: annex.
68. F. Balestra, S. Cristoloveanu, M. Bénachir, J. Brini, and T. Elewa, "Double-gate silicon on insulator transistor with volume inversion: a new device with greatly enhanced performance", *IEEE Electron Device Lett.* **8** (1987) 410–412.
69. F. Gámiz, J.B. Roldán, J.A. López–Villanueva, P. Cartujo–Cassinello, J.E. Carcellar, and P. Cartujo, "Monte Carlo simulation of electron transport in silicon-on-insulator devices", in *Silicon-On-Insulator Technology and Devices X*, eds. S. Cristoloveanu,

P.L.F. Hemment, K. Izumi, G.K. Celler, F. Assaderaghi, and Y–W. Kim, Electrochemical Soc., Pennington, vol. 2001–3, 2001, pp. 157–168.
70. D. Esseni, M. Mastrapasqua, C. Fiegna, G.K. Celler, L. Selmi, and E. Sangiorgi, "An experimental study of low field electron mobilityin in double-gate, ultra-thin SOI MOSFETs", in *IEDM Techn. Dig.*, 2001, pp. 19.7.1–19.7.4.
71. C.R. Cirba, S. Cristoloveanu, R.D. Schrimpf, L.C. Feldman, D.M. Fleetwood, and K.F. Galloway, "Total-dose radiation hardness of double-gate ultra-thin SOI MOSFETs", in *Silicon-On-Insulator Technology and Devices X*, eds. S. Cristoloveanu, G.K. Celler, J.G. Fossum, F. Gamiz, Y.W. Kim, Electrochemical Soc., Pennington, vol. 2003–05, pp. 493–498.
72. B. Yu, L. Chang, S. Ahmed *et al*, "FinFET scaling to 10nm gate length", in *IEDM Techn. Dig.*, 2002, pp. 251–254.
73. B.J. Blalock, S. Cristoloveanu, B. Dufrene, F. Allibert, and M.M. Mojarradi, "The multiple-gate MOS–JFET", *Int. J. High Speed Electronics and Systems* **12** (2002) 511–520.

International Journal of High Speed Electronics and Systems
Vol. 14, No. 2 (2004) 489–501

World Scientific
www.worldscientific.com

TOTAL-DOSE AND SINGLE-EVENT EFFECTS IN SILICON-GERMANIUM HETEROJUNCTION BIPOLAR TRANSISTORS

JOHN D. CRESSLER
School of Electrical and Computer Engineering
777 Atlantic Drive, N.W.
Georgia Institute of Technology
Atlanta, GA 30332-0250, USA
cressler@ece.gatech.edu

We present an overview of radiation effects in silicon-germanium heterojunction bipolar transistors (SiGe HBT). We begin by reviewing SiGe HBTs, and then examine the impact of ionizing radiation on both the *dc* and *ac* performance of SiGe HBTs, the circuit-level impact of radiation-induced changes in the transistors, followed by single-event phenomena in SiGe HBT circuits. While ionizing radiation degrades both the *dc* and *ac* properties of SiGe HBTs, this degradation is remarkably minor, and is far better than that observed in even radiation-hardened conventional Si BJT technologies. This fact is particularly significant given that no intentional radiation hardening is needed to ensure this level of both device-level and circuit-level tolerance (typically multi-Mrad TID). SEU effects are pronounced in SiGe HBT circuits, as expected, but circuit-level mitigation schemes will likely be suitable to ensure adequate tolerance for many orbital missions. SiGe HBT technology thus offers many interesting possibilities for space-borne electronic systems.

Keywords: SiGe; silicon-germanium; HBT; heterojunction bipolar transistor; radiation effects; total-dose; single event effects; SEU.

1. Introduction

At present, there are two recent but rapidly growing thrusts within the space community: 1) the use of commercial-off-the-shelf (COTS) parts whenever possible for space-borne systems as a cost-saving measure; and 2) the use of system-on-a-chip integration to lower chip counts and system costs, as well as simplify packaging and lower total system launch weight. The "holy-grail" in the realm of space electronics can thus be viewed as a conventional terrestrial IC technology with a system-on-a-chip capability, which is also radiation-hard as fabricated, without requiring any additional process modifications or layout changes. It is within this context that we discuss SiGe HBT technology as potentially such a "radiation-hard-as-fabricated" IC technology with possibly far-ranging implications for the space community.

2. The SiGe HBT

While the basic idea of using SiGe alloys to bandgap-engineer Si devices dates to the 1950s, the synthesis of defect-free SiGe films proved surprisingly difficult, and device-quality SiGe films were not successfully produced until the mid-1980s. Because of the difference in lattice constants between Si and Ge, SiGe strained layers are subject to a fundamental stability criterion limiting their thickness for a given Ge concentration. Introducing Ge into Si has a number of consequences for devices. First and most important, because Ge has a larger lattice constant than Si, the energy bandgap of Ge is smaller than that of Si (0.66 eV vs 1.12 eV), and thus SiGe will have a bandgap smaller than that of Si, making it a suitable candidate for bandgap engineering in Si. The compressive strain associated with SiGe alloys produces an additional bandgap shrinkage, and the net result is a bandgap reduction of approximately 75 meV for each 10% of Ge introduced. This Ge-induced "band offset" occurs predominantly in the valence band, making it conducive for use in tailoring *npn* bipolar transistors. Because a practical SiGe film must be very thin if it is to remain stable and hence defect free, it is a natural candidate for use in the base region of a bipolar transistor (which by definition must be thin to achieve high-frequency operation). The SiGe HBT represents the first practical bandgap-engineered transistor in the Si material system. [1]

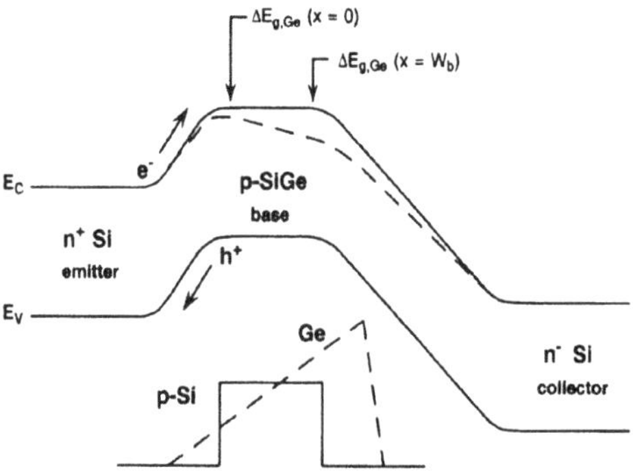

Fig. 1. Energy band diagram for a Si BJT and graded-base SiGe HBT, biased in forward active mode.

To intuitively understand how these band edge changes affect the operation of the SiGe HBT, first consider the band diagram of the device (Fig 1.). The introduction of Ge into the base region has two tangible *dc* consequences: 1) the potential barrier to injection of electrons from emitter into the base is decreased. Intuitively, this will yield exponentially more electron injection for the same applied V_{BE}, translating into higher collector current and hence higher current gain, provided the base current remains unchanged. Given that band edge effects generally couple strongly to transistor properties, we naively expect a strong dependence of J_C on Ge content. Of practical consequence, the introduction of Ge effectively decouples the base doping from the current gain, thereby providing device designers with much greater flexibility than in Si BJT design. If, for instance, the intended circuit application does not require high current gain, we can effectively trade the higher gain induced by the Ge band offset for a higher base doping level, leading to lower net base resistance, and hence better dynamic switching and noise characteristics. 2) The presence of a finite Ge content in the CB junction will positively influence the output conductance of the transistor, yielding higher Early voltage, even at constant base doping.

The introduction of Ge into the base region also has important consequences for *ac* device operation, since the Ge-gradient-induced drift field across the neutral base is aligned in a direction (from collector to emitter) such that it will accelerate the injected minority electrons across the base. We are thus able to add a large drift field component to the electron transport, effectively speeding up the diffusive transport of the minority carriers and thereby decreasing the base transit time. Even though the band offsets in SiGe HBTs are typically small by III-V technology standards, the Ge grading over the short distance of the neutral base can translate into large electric fields. For instance, a linearly graded Ge profile with a modest peak Ge content of 10%, graded over a 50-nm neutral base width, yields 75 mV / 50 nm = 15 kV/cm electric field, sufficient to accelerate the electrons to near saturation velocity. Because the base transit time typically limits the frequency response of a Si BJT, we would expect that the frequency response should be significantly improved by introducing this Ge-induced drift field.

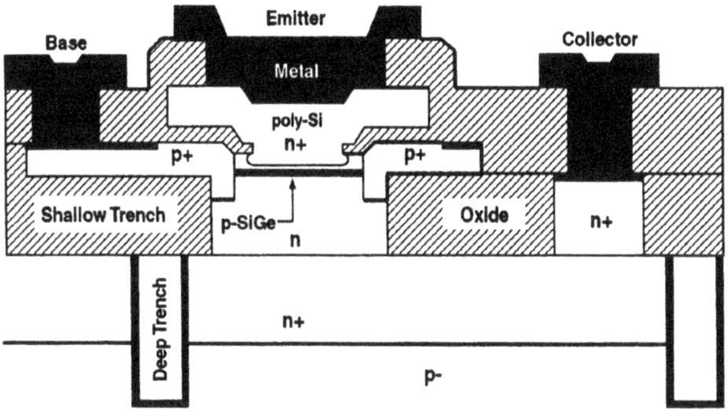

Fig. 2. Schematic cross-section of a representative first generation SiGe HBT, drawn through first metal.

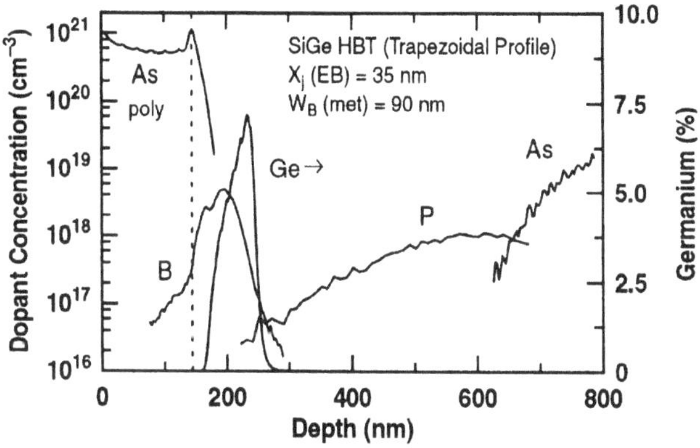

Fig. 3. Measured SIMS doping profile of a first generation SiGe HBT.

Perhaps most importantly, SiGe HBTs can be quite easily teamed with best-of-breed Si CMOS to form a monolithic SiGe HBT BiCMOS technology. While this might seem at first glance to be a mundane advantage, it is in fact a fundamental enabler for SiGe's long-term success, provided SiGe HBTs can be realized without an excessive cost penalty compared to standard Si ICs. The integration of SiGe HBTs with Si CMOS is also the fundamental departure point between SiGe technology and III-V technologies. If SiGe technology is to be successful in the long haul, it must bring to the table the RF and analog performance advantages of the SiGe HBT, and the low-power logic, integration level, and memory density of Si CMOS, into a single cost-effective IC that enables SoC integration (i.e., SiGe HBT BiCMOS). Typically, SiGe HBTs (often with multiple breakdown voltages) exist as an "adder" to a basic CMOS IC building-block core, to be swapped in or out as the application demands, without excessive cost burden. Typical state-of-the-art SiGe HBT BiCMOS technologies generally have a roughly 20% adder in mask count compared to generic digital CMOS, and are viewed by many as an acceptable compromise between performance benefit and cost, depending on the application. Fig. 2 and Fig. 3 show a schematic cross-section and doping profile of a first generation SiGe HBT.

3. Radiation Response of SiGe HBTs

The typical response of a SiGe HBT to irradiation can be seen in Fig. 4, which shows typical measured Gummel characteristics of a SiGe HBT, both before and after exposure to protons. [2-6] As expected, the base current increases after a sufficiently high proton fluence due to the production of G/R trapping centers, and hence the current gain of the device degrades. There are two main physical origins of this degradation. The base current density is inversely proportional to the minority carrier lifetime in the emitter, so that a degradation of the hole lifetime will induce an increase in the base current. In addition, ionization damage due to the charged nature of the proton fluence produces interface states and oxide trapped charges in the spacer layer at the emitter-base junction. These G/R centers also degrade I_B, particularly if they are placed inside the EB space charge region, where they will yield an additional non-ideal base current component (non-kT/q exponential voltage dependence). By analyzing a variety of device geometries, it can be shown that the radiation-induced excess base current is primarily associated with the EB spacer oxide at the periphery of the transistor, as naively expected, and is hence the radiation response is dominated by ionization damage rather than displacement damage. [1] The degradation of the current gain as a function of collector current for the pre-irradiated sample, and after exposure to three proton fluences, is shown in Fig. 5. For fluences up to 1×10^{13} p/cm^2 the peak current gain at 10 μA does not show a visible degradation, and at 1×10^{14} p/cm^2, a degradation of only about 8% compared to the pre-irradiated device is observed. This suggests that these SiGe HBTs are robust to TID for typical orbital proton fluences for realistic circuit operating currents above roughly 100 μA without any additional radiation hardening. These results are significantly better than for conventional diffused or even ion-implanted Si BJT technologies (even radiation-hardened ones).

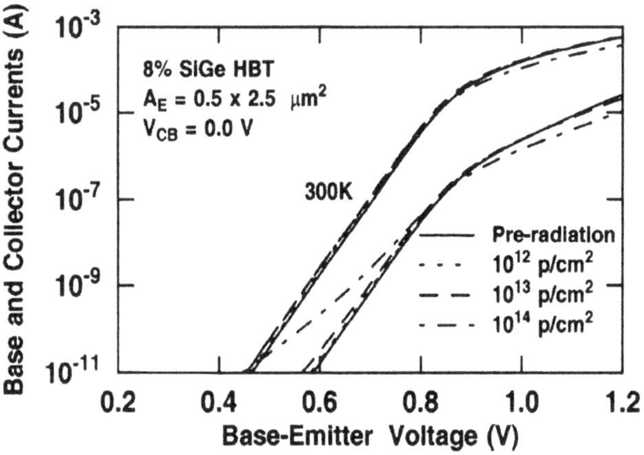

Fig. 4. Measured current-voltage characteristics of a SiGe HBT before and after proton exposure.

Of particular interest is the inference of the spatial location of the proton-induced traps in these devices. [4] The existence of proton-induced traps in the EB space charge region is clearly demonstrated by the G/R-induced increase in the non-ideal base current component shown in the Gummel characteristics. The existence of radiation-induced traps in the collector-base space charge region was verified by measuring the inverse mode Gummel characteristics of the device (emitter and collector leads swapped). In this case the radiation-induced traps in the CB junction now act as G/R centers in the inverse EB junction, with a signature non-kT/q exponential slope. 2-D simulations were calibrated to both measured data for the pre- and post-irradiated devices at a collector-base voltage of 0.0 V. In order to obtain quantitative agreement between the simulated and measured irradiated results, traps must be located uniformly throughout the device, and additional interface traps must be located around the emitter-base spacer oxide edge. Most of the radiation-induced recombination occurs inside the EB space charge region, leading to a non-ideal base current, as expected.

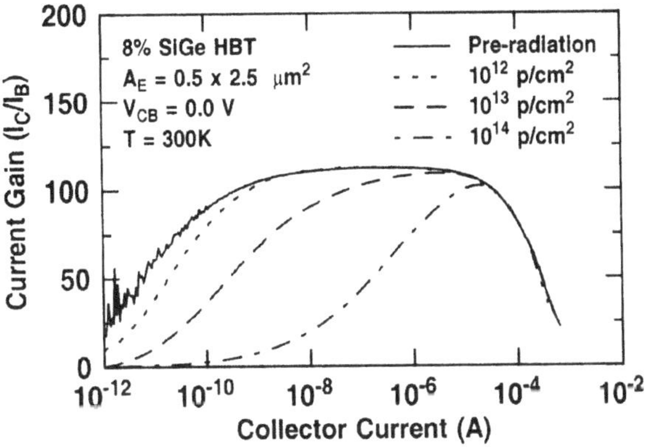

Fig. 5. Measured current gain versus bias current of a SiGe HBT as a function of proton fluence.

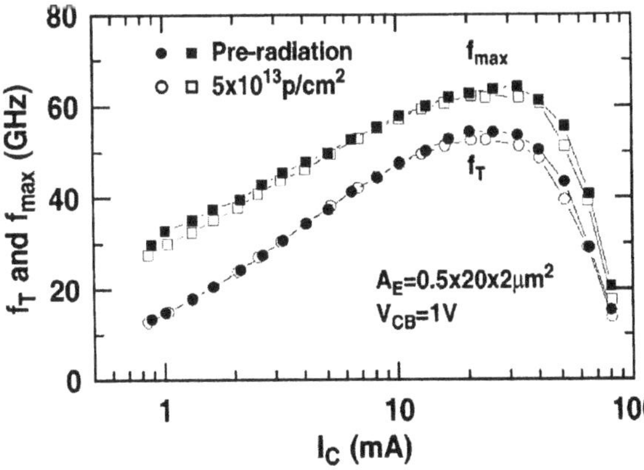

Fig. 6. Measured cutoff frequency versus bias current of a SiGe HBT after proton irradiation.

To assess the impact of radiation on the *ac* performance of the transistors, the S-parameters were measured to 40 GHz both before and after proton exposure. [7] From the measured S-parameters, the transistor cutoff frequency and maximum oscillation frequency were extracted (Fig. 6). Only a slight degradation in f_T and f_{max} is observed, the latter being expected from the minor increase of the base resistance with irradiation, due to either carrier removal, mobility / lifetime changes, or both. The broadband noise performance of SiGe HBTs is critical for space-borne transceivers and communications platforms. As shown in Fig. 7 the minimum noise figure (NF_{min}) degrades only slightly at 2.0 GHz after an extreme proton fluence of $5x10^{13}$ p/cm^2 (from 0.95 dB to a still-excellent value of 1.07 dB, a 12.6% degradation).

We note that careful comparisons between identically fabricated SiGe HBTs and Si BJTs (same device geometry and wafer lot, but without Ge in the base for the epitaxial-base Si BJT), show that the extreme level of total dose tolerance of SiGe HBTs is not *per se* due to the presence of Ge. [4] That is, the proton response of both the epitaxial base SiGe HBT and Si BJT are nearly identical. We thus attribute the observed radiation hardness to the unique and inherent structural features of the device itself, which from a radiation standpoint can be divided into three major aspects: 1) in these epitaxial base structures, the extrinsic base region is very heavily doped (> $5x10^{19}$ cm^{-3}) and located immediately below the EB spacer oxide region, effectively confining any radiation-induced damage, and its effects on the EB junction; 2) the EB spacer, known to be the most vulnerable damage point in conventional BJT technologies, is thin (< 0.20 μm wide) and composed of an oxide/nitride composite, the latter of which is known to produce an increased level of radiation immunity; 3) the active volume of these transistors is very small (W_E = 0.5 μm, and W_b < 150 nm), and the emitter, base, and collector doping profiles are quite heavily doped, effectively lessening the impact of displacement damage. We also note that these SiGe HBTs compare very favorably in both performance and radiation hardness with (more expensive) GaAs HBT technologies that are often employed in space applications requiring both very high speed and an extreme level of radiation immunity. [8]

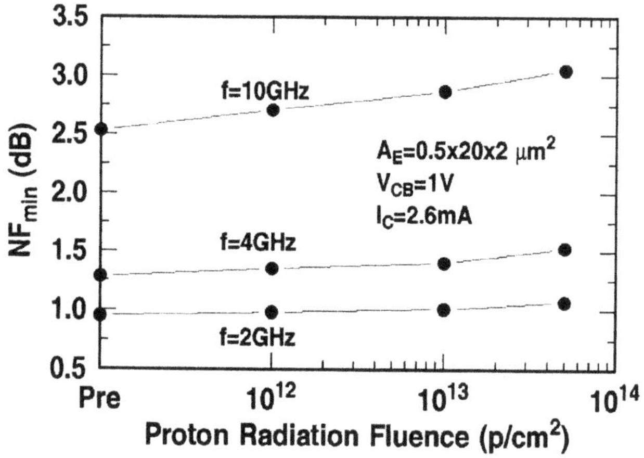

Fig. 7. Measured broadband noise performance of a SiGe HBT as a function of proton fluence.

Within the past few years, a pronounced low-dose-rate sensitivity to gamma irradiation that is not screened by the current test methods for ionizing radiation has been observed in Si bipolar technologies. The enhancement in device and circuit degradation at low gamma dose rates has come to be known as "Enhanced Low Dose Rate Sensitivity" (ELDRS). [9-10] The ELDRS effect was first reported in 1991, which demonstrated that existing radiation hardness test assurance methodologies were not appropriately considering worst case conditions. The physical origins underlying ELDRS have been hotly debated for years, and numerous mechanisms proposed. Recent attempts to understand ELDRS include a model suggesting that the lower net radiation induced trapped charge density at high-dose-rates is a result of a space charge phenomenon, caused by delocalized hole traps which occur in heavily damaged oxides such as bipolar base oxides. These traps can retain holes on a timescale of seconds to minutes, causing a buildup of positive charge in the oxide bulk during high-dose-rate irradiation. This is in contrast to low-dose-rate irradiation, where the irradiation time is much longer, effectively allowing the holes in the trap centers to be detrapped. Thus, in the high-dose-rate case, the larger total trapped hole density forces holes near the interface to be trapped closer to the interface, where they can be compensated by electrons from the silicon. This lowers the resultant net trapped charge density. [11]

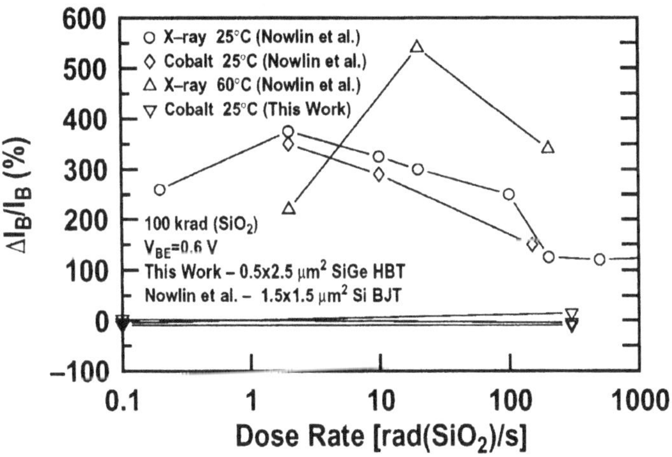

Fig. 8. Low dose gamma data comparing SiGe HBT ELDRS response with other Si BJT technologies.

To assess ELDRS in SiGe technology, low-dose-rate (0.1 rad(Si)/sec) and high-dose-rate (300 rad(Si)/sec) experiments were conducted using Cobalt-60. [12] As can be seen in Fig. 8, low-dose-rate effects in these SiGe HBTs were found to be nearly nonexistent, in striking contrast to reports of strong ELDRS in conventional Si bipolar technologies. We attribute this observed hardness to ELDRS to the same mechanisms responsible for the overall radiation hardness of the technology, and is likely more structural in nature than due to any unique advantage afforded by the SiGe base. Interestingly, an anomalous decrease in base current was also found in these devices at low-dose-rates, suggesting that a new physical phenomenon is present at low-dose-rates in these devices.

4. SiGe HBT Circuit Tolerance

For the successful deployment of SiGe technology into space-based systems, circuit-level radiation hardness is clearly more important than device-level hardness. As presented above, the TID device degradation is minor in the bias range of interest to most actual circuits (typically $I_C > 100\ \mu A$). In order to assess the impact of radiation exposure on actual SiGe HBT circuits, we have compared two very important, yet very different circuit types, one heavily used in analog ICs (the bandgap reference circuit), and one heavily used in RFICs (the voltage controlled oscillator).[13-14] Each circuit represents a key building block for realistic SiGe ICs that might be flown in space. Each of these SiGe HBT circuits was designed using fully calibrated SPICE models, layed-out, and then fabricated on the same wafer to facilitate unambiguous comparisons. In addition, because any realistic RF IC must also necessarily include passive elements such as monolithic inductors and capacitors, we have also investigated the effects of proton exposure on an RF LC bandpass filter. As can be seen from the data (Table 1), the impact of even extreme proton fluences has minimal effect on either the output voltage or temperature sensitivity of BGRs, the phase noise or tuning range of VCO's or passive elements, and is indicative of the overall robustness of this SiGe technology for analog and RF circuit applications.

Table 1. Comparison of SiGe HBT Circuit Performance Before and After Proton Exposure.

SiGe HBT Circuit	Parameters	Pre-radiation	After 5×10^{13} p/cm^2	Units
Bandgap Reference	V_{cc}	3.0	3.0	V
	I_{cc}	0.773	0.767	mA
	V_{out} at 300K	1.37416	1.372096	V
	Stability (-55 to 85°C)	81.2	81.7	ppm/°C
Voltage Controlled Oscillator	Frequency	5.0	5.0	GHz
	V_{cc}	3.3	3.3	V
	I_{cc}	22.5	22.5	mA
	Output Power	-5.0	-5.5	dBm
	Phase Noise	-112.5	-111.8	dBc/Hz
	Tuning Range	4,595-5,452	4,623-5,470	MHz
LC Bandpass Filter	Frequency	1.9	1.9	GHz
	Filter Q (@ 3dB BW)	7.6	7.6	-
	Insertion Loss	16.8	16.8	dB
	L	2.5	2.5	nH
	Inductor Q	7.4	7.4	-
	C	6.0	6.0	pF
	Capacitor Q	58	58	-

5. Single Event Effects in SiGe HBTs

Clearly, a space-qualified IC technology must demonstrate sufficient SEU immunity to support high-speed circuit applications as well as possess TID tolerance. It is well known that even III-V technologies that have significant TID tolerance often suffer from poor SEU immunity, particularly at high data rates. Recently, high-speed SiGe HBT digital logic circuits were found to be vulnerable to SEU at even low LET values.[15-16] In addition, successfully employed III-V HBT circuit-level hardening schemes using the current-sharing hardening (CSH) technique were found to be ineffective for these SiGe HBT logic circuits (Fig. 9). To understand single event effects in SiGe HBTs, one must use calibrated 2-D/3-D device simulation to assess the charge collection characteristics of SiGe HBTs. These device-level simulation results can then be coupled to circuit-level modeling to better understand circuit-level mitigation approaches.

From a device perspective, it is important to first assess the transistor charge collection characteristics as a function of terminal bias, load condition, substrate doping, and ion strike depth. [17-18] Bias and loading conditions were chosen to mimic representative circuit conditions within an actual ECL/CML digital circuit. Fig. 10 shows the charge collected by the collector versus time for different RC loads. The base and emitter terminals were grounded, the substrate bias was -5.2 V, the collector was connected to ground through an RC load, and the substrate doping was $5x10^{15}$ cm^{-3}. A uniform LET of 0.1 pC/μm (equivalent to 10 MeV-cm^2/mg) over 10 μm depth was used, which generates a total charge of 1.0 pC. The results clearly show that charge collection is highly dependent on the transistor load condition (i.e., circuit topology). As the load resistance increases, the collector-collected charge decreases. Note, however, that the emitter-collected charge increases correspondingly. The underlying physics is that more electrons exit through the emitter, instead of the collector. A larger load resistance presents a higher impedance to the electrons at the collector, and thus more electrons exit through the emitter. The collector of the adjacent device only collects a negligible amount of charge, despite the transient current spikes of the strike. Nearly all of the electrons deposited are collected by the collector and the emitter, although the partition between emitter and collector collection varies with the load condition. The impact on the SiGe base layer on the charge collection properties is a secondary effect.

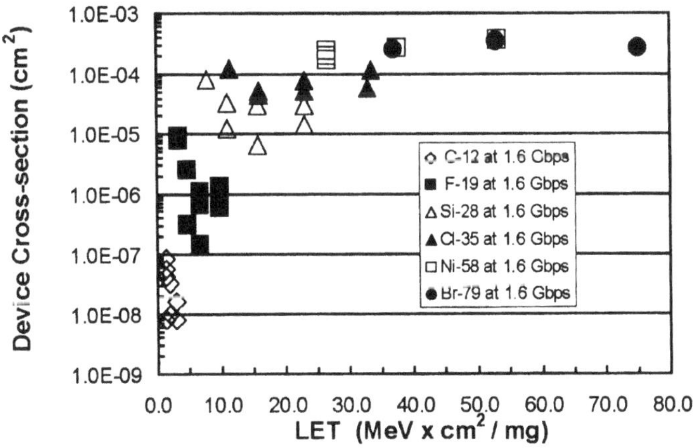

Fig. 9. Measured single event upset tolerance for first generation SiGe HBT digital circuits.

To better understand circuit-level SEU response, we combined these simulated charge-time profiles with circuit-level modeling in three different SiGe circuit architectures. [19] Circuit A is a straightforward ECL implementation of the standard rising edge-triggered flip-flop logic. Circuit B is the unhardened version of the D flip-flop used in the shift register results shown in Fig. 10. Circuit B uses fewer transistors and thus less power than circuit A, and is also faster than circuit A, allowing operation at higher clock rates. Because of these advantages, circuit B is very popular in high-speed bipolar digital circuit design. The circuit consists of a master stage and a slave stage. The master stage consists of a pass cell, a storage cell, a clocking stage, and a biasing control. The slave stage has a similar circuit configuration. Circuit C is the current-sharing hardened version of circuit B. (Refer to reference [1] for circuit level schematics for A, B, and C.) The circuit was used as a basic building block of the 32-stage shift-register data shown in Fig. 9. In this case, the current source transistor is divided into 5 paths, and these paths are maintained separately through the clocking stage and through the pass and storage cells. In essence, the input and output nodes of five copies of the switching circuits, including the controlling switch, clock, master and storage cells, are connected in parallel. The load resistance is shared by all the current paths.

The quasi-3D simulated SEU-induced transient currents were activated on one of the sensitive transistors in the respective circuits. The SEU currents were activated at 5.46 nsec (within the circuit hold time), immediately after the clock goes from low to high, a sensitive time instant for SEU-induced transient currents to produce an upset at the output. The input data is an alternating "0" and "1" series with a data rate of 2 Gbit/sec. Under these conditions, circuit A shows no upset at all, while circuits B and C show 5 and 3 continuous bits of data upset, respectively (Fig. 11). These results suggest that circuit A has the best SEU tolerance, while circuit C, the CSH hardened version, has better SEU tolerance than its unhardened companion version, circuit B. Circuit A, which shows no data upset at a switching current of 1.5 mA, does in fact show an upset when the switching current is lowered to 0.6 mA. This is consistent with our earlier observation that increasing switching current is effective in improving SEU performance for circuit C.

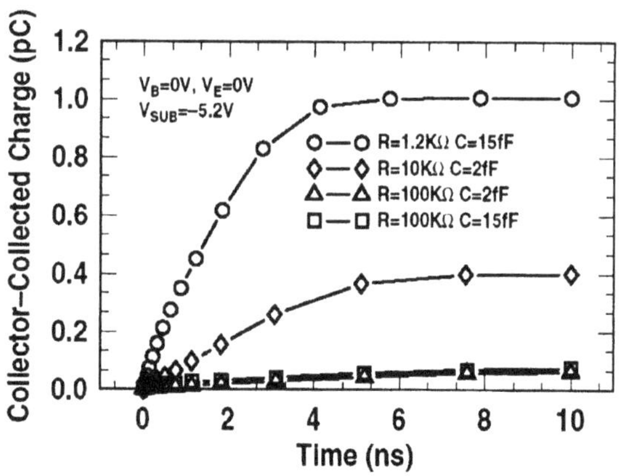

Fig. 10. Simulated charge collection characteristics of a SiGe HBT.

The fundamental reason for the observed better SEU tolerance of circuit A than for circuits B and C is that only one of the two outputs of the emitter-coupled pair being hit is affected by the ion-strike SEU current transients. As long as the differential output is above the logic switching threshold, the output remains unaffected, and no upset occurs. The collector voltage of the switching transistor decreases upon ion strike (compared to without SEU), however, and no upset is observed at the output, simply because the differential output remains above or below the relevant switching threshold (Fig. 12). These results suggest that circuit-level mitigation techniques can be used in SEU hardening of SiGe HBT logic, albeit at some level of additional power dissipation and circuit complexity.

6. Summary

While ionizing radiation degrades both the *dc* and *ac* performance of SiGe HBTs, this degradation is remarkably minor, even for extreme levels of radiation exposure, and is far better than that observed in even radiation-hardened conventional Si BJT technologies. This fact is particularly significant given that no intentional radiation hardening is needed to ensure this level of both device-level and circuit-level tolerance (typically multi-Mrad TID). SEU effects are pronounced in SiGe HBT circuits, as expected, but circuit-level mitigation schemes will likely be suitable to ensure adequate tolerance for many orbital missions. SiGe HBT technology thus offers many interesting possibilities for space-borne electronic systems.

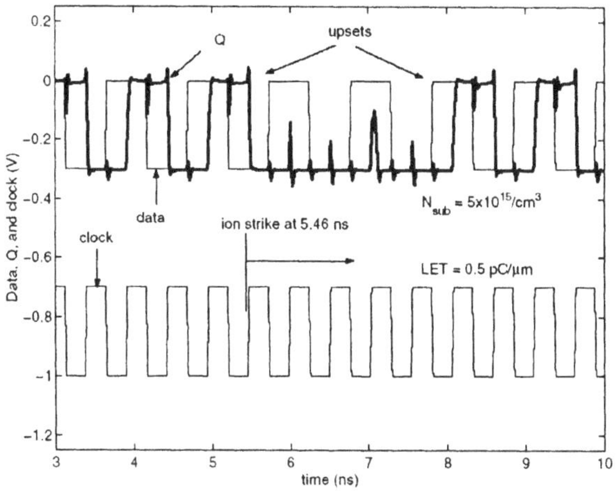

Fig. 11. Simulated SEU response of a SiGe HBT using circuit B architecture, showing upsets

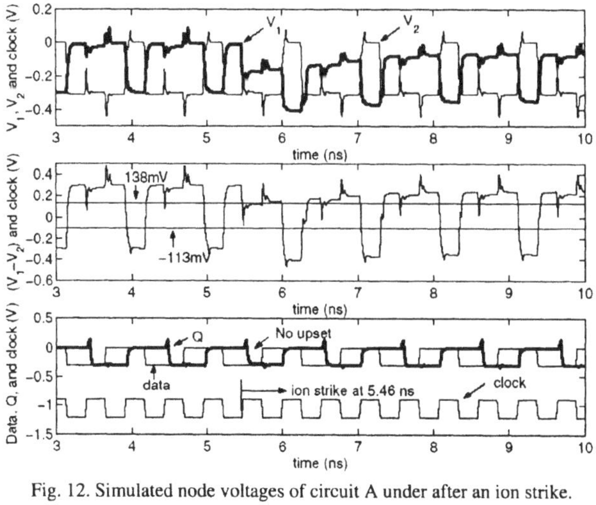

Fig. 12. Simulated node voltages of circuit A under after an ion strike.

Acknowledgement

This work was supported by DTRA under the Radiation Tolerant Microelectronics Program, NASA-GSFC under the Electronics Radiation Characterization (ERC) Program, the Auburn University CSPAE, NAVSEA Crane, and an IBM University Partner Award. I would like to thank R. Reed, P. Marshall, C. Marshall, L. Cohn, B. Kauffman, K. LaBel, H. Brandhorst, S. Clark, D. Emily, D. Harame, A. Joseph, S. Subbanna, D. Ahlgren, G. Freeman, B. Meyerson, D. Herman, and the IBM SiGe team for their support of this work, as well as H. Kim, T. Sanders and D. Hawkins for experimental support. I am indebted to the contributions of my students (J. Roldán, J. Babcock, W. Ansley, L. Vempati, G. Banerjee, S. Zhang, Z. Jin, M. Hamilton, G. Mullinax, and R. Krithivasan), as well as my colleague Guofu Niu.

References

1. J.D. Cressler and G. Niu, *Silicon-Germanium Heterojunction Bipolar Transistors*, Artech House, Boston, 2003.
2. J.A. Babcock, J.D. Cressler, L.S. Vempati, S.D. Clark, R.C. Jaeger, and D.L. Harame, "Ionizing radiation tolerance of high performance SiGe HBTs grown by UHV/CVD," *IEEE Trans. Nucl. Sci.* **42** (1995) 1558-1566.
3. J. Roldán, W.E. Ansley, J.D. Cressler, S.D. Clark, and D. Nguyen-Ngoc, "Neutron radiation tolerance of advanced UHV/CVD SiGe HBTs," *IEEE Trans. Nucl. Sci.* **44** (1997) 1965-1973.
4. J. Roldán, G. Niu, W.E. Ansley, J.D. Cressler, and S.D. Clark, "An investigation of the spatial location of proton-induced traps in SiGe HBTs," *IEEE Trans. Nucl. Sci.* **45** (1998) 2424-2430.
5. J.D. Cressler, M. Hamilton, G. Mullinax, Y. Li, G. Niu, C. Marshall, P. Marshall, H. Kim, M. Palmer, A. Joseph, and G. Freeman, "The Effects of proton irradiation on the lateral and vertical scaling of UHV/CVD SiGe HBT BiCMOS Technology," *IEEE Trans. Nucl. Sci.* **47** (2000) 2515-2520.
6. J.D. Cressler, R. Krithivasan, G. Zhang, G. Niu, P. Marshall, H. Kim, R. Reed, M. Palmer, and A. Joseph, "An investigation of the origins of the variable proton tolerance in multiple SiGe HBT BiCMOS technology generations," *IEEE Trans. Nucl. Sci.*, **49** (2002) 3203-3207.
7. S. Zhang, G. Niu, S.D. Clark, J.D. Cressler, M. Palmer, "The effects of proton irradiation on the RF performance of SiGe HBTs," *IEEE Trans. Nucl. Sci.* **46** (1999) 1716-1721.
8. S. Zhang, G. Niu, J.D. Cressler, S.J. Mathew, S.D. Clark, P. Zampardi, and R.L. Pierson, "A comparison of the effects of gamma irradiation on SiGe HBT and GaAs HBT technologies," *IEEE Trans. Nucl. Sci.* **47** (2000) 2521-2527.
9. R.L. Pease, "Total-dose issues for microelectronics in space systems," *IEEE Trans. Nucl. Sci.* **43** (1996) 442-452.
10. E.W. Enlow, R.L. Pease, W. Combs, R.D. Schrimpf and R.N. Nowlin, "Response of advanced bipolar processes to ionizing radiation," *IEEE Trans. Nucl. Sci.* **38** (1991) 1342-1351.
11. D.M. Fleetwood, S.L. Kosier, R.N. Nowlin, R.D. Schrimpf, R.A. Reber Jr., M. Delaus, P.S. Winokur, A. Wei, W.E. Combs, and R.L. Pease, "Physical mechanisms contributing to enhanced bipolar gain degradation at low dose rates," *IEEE Trans. Nucl. Sci.* **41** (1994) 1871-1883.
12. G. Banerjee, G. Niu, J.D. Cressler, S.D. Clark, M.J. Palmer, and D.C. Ahlgren, "Anomalous dose rate effects in gamma irradiated SiGe heterojunction bipolar transistors," *IEEE Trans. Nucl. Sci.* **46** (1999) 1620-1626.
13. J.D. Cressler, M.C. Hamilton, R. Krithivasan, H. Ainspan, R. Groves, G. Niu, S. Zhang, Z. Jin, C.J. Marshall, P.W. Marshall, H.S. Kim, R.A. Reed, M.J. Palmer, A.J. Joseph, and D.L.

Harame, "Proton radiation response of SiGe HBT analog and RF circuits and passives," *IEEE Trans. Nucl. Sci.* **48** (2001) 2238-2243.

14. S. Zhang, J.D. Cressler, G. Niu, C. Marshall, P. Marshall, H. Kim, R. Reed, M. Palmer, A. Joseph, and D. Harame, "The effects of operating bias conditions on the proton tolerance of SiGe HBTs," *Solid-State Elect.*, in press.
15. P. Marshall, M.A. Carts, A. Campbell, D. McMorrow, S. Buchner, R. Stewart, B. Randall, B. Gilbert, and R. Reed, "Single event effects in circuit hardened SiGe HBT logic at gigabit per second data rate," *IEEE Trans. Nucl. Sci.* **47** (2000) 2669-2674.
16. R. Reed, P. Marshall, H. Ainspan, C. Marshall, H. Kim, J.D Cressler, and G. Niu, "Single event upset test results on an IBM prescalar fabricated in IBM's 5HP germanium doped silicon process," *Proc. IEEE Nucl. Space Rad. Eff. Conf. Data Workshop*, Vancouver, Canada, July 2001, 172-176.
17. G. Niu, J.D. Cressler, M. Shoga, K. Jobe, P. Chu, and D.L. Harame, "Simulation of SEE-Induced Charge Collection in UHV/CVD SiGe HBTs," *IEEE Trans. Nucl. Sci.* **47** (2000) 2682-2689.
18. G. Niu, R. Krithivasan, J.D. Cressler, P. Marshall, C. Marshall, R. Reed, and D. Harame, "Modeling of single event effects in circuit-hardened high-speed SiGe HBT logic," *IEEE Trans. Nucl. Sci.* **48** (2001) 1849-1854.
19. G. Niu, R. Krithivasan, J.D. Cressler, P.A. Riggs, B.A. Randall, P. Marshall, and R. Reed, "A Comparison of SEU tolerance in high-speed SiGe HBT digital logic designed with multiple circuit architectures," *IEEE Trans. Nucl. Sci.* **49** (2002) 3107-3114.

International Journal of High Speed Electronics and Systems
Vol. 14, No. 2 (2004) 503–517

GAIN DEGRADATION AND ENHANCED LOW-DOSE-RATE SENSITIVITY IN BIPOLAR JUNCTION TRANSISTORS

R. D. SCHRIMPF

Vanderbilt University
Electrical Engineering and Computer Science Department
Nashville, TN 37235, USA

The current gain of irradiated bipolar junction transistors decreases due to increased recombination current in the emitter-base depletion region and the neutral base. This recombination current depends on the interaction of two factors: (1) decreased minority-carrier lifetime at the Si/SiO_2 interface or in the bulk Si and (2) changes in surface potential caused by charge in the oxide. In *npn* transistors, these two factors both result in increased base current, while in *pnp* devices, positive charge in the oxide moderates the increase in base current due to surface recombination. In some technologies, the amount of degradation that occurs at a given total dose increases as the dose rate decreases. This enhanced low-dose-rate sensitivity results from space-charge effects produced by slowly transporting holes and protons in the oxide that covers the emitter-base junction.

Keywords: Gain degradation, excess base current, total dose, bipolar transistors.

1. Introduction

Bipolar junction transistors (BJTs) have important applications in which they may be exposed to radiation, including space systems, defense systems, and high-energy particle accelerators. The principle effect of total ionizing dose and displacement damage on bipolar transistors is an increase in the number of defects that participate in Shockley-Read-Hall recombination (either at the Si/SiO_2 interface or in the Si bulk), resulting in increased base current and decreased current gain.[1, 2] In addition to gain degradation, bipolar integrated circuits also may suffer from device-to-device or collector-to-emitter leakage current caused by radiation-induced inversion under isolation oxides.[3-7] In circuits in which the absolute value of the current gain is not important as long as it exceeds some minimum value, radiation-induced leakage current may be the dominant failure mechanism. In this paper, the mechanisms responsible for gain degradation are described. In addition, possible explanations of the Enhanced Low-Dose-Rate Sensitivity (ELDRS) of gain degradation that is observed in many bipolar technologies are explored.

Bipolar and BiCMOS technologies are particularly important for analog, radio-frequency (RF), and mixed-signal integrated circuits. Much of the recent work on BJTs has focused on development of high-speed transistors (f_T > 50 GHz) for RF integrated circuits. In particular, BJTs based on silicon-germanium (SiGe) and gallium arsenide (GaAs) technologies show much promise for use in the fast-growing wireless communications market.[8] However, much of the market for BJTs remains in the traditional silicon-based analog integrated circuit area.

Bipolar transistors have advantages over MOSFETs in the areas of speed,[9] linearity, transconductance,[10] and current drive.

Bipolar and MOS (metal oxide semiconductor) technologies each have important roles in the microelectronics industry. For digital integrated circuits, MOS technology is favored because of its high packing density, low power dissipation, and flexible logic. The most frequently used bipolar analog integrated circuits include operational amplifiers, voltage regulators, phase-locked loops, analog-to-digital converters (ADCs), and digital-to-analog converters (DACs).[11]

BiCMOS integrated circuits combine bipolar and MOS transistors on a single substrate. In principle, each device type can be used for the applications at which it excels. In the case of bipolar transistors, the most desirable features are switching speed, current drive per unit area, noise performance, analog capability, and I/O speed.[12] In contrast, CMOS (complementary MOS) offers better power dissipation, noise margins, packing density, and digital-circuit complexity.[12] The bipolar and MOS transistors in BiCMOS technologies are both affected by radiation and thus the unique radiation-induced degradation characteristics of both device types (gain degradation, leakage currents, threshold-voltage shifts, surface-mobility degradation, breakdown-voltage changes, etc.) must be considered. However, BiCMOS technologies exhibiting very high levels of radiation tolerance have been developed.[13, 14]

2. Device Structures

2.1. Conventional npn BJTs

A cross-sectional view of a traditional *npn* BJT is shown in Figure 1. The dashed line in this figure indicates the portion of the device that is most sensitive to ionizing radiation. This example illustrates a junction-isolated device, although recessed oxides and trenches are often used in modern technologies.

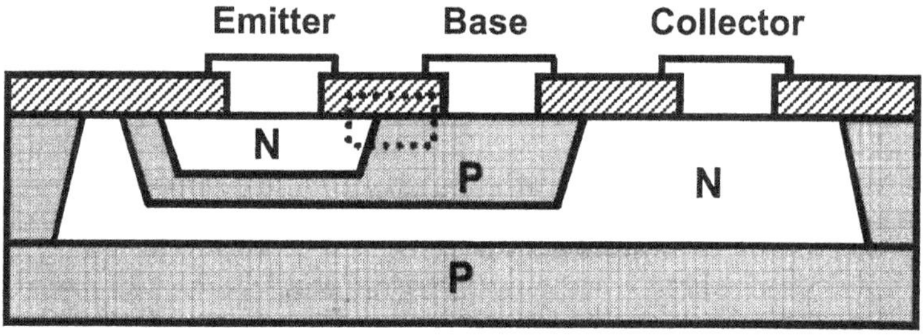

Figure 1 Cross-sectional view of a conventional *npn* BJT. The dashed line indicates the portion of the device that is most sensitive to ionizing radiation.

Polysilicon-emitter *npn* BJTs have significant advantages over crystalline-emitter devices,[15, 16] including smaller device size, reduced parasitics, and high current gain. Transistors with f_t = 30 GHz and no special hardening techniques maintained good performance to a total dose of 100 krad or a neutron fluence of 5×10^{13} cm^{-2}.[17]

2.2. Vertical, lateral, and substrate pnp BJTs

There are three basic types of *pnp* bipolar transistors that are used in integrated circuits. Vertical PNP transistors are similar in structure to vertical *npn* devices. Lateral *pnp* (LPNP) transistors have the active region of the device at the silicon surface and the current flows laterally between emitter and collector regions that are both at the surface. Substrate *pnp* devices have a vertical current-flow pattern, but the substrate serves as the collector for the device. Figure 2 shows qualitative cross sections of lateral and substrate *pnp* transistors. The lightly-doped *n*-type base can be quite sensitive to radiation, particularly at the oxide interface.

Vertical *pnp* transistors are usually relatively radiation-hard compared to vertical *npn* devices.[18, 19] In contrast, lateral and substrate *pnp* transistors tend to be relatively soft.[20-22] The total dose required to degrade lateral devices to half of their normalized current gain may be as much as 50 times less than that needed to observe current gain degradation of the same extent in vertical devices.[23]

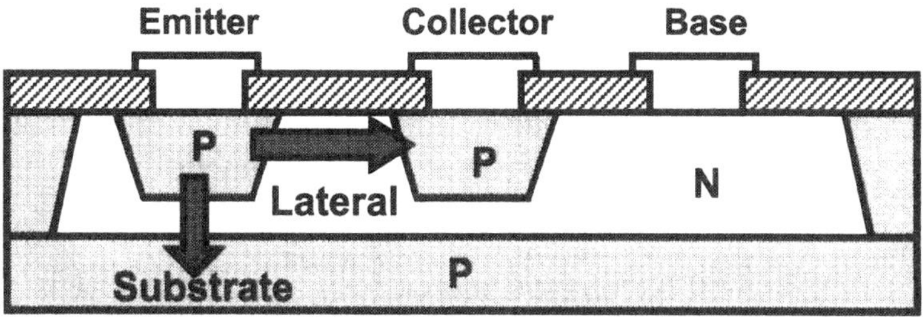

Figure 2 Representative cross-sections of lateral and substrate *pnp* transistors. The arrows indicate the current-flow paths for each type of device. The lightly-doped *n*-type base can be quite sensitive to radiation, particularly at the oxide interface.

3. Transistor-Level Degradation

3.1. Current components

The collector current in an *npn* BJT operating in the forward-active region consists of electrons that are injected from the forward-biased emitter-base junction, diffuse across the neutral base, and are swept through the collector depletion region by the electric field. If all of the electrons injected into the base reach the collector, the collector current density is equal to the diffusion current of electrons in the neutral base:

$$J_C = \frac{qD_{nB}n_{B0}}{W_B}\exp\left(\frac{qV_{BE}}{kT}\right), \tag{1}$$

where J_C is the collector current density, D_{nB} is the electron diffusivity in the base, n_{B0} is the equilibrium electron concentration in the base, W_B is the neutral base width, V_{BE} is the base-emitter voltage, and kT/q is the thermal voltage.

The main components of base current in an *npn* BJT are back-injection of holes from the base to the emitter (J_{B1}), recombination in the emitter-base depletion region (J_{B2}), and recombination in the neutral base (J_{B3}). In most situations, the back-injection component of the base current dominates. However, for devices that have been exposed to ionizing radiation, both J_{B2} and J_{B3} increase, with J_{B2} normally dominating. When the degradation is dominated by displacement damage, J_{B2} and J_{B3} both may increase significantly. The back-injected hole-current density from the base to the emitter, J_{B1}, is a diffusion current:

$$J_{B1} = \frac{qD_{pE}p_{E0}}{L_{pE}} \exp\left(\frac{qV_{BE}}{kT}\right), \tag{2}$$

where D_{pe} is the hole diffusivity in the emitter, p_{E0} is the equilibrium hole concentration in the emitter, and L_{pE} is the diffusion length for holes in the emitter. This assumes that the back-injected hole concentration decays exponentially with distance from the edge of the emitter-base depletion region. Analogous equations describe the operation of *pnp* transistors, with the roles of holes and electrons interchanged.

3.2. Recombination in the emitter-base depletion region

The primary effect of ionizing radiation-induced oxide trapped charge and interface traps is usually an increase in the base current resulting from enhanced recombination in the emitter-base depletion region. The amount by which the base current increases above its pre-irradiation value is called the excess base current (defined as $\Delta I_B = I_B - I_{B0}$, where I_{B0} is the pre-irradiation base current). The recombination-rate increase occurs mainly where the depletion region intersects the Si/SiO_2 interface, due to formation of interface traps that serve as recombination centers (the surface recombination velocity increases with the interface-trap density). When a BJT is exposed to energetic particles, displacement damage occurs in the bulk Si. The resulting defects reduce the minority-carrier lifetime.

The recombination rate is a function of position within the depletion region, exhibiting a strong peak where $n = p$. The ideality factor (defined as n_B in $\exp(qV/n_BkT)$) is 2 for recombination occurring at this peak and 1 for the ideal component of the base current. However, the recombination rate must be integrated throughout the depletion region to obtain the contribution of recombination to the base current. The excess base current due to surface recombination thus has an ideality factor between 1 and 2, which combines the effects of the different spatial locations at which the recombination takes place. However, it is a reasonable approximation in many cases to assume that the radiation-induced excess base current is dominated by the peak recombination rate, which leads to the results:[2]

$$J_{B2,surf} \propto v_{surf} \exp\left(\frac{qV}{2kT}\right) \tag{3}$$

and

$$J_{B2,bulk} \propto \frac{1}{\tau_d} \exp\left(\frac{qV}{2kT}\right), \tag{4}$$

where $J_{B2,surf}$ and $J_{B2,bulk}$ are the base current densities due to recombination in the depletion region where it intersects the Si/SiO_2 interface and in the depletion region in the bulk silicon, respectively, v_{surf} is the surface recombination velocity, and τ_d is the minority-carrier lifetime in the depletion region. Since the base current due to recombination in the depletion region increases more slowly with voltage than the back-injected (J_{B1}) component of the base current, its effect is greatest at low emitter-base voltages.

3.3. Recombination in the neutral base

When minority carriers are injected from the emitter into the neutral base, some of them recombine before they reach the collector junction. This neutral-base recombination may take place in either the bulk Si or at the Si/SiO_2 interface (if the Si surface is neutral, as determined by the charge in the oxide or the bias on any electrodes lying over the base region). The fraction of carriers that recombine is small in most unirradiated devices, but it may become significant following irradiation. In a BJT operating in the forward-active region, the distribution of minority carriers in the neutral base is approximately triangular and the total amount of minority-carrier charge (electrons, in this case) stored in the base is:

$$Q_B = \frac{1}{2} q W_B n_B(0) = \frac{q W_B n_{B0} \exp\left(\frac{qV_{BE}}{kT}\right)}{2} \tag{5}$$

If the average lifetime of minority carriers in the neutral base is τ_B, the base current density required to supply the recombination process is:

$$J_{B3} = \frac{Q_B}{\tau_B} = \frac{q W_B n_{B0} \exp\left(\frac{qV_{BE}}{kT}\right)}{2\tau_B}. \tag{6}$$

Unlike recombination in the emitter-base depletion region, which exhibits an ideality factor close to 2, the ideality factor associated with recombination in the neutral base is 1. When the device is irradiated, this component of the current increases since the lifetime in the base decreases. For ionizing radiation, the surface lifetime decreases due to interface-trap formation, but the bulk lifetime remains approximately constant. For displacement damage, the bulk lifetime decreases.

3.4. Current gain

The common-emitter current gain (β) is defined as the ratio of the collector current to the base current:

$$\beta = \frac{I_C}{I_B}. \tag{7}$$

In a simple one dimensional BJT, the peak pre-irradiation current gain can be estimated as the ratio of Eq. (1) to Eq. (2):

$$\beta = \frac{D_{nB} n_{B0} L_{pE}}{D_{pE} p_{E0} W_B} = \frac{D_{nB} N_{DE} L_{pE}}{D_{pE} N_{AB} W_B}, \tag{8}$$

where N_{DE} is the emitter doping and N_{AB} is the base doping. This simple estimate of the current gain does not include effects associated with either high or low bias levels. At high bias levels, the current gain is lower than the peak value due to high-level injection effects[24, 25] and the Kirk effect.[26] At low levels, the current gain is lower because of recombination in the emitter-base depletion region. When a BJT is irradiated, the base current increases because of increased recombination in the emitter-base depletion region and the neutral base. However, the collector current typically remains relatively constant, causing the current gain to decrease.[27, 28] An example Gummel plot (log I_C and I_B vs. V_{BE}) is shown in Figure 3. For this device, the collector current remains virtually unchanged except at very low bias levels, while the base current increases significantly. This causes a large reduction in the current gain, especially at low bias levels where the base current increases most rapidly. The current gain of an irradiated *npn* BJT is plotted vs. V_{BE} in Figure 4. The actual total dose levels that correspond to a given amount of degradation in a particular device type depend strongly on the technology used to fabricate the device; significant degradation may occur at total doses as low as 10 krad(SiO_2) in very soft technologies, while other technologies may exhibit relatively little degradation at 1 Mrad(SiO_2).

Increased recombination in the emitter-base depletion region does not reduce the collector current at a given bias level because the number of carriers injected into the base depends only on the doping of the base and the applied bias. If recombination increases in the depletion region, the emitter and base currents increase, but the collector current remains constant. However, when injected carriers recombine in the neutral base, they do not reach the collector junction and the collector current decreases.

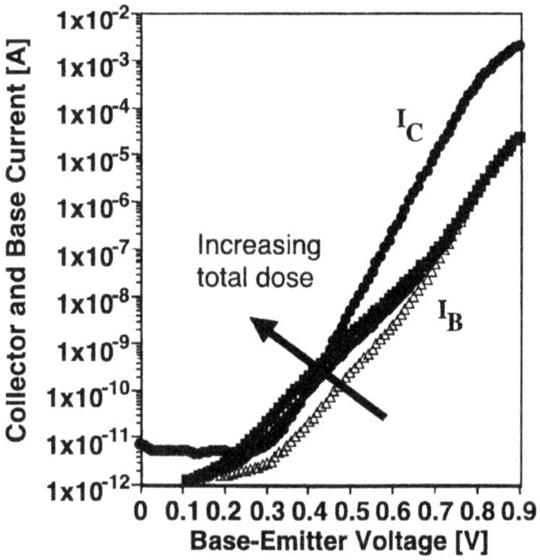

Figure 3 Collector and base current vs. base-emitter voltage for an irradiated *npn* BJT.

Early studies of gain degradation in BJTs focused on displacement damage. In these studies, it was shown that the change in the current gain due to neutron irradiation could be described by:[29-31]

$$\frac{1}{\beta} = \frac{1}{\beta_0} + \frac{K_1 \Phi_n}{2\pi f_T} \tag{9}$$

where K_1 is a constant of proportionality and Φ_n is the neutron fluence. This equation is equivalent to saying that the base current is inversely proportional to the minority-carrier lifetime in the emitter-base depletion region or the neutral base. The same type of relationship holds for displacement damage introduced by other particles, including heavy ions, protons, and electrons.[32]

In space systems, high-energy protons constitute a large part of the radiation to which devices are subjected. In addition to ionizing-dose effects, protons also cause displacement damage. The device-level effects due to these two types of damage are not simply additive; they interact in a complicated fashion.[33] It has been shown that proton irradiation can cause linear integrated circuits to fail at much lower total-dose levels than can be tolerated for γ irradiation.[34] In some parts, the proton irradiation produced different failure mechanisms than those that occurred for γ irradiation, including catastrophic failures.

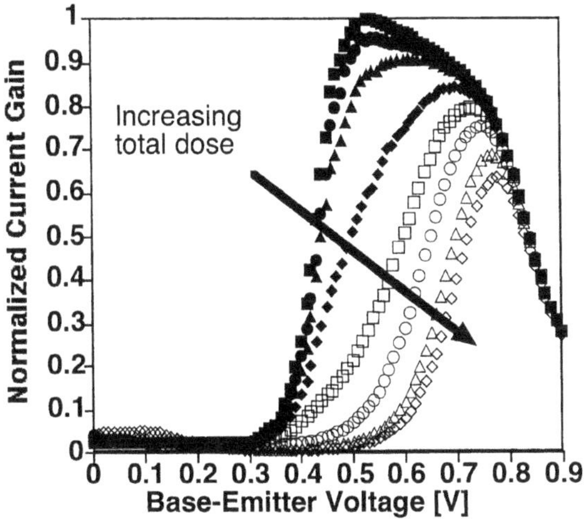

Figure 4 Normalized current gain vs. base-emitter voltage for an *npn* BJT irradiated to various total doses.

The increase in surface recombination velocity is approximately proportional to the density of recombination centers at the silicon/silicon dioxide interface that covers the emitter-base junction. Traps with energies near the middle of the silicon bandgap are the most effective. Thus, the recombination centers are not precisely the same as the interface traps (N_{it}) that are used to describe threshold-voltage shifts in MOS technologies. In discussions of irradiated MOSFETs, N_{it} usually refers to the total areal density of interface traps with energies between midgap and threshold. The increase in surface recombination velocity is often the dominant factor in determining the amount of radiation-induced gain degradation.

Since the net charge introduced into the oxide by ionizing radiation (N_{ox}) is positive, the depletion region spreads on the *p*-side of a *pn* junction. For *npn* transistors, the surface of the relatively lightly doped *p*-type base region becomes depleted. The recombination rate is maximized when the electron and hole populations are equal; this condition occurs at a point within the depletion region.

The increased recombination occurs around the edge of the emitter, so the amount of excess base current caused by surface recombination is proportional to the perimeter of the emitter.[35] Device layouts with large perimeter-to-area ratios increase the sensitivity to ionizing radiation because the excess base current (proportional to emitter perimeter) is large relative to the ideal component of the base current (proportional to emitter area).

Ionizing radiation also may change the effective doping concentration in *p*-type regions due to deactivation of acceptor atoms. This can affect the gain of *npn* transistors by changing the number of electrons injected into the base or increasing the sensitivity to radiation-induced charge in the oxide.[36]

Vertical *pnp* transistors are relatively radiation hard compared to vertical *npn* devices because positive radiation-induced trapped charge tends to accumulate the *n*-type base, reducing the width of the emitter-base depletion region. The positive charge depletes the *p*-type emitter, but the effect is typically small because of the high emitter doping.

In lateral *pnp* BJTs, the excess base current scales with the surface recombination velocity, which increases approximately in proportion to the number of interface traps. However, this increase is moderated by the effects of positive charge in the oxide over the base. This positive charge accumulates the surface of the base, suppressing surface recombination. Thus, the effects of radiation-induced positive oxide charge and increased surface recombination velocity oppose each other. This may lead to a sublinear dependence of excess base current on total dose. In addition, lateral *pnp* BJTs with lightly doped emitters are more sensitive to total ionizing dose than those with more heavily doped emitters.[37] In the lightly doped devices, the depletion region is able to spread into the emitter, increasing the surface area where recombination may occur in the same fashion as occurs in *npn* BJTs.

Lateral *pnp* transistors are affected more significantly by ionizing radiation than substrate *pnp* transistors, since the current flow pattern in the LPNP devices is lateral and directly under the oxide where the recombination centers occur, while the current flow path in the SPNP devices is vertical. The current flow path in a lateral *pnp* BJT is illustrated in Figure 2.

If the surface of an LPNP BJT is depleted, then the radiation-induced recombination centers are most effective and the increase in base current is large. However, the surface potential in the base region can be controlled by placing a gate over the neutral base region. The surface recombination rate can be reduced by biasing the gate so that the surface of the base is accumulated, rather than depleted.[28, 38, 39] An emitter-tied field plate can be obtained by simply extending the emitter metallization over the base;[28] this can be accomplished at no increase in device or process complexity. If the gate can be biased independently, this offers much more flexibility in controlling the amount of radiation-induced degradation. If the gate bias is negative during irradiation, the electric field in the oxide is directed from the silicon to the gate, minimizing the degradation due to positively charged holes and protons. However, this bias condition maximizes the recombination rate for a given defect density. The ideal situation is to adjust the gate bias dynamically so that it is negative when the device is not operating and positive when the device is operating.[39] This reduces the amount of defect buildup during non-operational periods and reduces the effects of the defects when the device operates. This technique is most effective for devices that have long periods of time when they are exposed to radiation but are not operating. There also is additional layout complexity due to the independent gate electrodes.

Since the gain degradation in lateral *pnp* transistors is caused by radiation-induced interface traps, the radiation tolerance can be improved by reducing interface-trap buildup. This was demonstrated by implanting high doses of arsenic into the oxide that covered the neutral base of *pnp* transistors.[40] Both interface-trap density and radiation-induced current-gain degradation were smaller in devices that were fabricated with the implanted oxide.

4. Enhanced low-dose-rate sensitivity (ELDRS)

4.1. Overview

For some bipolar transistors, the amount of degradation at a given total dose depends on the rate at which the dose is accumulated, with more degradation occurring at lower dose rates.[18, 41-49] This phenomenon is referred to as Enhanced Low-Dose-Rate Sensitivity (ELDRS). The amount of degradation typically varies most rapidly with dose rate somewhere between 0.1 and 10 rad(SiO_2)/s and is often approximately independent of dose rate for low and high dose rates. However, dose-rate sensitivity is very technology-dependent, so whether or not a given technology exhibits ELDRS and if so, the specific range of dose rates over which it occurs, are difficult to predict.

Laboratory testing is frequently conducted at dose rates of 50 rad(SiO_2)/s or higher in order to complete the tests in a reasonable amount of time. However, this is significantly above the dose rates at which ELDRS typically occurs. In contrast, dose rates in space systems vary dramatically depending on orbit and shielding, but values below 0.01 rad(SiO_2)/s are typical. This can lead to overestimation of device lifetime if high dose rate tests are used to predict space-system behavior. Figure 5 is an illustrative plot of the normalized current gain vs. dose rate for a particular lateral *pnp* BJT, showing that the device response varies significantly in the range of dose rates between the space environment and typical laboratory tests.

ELDRS is typically observed only in relatively low quality oxides irradiated at low electric fields. Oxides covering the base region of BJTs frequently satisfy these conditions because they may be thick, may not have an electrode covering the oxide, and may have been used as an implant mask.

Developing hardness-assurance and test methods that are able to deal with the ELDRS effect has been a major challenge for radiation-effects researchers in recent years.[45, 47, 48, 50-52] Although accelerated test methods have been proposed,[53, 54] low-dose-rate testing is still advisable in critical applications.

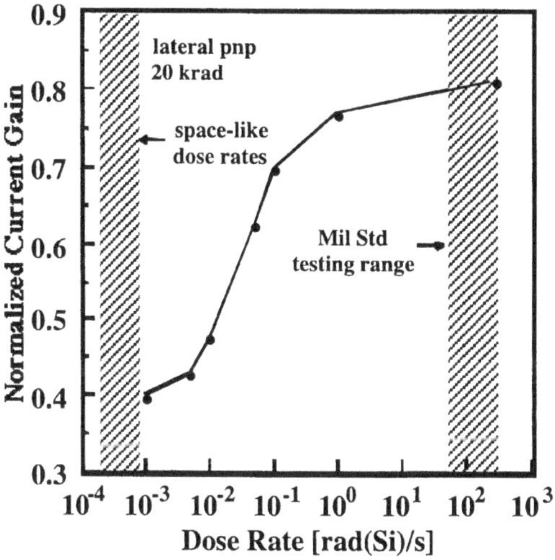

Figure 5 Normalized current gain vs. dose rate for a lateral *pnp* BJT irradiated to a total dose of 20 krad(Si) as a function of dose rate. The amount of degradation increases at low dose rates approaching those encountered in space (illustration courtesy of S. Witczak).

4.2. True dose-rate effects

Although the total-dose response of MOS integrated circuits frequently depends on dose rate, the effects in MOS ICs are usually time-dependent effects rather than true dose-rate effects.[55-61] A true dose-rate effect occurs when a physical process responsible for the degradation depends on a quantity that is related to the dose rate. An example is a space-charge effect in which the presence of more charge at high dose rates changes the electric field, affecting the number of carriers surviving initial recombination and their subsequent transport. A time-dependent effect is one in which the same response can be obtained by waiting for a period of time following irradiation at the higher dose rate equal to the amount of time that is required to reach the same total dose by irradiation at the lower dose rate. The densities of oxide-trapped charge (N_{ot}) and interface-trapped charge (N_{it}) both change following irradiation in many devices; typically, N_{ot} decreases due to annealing and N_{it} increases as slowly-transporting protons reach the interface. Both true dose-rate and time-dependent effects have been observed in bipolar devices, but it was the identification of a true dose-rate effect that was initially surprising.[41]

The ELDRS effect has been observed in both *npn* and *pnp* BJTs. An illustration of the excess base current vs. dose rate for an *npn* BJT is shown in Fig. 6.[48] This device was irradiated to a total dose of 40 krad(Si) and the base current was measured at V_{BE} = 0.6 V. For this device type, the ELDRS appears to saturate near 0.005 rad(Si)/s. The degradation at this dose rate is approximately five times that which occurs at higher dose rates.

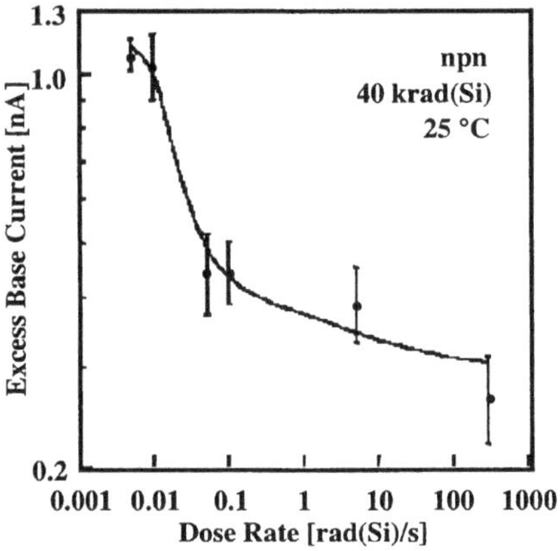

Figure 6 Excess base current vs. dose rate for an *npn* BJT measured at a total dose of 40 krad(Si) and V_{BE} = 0.6 V (after[48]).

4.3. *Physical models*

It has been suggested that ELDRS in BJTs is related to space-charge effects due to slowly-transporting holes.[49, 50, 62-65] The space charge that exists at higher dose rates retards the transport of holes (and also protons) to the interface. A critical dose rate exists at which the space charge has a significant effect on the electric field in the oxide. At dose rates far above or far below this, the charge buildup is nearly independent of dose rate. The transition between the two regions results in an S-shaped curve of excess base current vs. dose rate, as observed experimentally. Since both holes and hydrogen ions are affected by the space charge, this model applies to the dose-rate dependence observed experimentally in the buildup of both oxide charge and interface traps.[65]

At low dose rates, the net positive oxide charge and the interface-trap density are both found to be greater in some bipolar base oxides.[62] Thermally-stimulated current (TSC) measurements showed that the increased net positive oxide charge was due to increased electron compensation, rather than reduced hole trapping.

It also has been suggested that ELDRS may be a delayed reaction rate effect.[66, 67] This has been described by a model that assumes that the degradation results from interaction of two species, with different times required for the species to reach the Si/SiO_2 interface.[66, 67]

The ELDRS phenomenon also may depend on the thermal history of the devices[68] and the presence (or absence) of passivation layers on the die.[69] Both of these issues may be related to the packaging process. Although the physical mechanisms responsible for these effects have not been identified conclusively, hydrogen and mechanical strain induced by the passivation layer both may play important roles.

5. Conclusions

Bipolar transistors play a variety of important roles in systems where they are exposed to radiation. Although most of the parts currently deployed are fabricated in traditional Si technologies with vertical *npn* and lateral *pnp* transistors, there are exciting new applications for high-speed technologies utilizing heterojunction bipolar transistors. Understanding radiation effects at both the device and circuit levels is essential for fielding radiation-tolerant systems.

When BJTs are exposed to radiation, the base current increases and the current gain decreases. Increased recombination in the emitter-base depletion region is the main mechanism responsible for the increased base current. In *npn* BJTs, the degradation is due to the combined effects of increased surface recombination velocity due to radiation-induced interface traps and changes in surface potential caused by oxide trapped charge. In *pnp* transistors, the base current increases due to the increased surface recombination velocity, but the increase is moderated by the oxide trapped charge since it moves the base surface toward accumulation. Typically lateral *pnp* transistors are softer than vertical *npn* transistors because the current path is near the Si/SiO_2 interface.

Many bipolar technologies degrade more at low dose rates than at high dose rates for a given total dose. This ELDRS phenomenon is a difficult hardness-assurance challenge because testing at standard laboratory dose rates overestimates the radiation hardness of the parts in space systems. Methods of predicting the low-dose-rate response using laboratory dose rates, including irradiation at high temperature, have been proposed. While these methods have been reasonably successful at identifying technologies that suffer from ELDRS, it is still difficult to identify a single test for all bipolar technologies.

Acknowledgements

I would like to express sincere appreciation to my many collaborators, students, former students, and sponsors in the bipolar total-dose area, including: Dennis Ball, Hugh Barnaby, Jérôme Boch, Philippe Calvel, Claude Cirba, Ryan Cizmarik, Lew Cohn, Pat Cole, Bill Combs, Mike DeLaus, Dave Emily, Ed Enlow, Dan Fleetwood, Pascal Fouillat, Ken Galloway, Russ Graves, Tim Holman, Allan Johnston, Steve Kosier, Jeff Krieg, Mike Maher, Andre Martinez, Lloyd Massengill, Randall Milanowski, Nathan Nowlin, Ron Pease, Dale Platteter, Dawn Schmidt, Marty Shaneyfelt, Mayrant Simons, Jeff Titus, Tom Turflinger, Diek van Nort, Peter Winokur, Steve Witczak, Andy Wei, and Andrew Wu. I would particularly like to thank the Defense Threat Reduction Agency and NAVSEA-Crane for their long-term support of research in this area.

References

[1] R. D. Schrimpf, "Recent Advances in Understanding Total-Dose Effects in Bipolar Transistors," *IEEE Trans. Nucl. Sci.*, vol. 43, pp. 787-796, 1996.

[2] S. L. Kosier, A. Wei, R. D. Schrimpf, D. M. Fleetwood, M. DeLaus, R. L. Pease, and W. E. Combs, "Physically Based Comparison of Hot-Carrier-Induced and Ionizing-Radiation-Induced Degradation in BJTs," *IEEE Trans. Electron Devices*, vol. 42, pp. 436-444, 1995.

[3] R. L. Pease, R. M. Turfler, D. Platteter, D. Emily, and R. Blice, "Total Dose Effects in Recessed Oxide Digital Bipolar Microcircuits," *IEEE Trans. Nucl. Sci.*, vol. NS-30, pp. 4216-4223, 1983.

[4] J. L. Titus and D. G. Platteter, "Wafer Mapping of Total Dose Failure Thresholds in a Bipolar Recessed Field Oxide Technology," *IEEE Trans. Nucl. Sci.*, vol. 34, pp. 1751-1756, 1987.

[5] E. W. Enlow, R. L. Pease, W. E. Combs, and D. G. Platteter, "Total dose induced hole trapping in trench oxides," *IEEE Trans. Nucl. Sci.*, vol. 36, pp. 2415-2422, 1989.

[6] J. P. Raymond, R. A. Gardner, and G. E. LaMar, "Characterization of radiation effects on trench-isolated bipolar analog microcircuit technology," *IEEE Trans. Nucl. Sci.*, vol. 39, pp. 405-412, 1992.

[7] W. C. Jenkins, "Dose-rate-independent total dose failure in 54F10 bipolar logic circuits," *IEEE Trans. Nucl. Sci.*, vol. 39, pp. 1899-1902, 1992.

[8] J. D. Cressler, "SiGe HBT Technology: A New Contender for Si-Based RF and Microwave Applications," *IEEE Trans. Microwave Theory and Techniques*, vol. 46, pp. 572-589, 1998.

[9] S. Subbanna, J. Johnson, G. Freeman, R. Volant, R. Groves, D. Herman, and B. Meyerson, "Prospects of silicon-germanium-based technology for very high-speed circuits," in *IEEE Microwave Symposium Digest*, pp. 361-364, 2000.

[10] R. M. Warner, Jr. and R. D. Schrimpf, "BJT-MOSFET Transconductance Comparisons," *IEEE Trans. Electron Devices*, vol. ED-34, pp. 1061-1065, 1987.

[11] P. R. Gray and R. G. Meyer, *Analysis and Design of Analog Integrated Circuits*. New York: John Wiley & Sons, 1993.

[12] A. R. Alvarez, *BiCMOS Technology and Applications*. Boston: Kluwer Academic Publishers, 1993.

[13] M. Dentan, E. Delagnes, N. Fourches, M. Rouger, M. C. Habrard, L. Blanquart, P. Delpierre, R. Potheau, R. Truche, J. P. Blanc, E. Delevoye, J. Gautier, J. L. Pelloie, d. Pontcharra, O. J.; Flament, J. L. Leray, J. L. Martin, J. Montaron, and O. Musseau, "Study of a CMOS-JFET-bipolar radiation hard analog-digital technology suitable for high energy physics electronics," *IEEE Trans. Nucl. Sci.*, vol. 40, pp. 1555-1560, 1993.

[14] M. Dentan, P. Abbon, E. Delagnes, N. Fourches, D. Lachartre, F. Lugiez, B. Paul, M. Rouger, R. Truche, J. P. Blanc, C. Leroux, E. Delevoye-Orsier, J. L. Pelloie, J. de Pontcharra, O. Flament, J. M. Guebhard, J. L. Leray, J. Montaron, and O. Musseau, "DMILL, a mixed analog-digital radiation-hard BICMOS technology for high energy physics electronics," *IEEE Trans. Nucl. Sci.*, vol. 43, pp. 1763-1767, 1996.

[15] I. R. C. Post, P. Ashburn, and G. R. Wolstenhome, "Polysilicon Emitters for Bipolar Transistors: A Review and Re-Evaluation of Theory and Experiment," *IEEE Trans. Electron Devices*, vol. 39, pp. 1717-1731, 1992.

[16] K. O, P. Garone, C. Tsai, B. Scharf, M. Higgins, D. Mai, C. Kermarrec, and J. Yasaitis, "A Double Polysilicon Self-Aligned npn Bipolar Process (ADRF) with Optional NMOS Transistors for RF and Microwave Applications," in *IEEE BCTM Proceedings*, pp. 221-224, 1994.

[17] O. Flament, S. Synold, J. de Pontcharra, and S. Niel, "Radiation tolerance of npn bipolar technology with 30 GHz F_t," *IEEE Trans. Nucl. Sci.*, vol. 47, pp. 654-658, 2000.

[18] R. N. Nowlin, E. W. Enlow, R. D. Schrimpf, and W. E. Combs, "Trends in the Total-Dose Response of Modern Bipolar Transistors," *IEEE Trans. Nucl. Sci.*, vol. 39, pp. 2026-2035, 1992.

[19] A. H. Pawlikiewicz, "High tolerance to radiation effects and low noise performance of ACUTE-a complementary bipolar SOI IC technology," *IEEE Trans. Nucl. Sci.*, vol. 43, pp. 761-768, 1996.

[20] J. Beaucour, T. Carriere, A. Gach, and P. Poirot, "Total Dose Effects on Negative Voltage Regulator," *IEEE Trans. Nucl. Sci.*, vol. 41, pp. 2420-2426, 1994.

[21] A. H. Johnston, B. G. Rax, and C. I. Lee, "Enhanced Damage in Linear Bipolar Integrated Circuits at Low Dose Rates," *IEEE Trans. Nucl. Sci.*, vol. 42, pp. 1650-1659, 1995.

[22] S. L. Kosier, R. D. Schrimpf, A. Wei, M. DeLaus, D. M. Fleetwood, and W. E. Combs, "Effects of Oxide Charge and Surface Recombination Velocity on the Excess Base Current of BJTs," in *Proc. IEEE Bipolar/BiCMOS Circuits and Technology Meeting*, pp. 211-214, 1993.

[23] D. M. Schmidt, D. M. Fleetwood, R. D. Schrimpf, R. L. Pease, R. J. Graves, G. H. Johnson, K. F. Galloway, and W. E. Combs, "Comparison of Ionizing-Radiation-Induced Gain Degradation in Lateral, Substrate, and Vertical PNP BJTs," *IEEE Trans. Nucl. Sci.*, vol. 42, pp. 1541-1549, 1995.

[24] E. S. Rittner, "Extension of the Theory of the Junction Transistor," *Phys. Rev.*, vol. 94, pp. 1161, 1954.

[25] W. M. Webster, "On the Variation of Junction-Transistor Current-Amplification Factor with Emitter Current," *Proc. IRE*, vol. 42, pp. 914-920, 1954.

[26] C. T. Kirk, Jr., "A Theory of Transistor Cutoff Frequency (f_t) Falloff at High Current Densities," *IRE Trans. Electron Devices*, vol. 9, pp. 164, 1962.

[27] A. Wei, S. L. Kosier, R. D. Schrimpf, W. E. Combs, and M. DeLaus, "Excess Collector Current Due to an Oxide-Trapped-Charge-Induced Emitter in Irradiated NPN BJTs," *IEEE Trans. Electron Devices*, vol. 42, pp. 923-927, 1995.

[28] H. J. Barnaby, R. D. Schrimpf, D. M. Fleetwood, and S. L. Kosier, "The Effects of Emitter-Tied Field Plates on Lateral PNP Ionizing Radiation Response," in *IEEE BCTM Proc.*, pp. 35-38, 1998.

[29] G. C. Messenger and J. P. Spratt, "The Effects of Neutron Irradiation on Germanium and Silicon," *Proc. IRE*, vol. 46, pp. 1038-1044, 1958.

[30] G. C. Messenger and R. A. Hubbs, "A General Proof of the Beta Degradation Equation for Bulk Displacement Damage," *IEEE Trans. Nucl. Sci.*, vol. 20, 1973.

[31] G. C. Messenger and M. S. Ash, *The Effects of Radiation on Electronic Systems*. New York: Van Nostrand-Reinhold, 1986.

[32] C. J. Dale, P. W. Marshall, E. A. Burke, G. P. Summers, and E. A. Wolicki, "High energy electron induced displacement damage in silicon," *IEEE Trans. Nucl. Sci.*, vol. 35, pp. 1208-1214, 1988.

[33] H. J. Barnaby, R. D. Schrimpf, A. L. Sternberg, V. Berthe, C. R. Cirba, and R. L. Pease, "Proton Radiation Response Mechanisms in Bipolar Analog Circuits," *IEEE Trans. Nucl. Sci.*, vol. 48, pp. 2074-2080, 2001.

[34] B. G. Rax, A. H. Johnston, and T. Miyahira, "Displacement damage in bipolar linear integrated circuits," *IEEE Trans. Nucl. Sci.*, vol. 46, pp. 1660-1665, 1999.

[35] R. N. Nowlin, R. D. Schrimpf, E. W. Enlow, W. E. Combs, and R. L. Pease, "Mechanisms of Ionizing-Radiation-Induced Gain Degradation in Modern Bipolar Devices," in *Proc. 1991 IEEE Bipolar Circuits and Tech. Mtg.*, pp. 174-177, 1991.

[36] S. C. Witczak, R. C. Lacoe, M. R. Shaneyfelt, D. C. Mayer, J. R. Schwank, and P. S. Winokur, "Implications of radiation-induced dopant deactivation for npn bipolar junction transistors," *IEEE Trans. Nucl. Sci.*, vol. 47, pp. 2281-2288, 2000.

[37] A. Wu, R. D. Schrimpf, H. J. Barnaby, D. M. Fleetwood, R. L. Pease, and S. L. Kosier, "Radiation-Induced Gain Degradation in Lateral PNP BJTs with Lightly and Heavily Doped Emitters," *IEEE Trans. Nucl. Sci.*, vol. 44, pp. 1914-1921, 1997.

[38] P. Cazenave, P. Fouillat, X. Montagner, H. Barnaby, R. D. Schrimpf, L. Bonora, J. P. David, A. Touboul, M.-C. Calvet, and P. Calvel, "Total Dose Effects on Gate Controlled Lateral PNP Bipolar Junction Transistors," *IEEE Trans. Nucl. Sci.*, vol. 45, pp. 2577-2583, 1998.

[39] H. Barnaby, C. Cirba, R. D. Schrimpf, S. Kosier, P. Fouillat, and X. Montagner, "Minimizing Gain Degradation in Lateral PNP Bipolar Junction Transistors Using Gate Control," *IEEE Trans. Nucl. Sci.*, vol. 46, pp. 1652-1659, 1999.

[40] M. Kato, K. Watanabe, and T. Okabe, "Radiation effects on ion-implanted silicon-dioxide films," *IEEE Trans. Nucl. Sci.*, vol. 36, pp. 2199-2204, 1989.

[41] E. W. Enlow, R. L. Pease, W. E. Combs, R. D. Schrimpf, and R. N. Nowlin, "Response of Advanced Bipolar Processes to Ionizing Radiation," *IEEE Trans. Nucl. Sci.*, vol. NS-38, pp. 1342-1351, 1991.

[42] S. L. Kosier, W. E. Combs, A. Wei, R. D. Schrimpf, D. M. Fleetwood, M. DeLaus, and R. L. Pease, "Bounding the Total-Dose Response of Modern Bipolar Transistors," *IEEE Trans. Nucl. Sci.*, vol. 41, pp. 1864-1870, 1994.

[43] R. N. Nowlin, D. M. Fleetwood, R. D. Schrimpf, R. L. Pease, and W. E. Combs, "Hardness-Assurance and Testing Issues for Bipolar/BiCMOS Devices," *IEEE Trans. Nucl. Sci.*, vol. 40, pp. 1686-1693, 1993.

[44] R. N. Nowlin, D. M. Fleetwood, and R. D. Schrimpf, "Saturation of the Dose-Rate Response of BJTs Below 10 rad(SiO_2)/s: Implications for Hardness Assurance," *IEEE Trans. Nucl. Sci.*, vol. 41, pp. 2637-2641, 1994.

[45] R. D. Schrimpf, R. J. Graves, D. M. Schmidt, D. M. Fleetwood, R. L. Pease, W. E. Combs, and M. DeLaus, "Hardness Assurance Issues for Lateral PNP Bipolar Junction Transistors," *IEEE Trans. Nucl. Sci.*, vol. 42, pp. 1641-1649, 1995.

[46] A. Wei, S. L. Kosier, R. D. Schrimpf, D. M. Fleetwood, and W. E. Combs, "Dose-Rate Effects on Radiation-Induced Bipolar Junction Transistor Gain Degradation," *Appl. Phys. Lett.*, vol. 65, pp. 1918-1920, 1994.

[47] S. C. Witczak, R. D. Schrimpf, K. F. Galloway, D. M. Fleetwood, R. L. Pease, J. M. Puhl, D. M. Schmidt, W. E. Combs, and J. S. Suehle, "Accelerated Tests for Simulating Low Dose Rate Gain Degradation of Lateral and Substrate PNP Bipolar Junction Transistors," *IEEE Trans. Nucl. Sci.*, vol. 43, pp. 3151-3160, 1996.

[48] S. C. Witczak, R. D. Schrimpf, D. M. Fleetwood, K. F. Galloway, R. C. Lacoe, D. C. Mayer, J. M. Puhl, R. L. Pease, and J. S. Suehle, "Hardness Assurance Testing of Bipolar Junction Transistors at Elevated Irradiation Temperatures," *IEEE Trans. Nucl. Sci.*, vol. 44, pp. 1989-2000, 1997.

[49] S. C. Witczak, R. C. Lacoe, D. C. Mayer, D. M. Fleetwood, R. D. Schrimpf, and K. F. Galloway, "Space Charge Limited Degradation of Bipolar Oxides at Low Electric Fields," *IEEE Trans. Nucl. Sci.*, vol. 45, pp. 2339-2351, 1998.

[50] D. M. Fleetwood, S. L. Kosier, R. N. Nowlin, R. D. Schrimpf, R. A. Reber, Jr., M. DeLaus, P. S. Winokur, A. Wei, W. E. Combs, and R. L. Pease, "Physical Mechanisms Contributing to Enhanced Bipolar Gain Degradation at Low Dose Rates," *IEEE Trans. Nucl. Sci.*, vol. 41, pp. 1871-1883, 1994.

[51] R. L. Pease, "Total-dose issues for microelectronics in space systems," *IEEE Trans. Nucl. Sci.*, vol. 43, pp. 442-452, 1996.

[52] J. Krieg, T. Turflinger, J. Titus, P. Cole, P. Baker, M. Gehlhausen, D. Emily, L. Yang, R. L. Pease, H. Barnaby, R. Schrimpf, and M. C. Maher, "Hardness Assurance Implications of Bimodal Total Dose Response in a Bipolar Linear Voltage Comparator," *IEEE Trans. Nucl. Sci.*, vol. 46, pp. 1627-1632, 1999.

[53] R. L. Pease, L. M. Cohn, D. M. Fleetwood, M. A. Gehlhausen, T. L. Turflinger, D. B. Brown, and A. H. Johnston, "A Proposed Hardness Assurance Test Methodology for Bipolar Linear Circuits and Devices in a Space Ionizing Radiation Environment," *IEEE Trans. Nucl. Sci.*, vol. 44, pp. 1981-1988, 1997.

[54] R. L. Pease, J. Krieg, M. Gehlhausen, D. Platteter, and J. Black, "Evaluation of Proposed Hardness Method for Bipolar Linear Circuits with Enhanced Low Dose Rate Sensitivity (ELDRS)," *IEEE Trans. Nucl. Sci.*, vol. 45, pp. 2665-2672, 1998.

[55] A. H. Johnston, "Super Recovery of Total Dose Damage in MOS Devices," *IEEE Trans. Nucl. Sci.*, vol. NS-31, pp. 1427-1433, 1984.

[56] J. R. Schwank, P. S. Winokur, P. J. McWhorter, F. W. Sexton, P. V. Dressendorfer, and D. C. Turpin, "Physical Mechanisms Contributing to Device 'Rebound'," *IEEE Trans Nucl. Sci.*, vol. 31, pp. 1434-1438, 1984.

[57] P. S. Winokur, F. W. Sexton, J. R. Schwank, D. M. Fleetwood, P. V. Dressendorfer, T. F. Wrobel, and D. C. Turpin, "Total-Dose Radiation and Annealing Studies: Implications for Hardness Assurance Testing," *IEEE Trans. Nucl. Sci.*, vol. NS-33, pp. 1343-1351, 1986.

[58] P. S. Winokur, F. W. Sexton, G. L. Hash, and D. C. Turpin, "Total-Dose Failure Mechanisms of Integrated Circuits in Laboratory and Space Environments," *IEEE Trans. Nucl. Sci.*, vol. NS-34, pp. 1448-1454, 1987.

[59] T. R. Oldham, A. J. Lelis, H. E. Boesch, Jr., J. M. Benedetto, F. B. McLean, and J. M. McGarrity, "Post-Irradiation Effects in Field-Oxide Isolation Structures," *IEEE Trans. Nucl. Sci.*, vol. 34, pp. 1184, 1987.

[60] D. M. Fleetwood, P. S. Winokur, and J. R. Schwank, "Using Laboratory X-Ray and Cobalt-60 Irradiations to Predict CMOS Device Response in Strategic and Space Environments," *IEEE Trans. Nucl.Sci.*, vol. NS-35, pp. 1497-1505, 1988.

[61] D. M. Fleetwood, P. S. Winokur, L. C. Riewe, and R. L. Pease, "An Improved Standard Total Dose Test for CMOS Space Electronics," *IEEE Trans. Nucl. Sci.*, vol. NS-36, pp. 1963-1970, 1989.

[62] D. M. Fleetwood, L. C. Riewe, J. R. Schwank, S. C. Witczak, and R. D. Schrimpf, "Radiation Effects at Low Electric Fields in Thermal, SIMOX, and Bipolar-Base Oxides," *IEEE Trans. Nucl. Sci.*, vol. 43, pp. 2537-2546, 1996.

[63] D. M. Fleetwood, M. J. Johnson, T. L. Meisenheimer, P. S. Winokur, W. L. Warren, and S. C. Witczak, "1/f noise, hydrogen transport, and latent interface-trap buildup in irradiated MOS devices," *IEEE Trans. Nucl. Sci.*, vol. 44, pp. 1810-1817, 1997.

[64] R. J. Graves, C. R. Cirba, R. D. Schrimpf, R. J. Milanowski, A. Michez, D. M. Fleetwood, S. C. Witczak, and F. Saigne, "Modeling Low-Dose-Rate Effects in Irradiated Bipolar-Base Oxides," *IEEE Trans. Nucl. Sci.*, vol. 45, pp. 2352-2360, 1998.

[65] S. N. Rashkeev, C. R. Cirba, D. M. Fleetwood, R. D. Schrimpf, S. C. Witczak, A. Michez, and S. T. Pantelides, "Physical Model for Enhanced Interface-Trap Formation at Low Dose Rates," *IEEE Trans. Nucl. Sci.*, vol. 49, pp. 2650-2655, 2002.

[66] R. K. Freitag and D. B. Brown, "Low dose rate effects on linear bipolar IC's: Experiments on the time dependence," *IEEE Trans. Nucl. Sci.*, vol. 44, pp. 1906-1913, 1997.

[67] R. K. Freitag and D. B. Brown, "Study of Low-Dose-Rate Radiation Effects on Commercial Linear Bipolar ICs," *IEEE Trans. Nucl. Sci.*, vol. 45, pp. 2649-2658, 1998.

[68] M. R. Shaneyfelt, J. R. Schwank, S. C. Witczak, D. M. Fleetwood, R. L. Pease, P. S. Winokur, L. C. Riewe, and G. L. Hash, "Thermal-stress effects and enhanced low dose rate sensitivity in linear bipolar ICs," *IEEE Trans. Nucl. Sci.*, vol. 47, pp. 2539-2545, 2000.

[69] M. R. Shaneyfelt, R. L. Pease, J. R. Schwank, M. C. Maher, G. L. Hash, D. M. Fleetwood, P. E. Dodd, C. A. Reber, S. C. Witczak, L. C. Riewe, H. P. Hjalmarson, J. C. Banks, B. L. Doyle, and J. A. Knapp, "Impact of passivation layers on enhanced low-dose-rate sensitivity and pre-irradiation elevated-temperature stress effects in bipolar linear ICs," *IEEE Trans. Nucl. Sci.*, vol. 49, pp. 3171-3179, 2002.

International Journal of High Speed Electronics and Systems
Vol. 14, No. 2 (2004) 519–541

World Scientific
www.worldscientific.com

TOTAL DOSE EFFECTS IN LINEAR BIPOLAR INTEGRATED CIRCUITS

H. J. BARNABY

The University of Arizona
Tucson, Arizona, USA, 85721

Abstract. Electronics systems that operate in space or strategic environments can be severely damaged by exposure to ionizing radiation. Space-based systems that utilize linear bipolar integrated circuits are particularly susceptible to radiation-induced damage because of the enhanced sensitivity of these circuits to the low rate of radiation exposure. The phenomenon of enhanced low-dose-rate sensitivity (ELDRS) demonstrates the need for a comprehensive understanding of the mechanisms of total dose effects in linear bipolar circuits. The majority of detailed bipolar total dose studies to date have focused on radiation effects mechanisms at either the process or transistor level. The goal of this text is to provide an overview of total dose mechanisms from the circuit perspective; in particular, the effects of transistor gain degradation on specific linear bipolar circuit parameters and the effects of circuit parameter degradation on select linear bipolar circuit applications.

Keywords: Bipolar junction transistor, linear bipolar circuit, operational amplifier, voltage comparator, voltage regulator, total ionizing dose, enhanced low-dose-rate sensitivity, current gain, base current, Co-60 γ-rays, irradiation bias, input transistor, input bias current, input offset voltage, input offset current, maximum output drive current

1. Introduction

In 1962, radiation induced degradation in a discrete bipolar junction transistor (BJT) was identified as the cause of failure in the newly-deployed communications satellite Telstar [1]. At that time the primary failure mechanism was determined to be related to the ionization of an unspecified gas within the encapsulated part's package [2]. The excess charge generated by ionization was transported to the device surface, causing deterioration in the transistor's electrical characteristics. Although gas-filled packages are still used today (most hermetic packages are back-filled with nitrogen), ionized gas has rarely been identified as the primary failure mechanism in irradiated bipolar parts. Instead, radiation-induced ionization in the oxide layers, which are used ubiquitously in conventional bipolar processes, is the primary source of concern for the long-term reliability of systems that operate in harsh radiation environments and utilize BJTs and linear bipolar circuits.

Modern satellite and strategic systems rely heavily on linear bipolar parts for numerous subsystem functions, including signal amplification, voltage regulation, and the conversion of information between analog and digital. Among the most widely used linear part types in these systems are: operational amplifiers, voltage comparators, and voltage regulators. Linear bipolar circuits manufactured in conventional processes are especially attractive to both system designers and program managers because of their versatility and low cost. Indeed, a standard commercial-off-the-shelf quad operational amplifier can be purchased for about $1.

Designers of systems operating in space and/or nuclear environments must balance these advantages with the sensitivity of bipolar components to several radiation threats. The class of significant radiation effects on linear bipolar circuits includes single event transients (time-dependent variations in the voltage or current at an analog circuit's external ports that are caused by the passage of heavy ions through a sensitive node) [3] and displacement damage (degradation to electrical characteristics caused by radiation induced damage to the semiconductor bulk) [4-6]. However, since its first observation on the Telstar satellite over four decades ago, radiation-induced ionization damage to bipolar components has generally been considered the most significant threat to linear bipolar circuits operating in radiation environments.

The numerous physical mechanisms of total ionizing dose (TID) damage in bipolar components have been investigated for over four decades [7]. The research has ranged from the study of the basic physics of radiation defect buildup and transistor response [8-21] to the identification, analysis, and characterization of circuit-level response mechanisms [22-32]. Research in the past decade has primarily focused on the phenomenon of enhanced low-dose-rate sensitivity (ELDRS). In 1991, it was reported that the amount of radiation-induced parametric degradation in a bipolar device was dependent on the dose rate at which the TID accumulated, with more degradation occurring at lower dose rates [12, 33]. Since there exist radiation environments (e.g., space) where electronic components are irradiated at low dose rates, ELDRS is a real problem for systems designed with linear bipolar circuits. Indeed, the costs in time and resources for qualifying bipolar parts to operate in low-dose-rate environments are often prohibitive. The data in Figure 1 illustrate ELDRS for various linear bipolar circuits. As the figure indicates, the relative amount of damage, normalized to a dose rate of 50 rad(Si)/s, increases for most of the parts as the dose rate is decreased. In this data set, all of the parts have been irradiated to 50 krad(Si). The data also show that ELDRS is a function of part type. The researchers who first reported these data in 1994 correlated the polarity of sensitive transistors in the circuit (i.e., pnp or npn BJTs) with the circuit's low-dose-rate sensitivity. It was suggested that linear bipolar parts designed with pnp devices in sensitive regions of the circuit topology were likely to show more ELDRS than other parts [25]. While transistor polarity continues to be considered a mechanism for TID effects and ELDRS in bipolar circuits, in the past few years the origins of low-dose-rate sensitivity have begun to be related more to fabrication variables, such as the post-processing thermal cycles and the type and quality of circuit passivation materials [34-36].

Concern regarding ELDRS in BJTs has been the driving force behind TID research in linear bipolar circuits. Understanding response mechanisms related to temperature stress and post-irradiation annealing have also been areas of active research. However, this paper will focus on describing TID mechanisms at circuit and application levels rather than investigating dose rate, thermal, or annealing effects in bipolar devices. These discussions will be supported by specific examples. The text is organized into three sections: 1) an overview of total dose effects in bipolar devices and circuits, 2) a description and analysis of total dose effects on the critical specification parameters of operational amplifiers and voltage comparators, and 3) a description and analysis of total dose effects on the critical specification parameters of voltage regulators. The latter two sections will consider how radiation-induced degradation in circuit parameters can affect the performance of linear circuit applications as well as examine how parametric response to TID is affected by circuit topology, bias conditions, and other design variables. The discussions will be supported by detailed analyses of the TID degradation mechanisms of three commercially available linear bipolar circuits: the LM101 operational amplifiers, the LM111 voltage comparator, and the LM117 voltage regulator. These circuits were selected for analysis primarily because the methods used to test these parts and their unique radiation responses provide insights into total dose effects in most conventional linear bipolar integrated circuits.

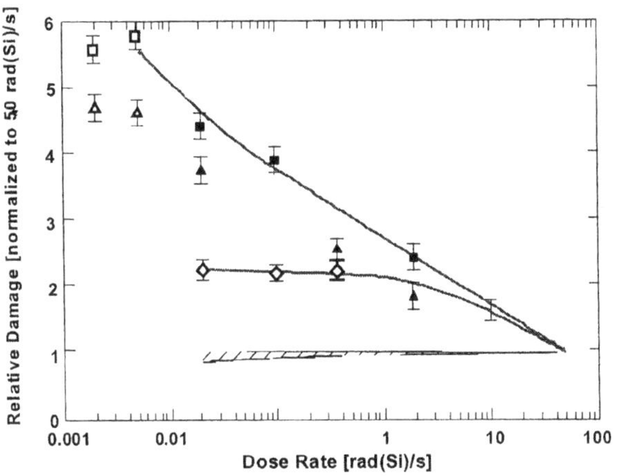

Figure 1. Relative damage vs. dose rate for irradiation to 50 krad(Si) for several linear ICs [25].

2. Overview of Total Ionizing Dose Effects

A. Bipolar Junction Transistors

Total ionizing dose effects in bipolar technologies are primarily characterized by the buildup of two types of defects within an integrated circuit's oxide layers (fixed net positive-oxide-trapped-charge) or at the semiconductor-oxide interface (interface traps) [2]. The principal result of these defects is a reduction of common emitter current gain (β) of the BJTs integrated into the linear bipolar circuit. Common emitter gain, the primary figure of merit for bipolar devices, is defined as the ratio of collector current (I_C) to base current (I_B), i.e., $\beta = I_C/I_B$.

The curves in Figure 2 exemplify the effects of ionization defects on transistor current gain. These data show the degradation in current gain vs. base-emitter voltage for a vertical NPN BJT exposed to various total dose levels [15]. The primary cause of reduced current gain in irradiated BJTs (of both polarities) is an increase in base current (I_B) due to increased recombination via the radiation-induced oxide defects [9, 15, 18]. In npn transistors, TID has a negligible impact on collector current [15]. This is indicated on the Gummel curves (I_C and I_B vs. base-emitter voltage) in Figure 3. These data correspond to those presented in Figure 2. Studies have demonstrated that the collector current in pnp transistors can show small but measurable reductions upon radiation exposure [21, 37]

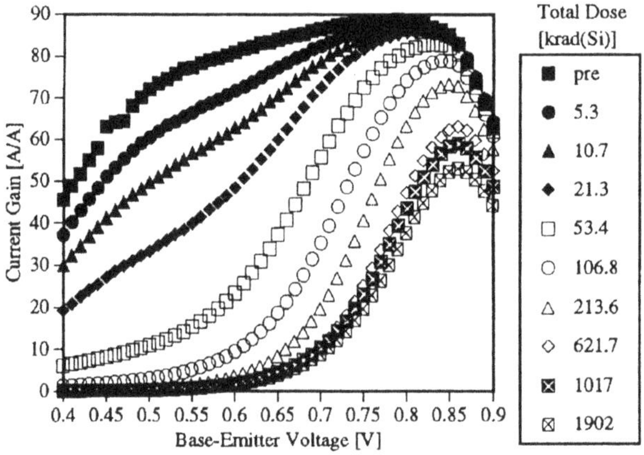

Figure 2. Current gain vs. base-emitter voltage for NPN BJTs exposed to various levels of TID [15].

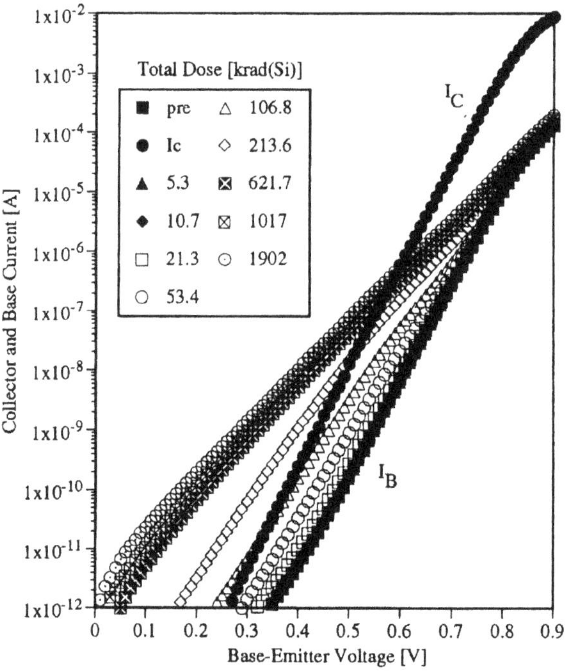

Figure 3. I_C and I_B vs. base-emitter voltage (Gummel curves). Data correspond to the gain curves in Figure 2 [15].

B. Operational Amplifiers and Voltage Comparators

Reduced current gain can have a significant impact on critical linear bipolar circuit parameters. For operational amplifiers and voltage comparators, the most sensitive specification parameters to TID are typically: input bias current (I_b^+ and I_b^-), input offset current (I_{os}), and input offset voltage (V_{os}). The two input bias currents and the offset voltage are shown as independent sources in the idealized amplifier/comparator model in Figure 4. I_b^+ and I_b^- are modeled as ideal current sources from the non-inverting and inverting input nodes (V^+ and V^-), respectively, and V_{os} is an ideal voltage source between the actual, external non-inverting node (V^+) and the ideal non-inverting node (V_I). The input offset current is not represented on the figure because it is simply the magnitude of the difference between I_b^+ and I_b^-.

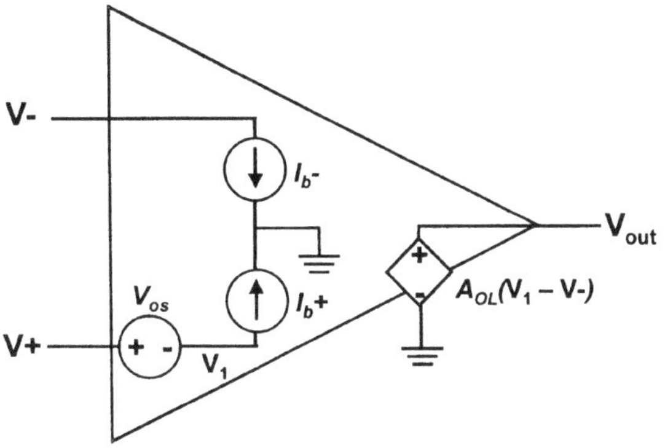

Figure 4. Circuit model of operational amplifier/voltage comparator. DC current sources represent references for input bias currents and dc voltage source represents reference for input offset voltage.

All linear components designed with bipolar differential amplifier input stages (e.g., operational amplifiers and voltage comparators) require some input bias current to operate. However, these currents are designed to be very low and matched ($I_{os} \approx 0$A) in order to achieve high circuit input impedance and prevent signal distortions (due to non-linearities) and loss of dynamic range (due to offsets). For standard commercial-off-the-shelf bipolar amplifier/comparator circuits, a nominal value for input bias current is 10 nA. Maximum specifications are rarely higher than 100 nA [38]. Exposure to ionizing radiation often causes significant increases in input bias current. This effect is observed in Figure 5 [29] in which the input bias current for LM124 operational amplifiers is plotted vs. total dose at various dose rates. In these tests, irradiations were performed with a Co-60 γ-ray source. The figure indicates that, even at a standard laboratory dose rate of 100 rad(Si)/s, input bias current increases above its maximum specification (50 nA) before the exposure level reaches 100 krad(Si). The data also demonstrate the LM124 exhibits enhanced, low-dose-rate sensitivity. Indeed, at the lowest dose rate (1 mrad(Si)/s), the part reaches its specification limit before 10 krad(Si).

A monotonic increase in bias current is typically a signature of gain degradation in a circuit's input transistors. In most bipolar operational amplifiers and voltage comparators, the input bias currents *are* the base currents of the respective input transistors. Thus, if other dc operating point voltages and currents remain fixed as a function of total dose, radiation-induced increases in input transistor base current will lead directly to a monotonic increase in input bias current at the circuit level. In a subsequent section of this paper, it will be shown that input bias current does not always increase monotonically with total dose. In some circuits, the topology is such that gain degradation in non-input BJTs can also have an impact on the bias current degradation.

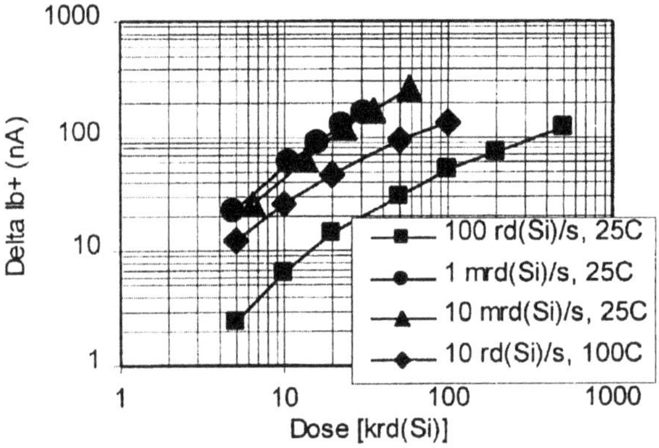

Figure 5. LM124 average increase in I_b+ vs. dose [29]

Ionizing radiation can also affect the offset parameters, V_{os} and I_{os}. Offset errors can further affect the functionality of circuit applications, particularly in mixed-signal applications that require large dynamic ranges (e.g., bit resolution in analog-to-digital converters). Radiation-induced drifts in offset parameters are often signatures of an imbalance in the rate of degradation in BJTs designed to operate as matched pairs. For example, if the current gain in an emitter-coupled input transistor pair (of an amplifier/comparator input stage) degrades unevenly such that there exists a non-negligible gain differential, $\Delta\beta$, then the input bias current offset will degrade to

$$I_{os} = I_b\left(\frac{\Delta\beta}{\beta}\right), \tag{1}$$

where I_b is the average input bias current $((I_b^- + I_b^+)/2)$ [39]. Non-uniform degradation arises when different dc bias conditions are applied to matched transistors during radiation exposure [29, 40]. This effect is illustrated in Figure 6, where the increase in the input offset current of a LM139 voltage comparator is plotted vs. total dose for different dose rates. In this experiment, the parts were exposed in Co-60 sources under large differential input voltage bias conditions. As with the I_b data of the LM124, I_{os} for the biased LM139 exhibits a low-dose-rate effect.

Degradation of input offset voltage is strongly dependent on irradiation bias conditions. However, unlike input offset current, a deterioration in V_{os} is most likely not dependent on gain degradation in the input transistors of amplifiers and comparators. Although the exact causes of radiation-induced offset voltage degradation have yet to be precisely identified, likely mechanisms include: 1) unbalanced gain degradation in transistors used to form the active loads for the input stage differential amplifier(s) [39] and 2) unbalanced degradation in the devices used to form the circuit's output stage. The susceptibility of the output stage amplifier to biased irradiations will be discussed in detail in section 3.

The effect of total dose irradiations on the V_{os} parameter is shown in Figure 7. The figure plots the increase in offset voltage vs. total dose for an LM111 voltage comparator, biased with a large differential input voltage during radiation exposure. As with the previous radiation responses shown in Figures 5 and 6, the V_{os} parameter of this circuit also demonstrates ELDRS.

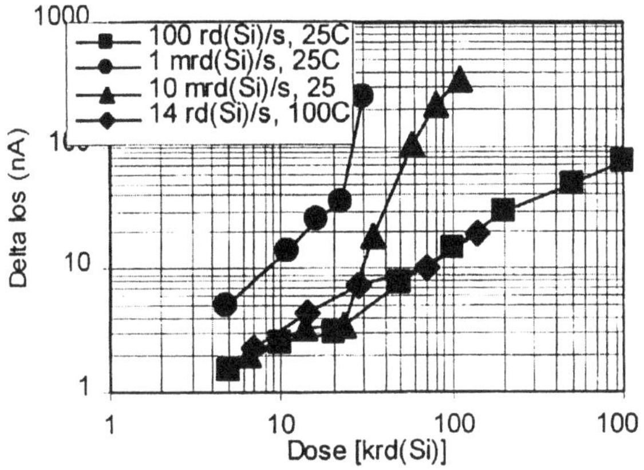

Figure 6. LM139 average increase in I_{os} vs. dose [29].

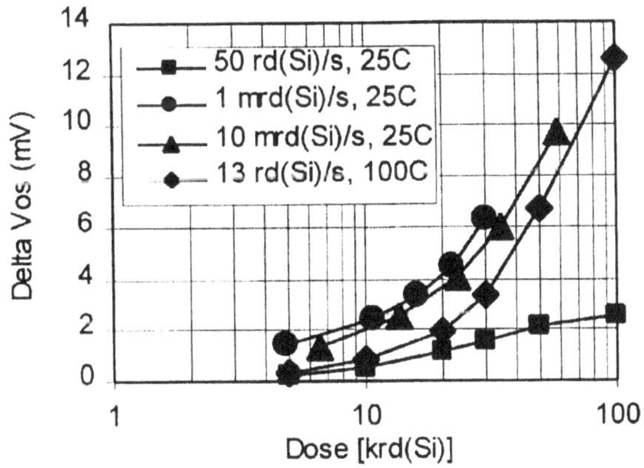

Figure 7. LM111 average increase in V_{os} vs. total dose [29].

C. Voltage Regulators

The function of a voltage regulator is to provide a constant voltage, V_{out}, to a load, I_o. In many regulator circuits, an unregulated, potentially variable voltage V_{in} serves as both the regulator input and one of the power supply rails. Voltage regulators are commonly used in ac-to-dc converter applications such as power supplies. Their function in these applications is to remove time-dependent ripples in voltage signals in order to generate dc voltages that are stable with input voltage, load, and temperature [39]. A typical voltage regulator application is shown in Figure 8. In this adjustable configuration, the output voltage is set by an internal band-gap voltage reference (V_{ref}) and an adjustable voltage divider determined by external resistors R_x and R_y [29].

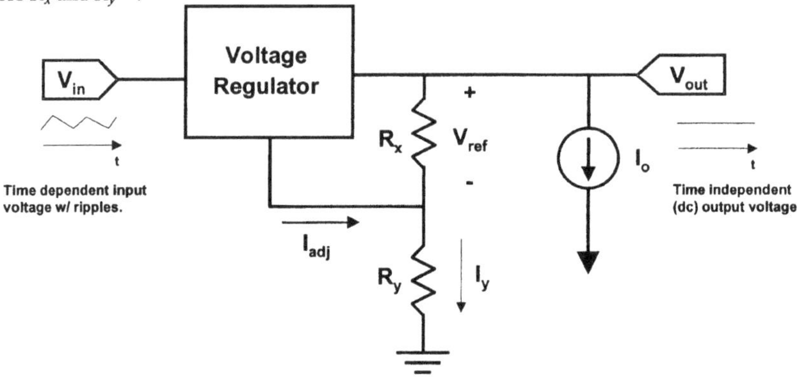

Figure 8. Typical adjustable voltage regulator configuration [38]

In normal operation, the adjust pin current is small ($I_{adj} \approx 50$ uA) relative to the current through resistor R_y (1 mA < I_y < 4 A), and thus the relationship between the output and reference voltage can be approximated as

$$V_{out} = V_{ref} \frac{R_x + R_y}{R_x}. \tag{2}$$

Degradation in the regulated output voltage is caused by radiation-induced alterations to the critical regulator parameters, I_{adj}, V_{ref}, and the maximum output drive current. Figure 9 plots the change in I_{adj} as a function of total dose for LM117 voltage regulators exposed to Co-60 γ-rays at a dose rate of 150 rad(Si)/s. The adjust current response to TID is non-monotonic with dose. In a 1996 study on the LM117, it was determined that the behavior of I_{adj} with dose was a function of multiple BJT degradation mechanisms in the four transistors that constitute the internal bandgap reference (i.e., gain degradation in the two lateral pnp devices and both gain degradation and collector-to-emitter leakage in the two vertical npn devices [22]).

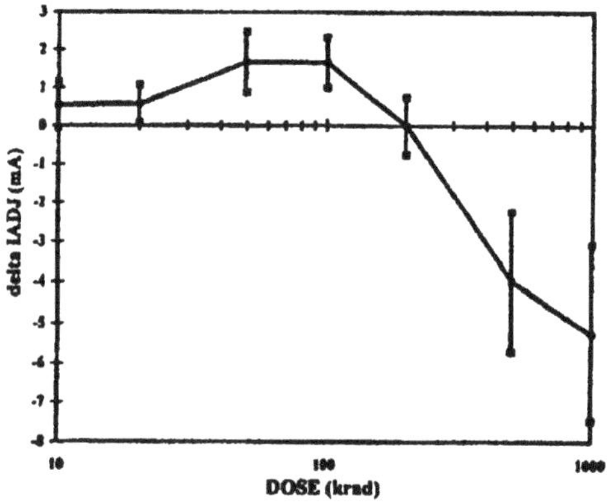

Figure 9. Adjust pin current vs. total dose for LM117 adjustable voltage regulator [22].

In Figure 10, the change in V_{ref} is plotted vs. total dose for an HS117 voltage regulator, for biased and unbiased parts, irradiated at two different dose rates. The parts were irradiated with Co60 γ-rays at dose rates of 61, 0.25 (high temperature), and 0.05 rad/s. These data, originally reported in 1998, indicate the HS117 circuit exhibits a significant ELDRS effect, but only when no bias is applied during radiation exposure [41]. This response is consistent with the prevailing model for low-dose-rate effects in bipolar parts, which suggests a low electric field across a BJT oxide is a necessary condition for ELDRS [14]. This condition exists when a part is unbiased since there exist no externally applied voltages across the any of the circuit's oxide layers that are ionized during radiation exposure.

Voltage regulators are designed to maintain stable output voltages while driving a range of load currents. Minimum load currents are nominally between of 1 and 10 mA. Maximum output drive currents ($I_o(max)$) range between 0.5 and 4 A. Loads above $I_o(max)$ will result in a drop in V_{out} (loss of regulation). The ability of a voltage regulator to supply current to large loads is dependent on the current gain of the regulator circuit's output transistor(s). Thus, if the output device(s) is susceptible to TID gain degradation, $I_o(max)$ will show significant deterioration, as illustrated in Figure 11 [42]. This data set, obtained from measurements on Co-60 irradiated 29372 low-dropout voltage regulators, shows $I_o(max)$ to be both total dose and dose rate sensitive. The output transistor of the 29372 circuit is a large multiple emitter lateral pnp transistor. Numerous studies on radiation effects on lateral pnp (lpnp) BJTs have shown these devices to be particularly vulnerable to gain degradation from TID [9, 18, 31]. This is due to the fact that, as they are lateral structures, the primary path for current in lpnp BJTs is near the semiconductor-oxide interface, in close proximity to radiation-induced defects [9].

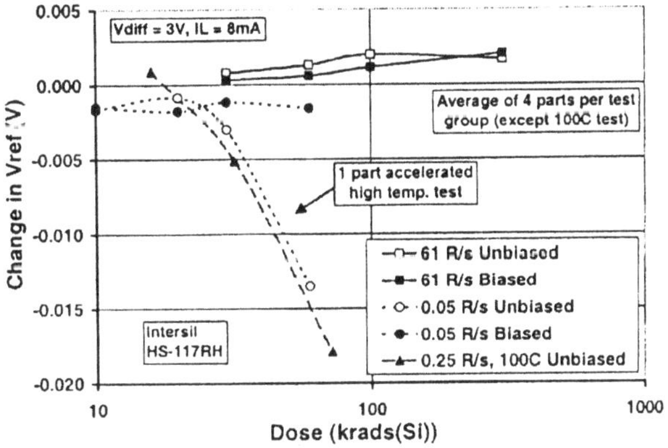

Figure 10. Change in V_{ref} vs. dose for an HS117 voltage regulator exposed at high and low-dose-rates at two bias conditions (biased and unbiased) [41].

In this section we have attempted to provide a general overview of TID effects in bipolar devices and circuits. In bipolar transistors, exposure to TID reduces the current gain primarily by increasing the base recombination current. Because of mechanisms related to the low electric fields across bipolar oxide layers, parametric degradation in BJTs is often enhanced when devices are exposed at low dose rates. Linear circuits fabricated with BJTs (e.g., operational amplifiers, voltage comparators, and voltage regulators) are, by extension, also susceptible to radiation-induced parametric degradation and ELDRS. These effects are illustrated in the data sets presented above.

In the next two sections, TID effects on critical circuit specification parameters for selected linear bipolar parts will be discussed. These discussions will include examples of how radiation-induced parametric degradation at the circuit-level may effect subsystem applications and approaches used in identifying TID degradation mechanisms at the circuit-level.

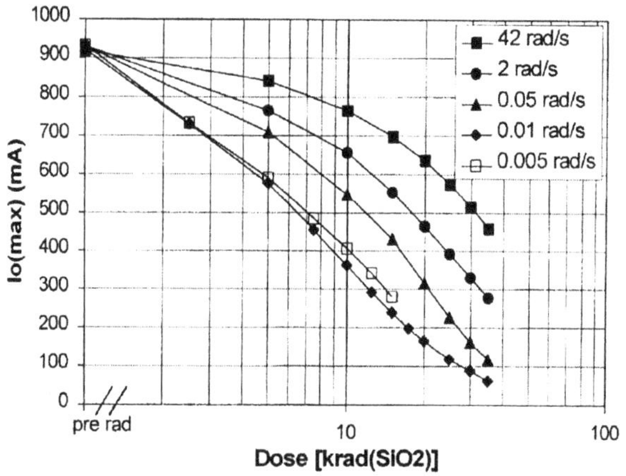

Figure 11. $I_o(max)$ vs. dose for a 29372 low-dropout regulator at several dose rates, showing low-dose-rate sensitivity [42].

3. Total Ionizing Dose Effects on Operational Amplifiers and Voltage Comparators

A. Effects of Input Bias Current Degradation on a Current-to-Voltage Converter

Figure 12 shows the schematic for a widely-used operational amplifier application: the current-to-voltage (IV) converter. Circuits similar to the IV converter can be found in many mixed signal applications, such as R2R ladder D/A converters. In this application, the current into or out of a device-under-test, I_{DUT}, is converted to a voltage, V_O, so that it may be sampled digitally. The ideal relationship between the I_{DUT} and V_O is

$$V_O = -I_{DUT} R_{fb}, \tag{3}$$

where R_{fb} is the feedback resistor. If the digital electronics of a system are capable of sampling voltages between 5 mV and 5V (e.g., a 12 bit A/D converter), a 100 kΩ feedback resistor will enable the measurement of currents between 50 nA (5 mV) and 50 uA (5 V). If I_{DUT} is equal to 100 nA, the converter will ideally generate an output voltage of 10 mV. This sample voltage can then be used to calculate I_{DUT} using Equation 3. This ideal operation is illustrated schematically in Figure 11a. If the input bias current of the operational amplifier increases significantly, it will cause the IV converter to have a measurable offset error:

$$\Delta V_O = \Delta I_b R_{fb}. \tag{4}$$

This will cause the measured output voltage to be reduced to $V_O - \Delta V_O$, and unless circuitry is built into the system to correct for this continuously degrading offset, the current will be computed as $I_{DUT} - \Delta I_b$. Input bias current degradation of over 50 nA, which is possible after 10 krad of exposure at a 1 mrad/s dose rate – or less than four months in space [30] – will make it impossible to be more than 50 % accurate in current measurements below 100 nA.

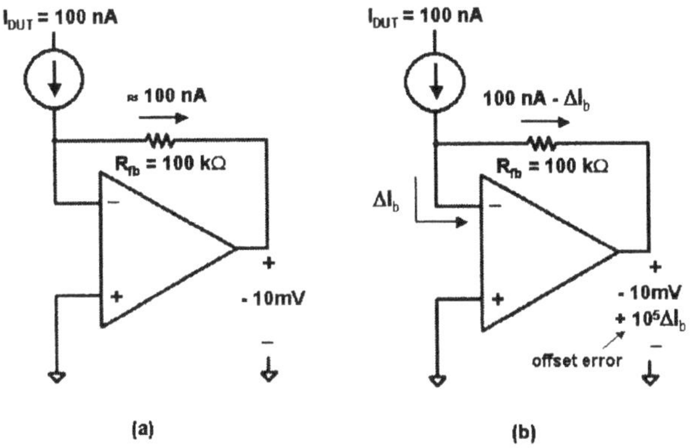

Figure 12. Schematics of current-to-voltage converter applications show a) ideal operation and b) operation with offset error which is increased by radiation-induced input bias current degradation.

Radiation-induced offset errors and loss of dynamic range are very common in mixed signal applications that rely on low input bias current in their operational amplifiers. Problems associated with input bias current degradation may be avoided by using linear bipolar circuits with JFET transistor inputs [40].

B) Radiation Response of Input Bias Current in an LM111 Voltage Comparator

Most radiation response measurements of bipolar operational amplifiers and voltage comparators have indicated input bias current increases monotonically with TID exposure [24, 30, 40, 43]. However, TID data on LM111 voltage comparators reported in 1999 showed that input bias current exhibited a downward trend or "recovery" for doses above 100 krad(SiO_2) [23]. Figure 13 shows the input bias current radiation response characteristics for Co-60-irradiated LM111 voltage comparators at two dose rates. Once again, this linear bipolar circuit exhibits enhanced low-dose-rate sensitivity.

The authors of this LM111 study demonstrated, through electrical measurements on isolated circuit transistors, parameter extraction, and SPICE simulations, that increased input bias current (I_{IB}) was due to degradation in the circuit's input transistors and that "recovery" was due to gain degradation in *non-input* devices integrated into the LM111 internal circuitry [23]. Figure 14a shows the equivalent, single-ended subcircuit schematic for the differential input stage of the LM111 comparator. Figure 14b shows, along with the high dose rate data, the simulated subcircuit I_{IB} characteristics with total dose. The figure demonstrates that the simulations reproduce the non-monotonic characteristics of the experimental data [23]. A detailed breakdown of the numerous circuit mechanisms underlying the radiation response of the LM111 input bias current was presented by Barnaby *et al.* in their 1999 paper. A brief summary of their conclusions is as follows:

Under some basic assumptions, the change in I_{IB} can be approximated as

$$\Delta I_{IB} \cong K\left[\frac{I_{SE1(pre)}\Delta V_{eb1}}{n_e V_T} + \Delta I_{SE1}\left(1 + \frac{\Delta V_{eb1}}{n_e V_T}\right)\right], \tag{5}$$

where K is a proportionality constant, ΔV_{ebl} is the radiation-induced change in the forward bias across the emitter base junction of Q1 (this is assumed to be 50 mV or less), n_e is the base-emitter leakage ideality factor of Q1, $I_{SE1(pre)}$ is the pre-irradiated value for the base-emitter leakage current of Q1, and ΔI_{SE1} is the radiation-induced change in the leakage current for Q1 [44]. ΔI_{SE1} increases monotonically with total dose. This is caused by increased carrier recombination within the emitter-base depletion region of the input device, Q1 [9, 15, 18]. As indicated by Equation 5, increasing ΔI_{SE1} in turn increases I_{IB}. Therefore, gain degradation in the LM111 input structures causes the monotonic increase in I_{IB} below 100 krad(SiO2).

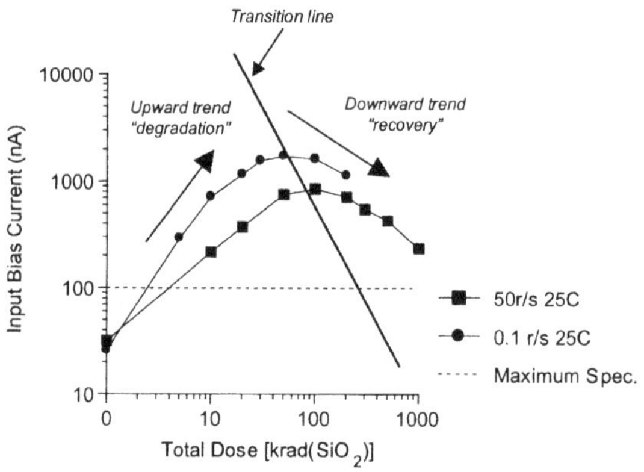

Figure 13. Input bias current radiation response for LM111 comparator at 50 and 0.1 rad(SiO_2)/s. These data indicate ELDRS and parametric "recovery" at high doses. The maximum specification is 100 nA [23].

The "recovery" in the response at doses above 100 krad(SiO_2) was determined to be caused by the effects of gain degradation in transistors Q2 and Q3. Degradation in Q3 leads to an increase in I_{b3}. Degradation in Q2 leads to a decrease in I_{c2}. The combined effect of radiation on these two dc current components steals collector current away for the input transistor Q1. This causes V_{ebl} to decrease and, as indicated in Equation 5, causes a reduction in I_{IB}. gain degradation in non-input devices thus compensates for increased base current in the LM111 input device and leads to a partial recovery in the input bias current response at high doses [23].

One of the most important conclusions that can be drawn from this analysis is that TID degradation mechanisms at the circuit level can be significantly more complex than the responses of individual transistors. Analysis and separation of these higher-order circuit mechanisms typically require pre- and post-irradiation transistor parameter extraction and SPICE simulation. The scanning electron microscope has also been used extensively to characterize circuit-level TID mechanisms.

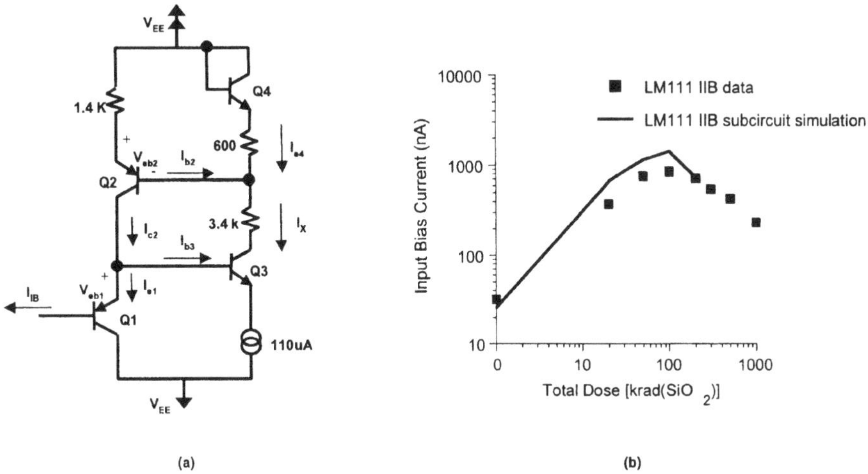

Figure 14. (a) Equivalent, single-ended subcircuit schematic for the differential input stage of the LM111 comparator and (b) simulated and experimental input bias current radiation response. The simulations model the upward ("degradation") trend at low doses and subsequent downward ("recovery") trend at high doses [23].

C. S.E.M. Irradiation of LM101 Operational Amplifier

As mentioned previously, irradiation bias can have a significant impact on the radiation response of bipolar devices and circuits (e.g., Figure 10). Moreover, as demonstrated above, parametric degradation in linear bipolar circuits may be the result of complex interactions between several degraded devices. Understanding how the radiation response of individual devices and subcircuits contribute to the overall circuit response is in many instances a difficult task that is further complicated by additional variables such as irradiation bias conditions. Several studies have used scanning electron microscopes (SEMs) to help separate and characterize circuit-level TID mechanisms [31, 45]. In this section, SEM irradiation data on LM101 operational amplifiers is discussed.

As long as the electron energy is set to an appropriate level, a SEM can cause ionization damage that may be correlated to the damage caused by high-energy photons (e.g., Co-60 γ-rays). In the SEM data presented below, the electron energy was set to 40 keV. The electron flux was set to $1.88\text{x}10^{10}$ e/cm^{2}s, which was equivalent to a dose rate of 2.5 krad(Si)/s (i.e., $7.5\text{x}10^6$ e/cm^{2}s – 1rad(Si)).

Figures 15 through 18 plot the increased input bias current (ΔI_{IB}) and offset voltage (ΔV_{OS}), for LM101 operational amplifiers irradiated by 40 keV SEM electrons. For the data shown in Figure 15, whole LM101 chips were exposed without an externally applied bias. Under these conditions, ΔI_{IB} increases linearly with total dose to a level well above the maximum specification of 75 nA after 1 Mrad equivalent of electron irradiations, while ΔV_{OS} does not increase significantly (less than 5% of the 2 mV specification). Figure 16 shows the radiation response of whole chips biased during SEM exposures. The results are similar to those shown in Figure 15 up to 100 krad(SiO2). However, above 100 krad(SiO_2), there is a sharp increase in ΔV_{OS} and a significant upward shift in ΔI_{IB}.

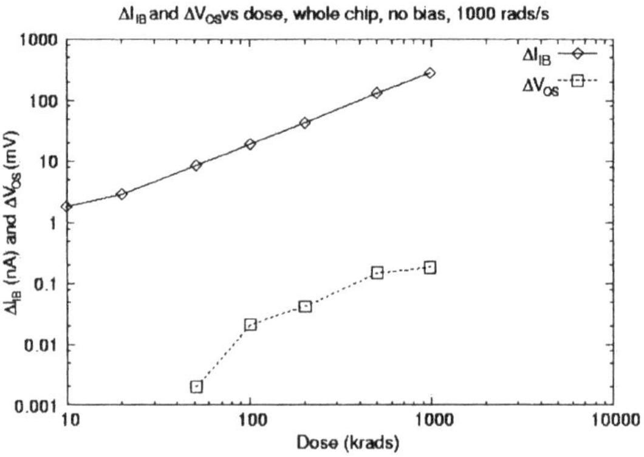

Figure 15. Increased input bias current and offset voltage in unbiased LM101 operational amplifiers irradiated by 40 keV SEM electrons. Data show bias current degrades and offset voltage does not degrade in response to exposure.

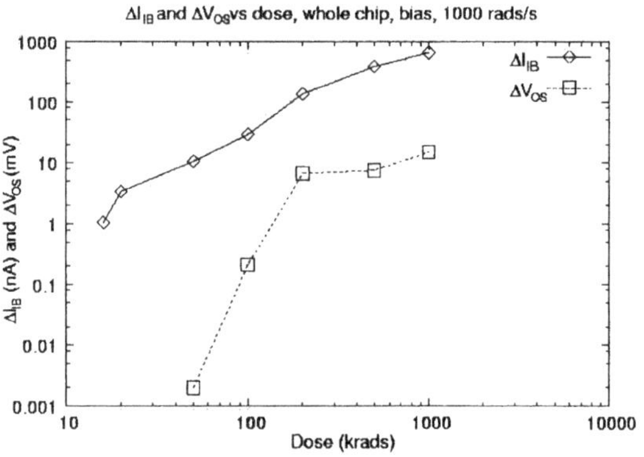

Figure 16. Increased input bias current and offset voltage in biased LM101 operational amplifier irradiated by 40 keV SEM electrons. Data show bias current degrades and offset voltage does not degrade in response to exposure.

In order to identify circuit-level mechanisms that underlie the bias dependence of the operational amplifier's radiation response, localized SEM irradiations were performed on two subcircuits of the LM101: the input npn transistor emitter-coupled-pair and the output stage. In Figure 17, the parametric responses are plotted for localized irradiations on the input transistors. In Figure 18, ΔI_{IB} and ΔV_{OS} are plotted for localized irradiations on the LM101 output stage.

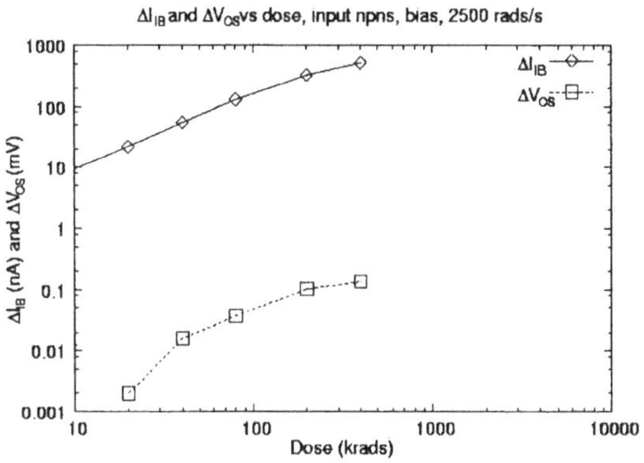

Figure 17. Increased input bias current and offset voltage in biased localized irradiations of the input transistors of LM101 operational amplifiers. Data show bias current degrades and offset voltage does not degrade in response to input transistor exposure.

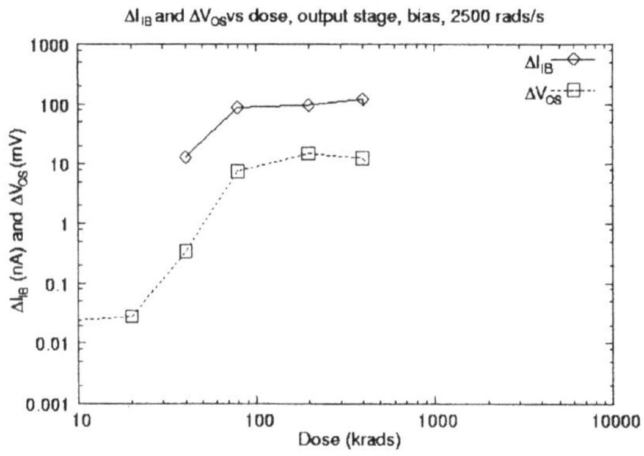

Figure 18. Increased input bias current and offset voltage in biased localized irradiations of the output stage of LM101 operational amplifiers. Data show both bias current degrades and offset voltage after 100 krad(SiO_2).

The results of these localized SEM irradiations effectively demonstrate numerous TID mechanisms for the LM101 circuit. First of all, the response to localized irradiations on the input transistors demonstrates gain degradation in these devices causes deterioration in ΔI_{IB} but not in ΔV_{OS}. Second, the fact that ΔI_{IB} degraded similarly when either the whole chip was irradiated under no bias, or when only the input devices were irradiated under bias, indicates that the gain degradation in input transistors does not depend significantly on bias. Lastly, above 100 krad(Si), gain degradation in the BJTs of the output stage caused an abrupt increase in ΔV_{OS} , as well as additional degradation in ΔI_{IB}. The results confirm offset voltage is dependent on irradiation bias and is only a function of degradation in *non-input* structures. An additional and surprising result is that input bias current shows additional degradation due to deterioration in output stage components. All of these results are consistent with claims made in section 2*B* regarding the circuit-level mechanisms of parametric degradation in operational amplifiers and voltage comparators. These experimental results demonstrate that the identification of the complex interactions between individual circuit-level radiation response mechanisms can be facilitated by the use of localized insitu SEM irradiations.

4. Total Ionizing Dose Effects on Voltage Regulator

A. Effects of $I_o(max)$ Degradation on Mixed Signal Regulator Application

Voltage regulators are used extensively in mixed signal applications. Figure 19 shows an application used for generating digitally selected regulated output voltages [38]. The regulator in this circuit is the 29372 low-dropout regulator. The output voltage, V_{out}, is fixed by inputs *d2*, *d1*, and *d0*. Small divider resistors are chosen to minimize output resistance ($R2 = 25\Omega$, $R1 = 6\Omega$). Figure 20 illustrates the ideal response of V_{out} and I_o to the four unique digital input sequences – (0,0,0), (0,0,1), (0,1,1), (1,1,1). When all of the inputs are low, the voltage output is 6.2V. One high input produces an output of 11.2V. Two high inputs generate 16.2 volts of regulated output. When all three inputs are switched high, the output is maximized at 21.2V.

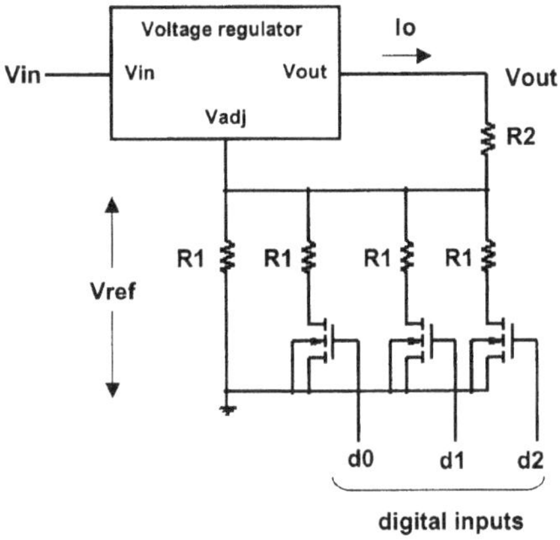

Figure 19. Voltage regulator application: a 3-bit digitally-controlled analog waveform generator [38].

For this application, the output voltage can be approximated as

$$V_{out} \cong V_{ref} \frac{R1 + NR2}{R1}, \tag{6}$$

where V_{ref} ($\approx$ 1.2V) is the fixed reference voltage and N is related to the sum of the digital inputs states (i.e., $N = 1 + d2 + d1 + d0$). Using Equation 6 and the relationship between V_{ref} and I_o, it can be shown that

$$V_{out} \leq I_o(max)\left(\frac{R1}{N} + R2\right). \tag{7}$$

The voltage output is therefore limited by the maximum output drive current of the regulator. The data presented in section 2*C* demonstrated how $I_o(max)$ of the 29372 low-dropout regulator degrades in response to ionizing radiation exposure. Indeed, Figure 11 shows $I_o(max)$ can drop as much as 90% after TID exposures of less than 40 krad(SiO_2). Figure 21 illustrates the simulated effect of $I_o(max)$ degradation on the operation of the circuit. Figure

21a indicates that if $I_o(max)$ drops to 0.6 mA, the voltage set by the (1,1,1) sequence (i.e., N = 4) will degrade to 15.9V. As Figure 21b demonstrates, if the maximum output drive current drops to 0.2 mA, aside from the trivial condition (0,0,0), no digital input sequences generate the desired regulated outputs. These simulated responses are consistent with the Equation 7 and clearly demonstrate how deterioration in circuit specification parameters can affect specific types of regulator applications.

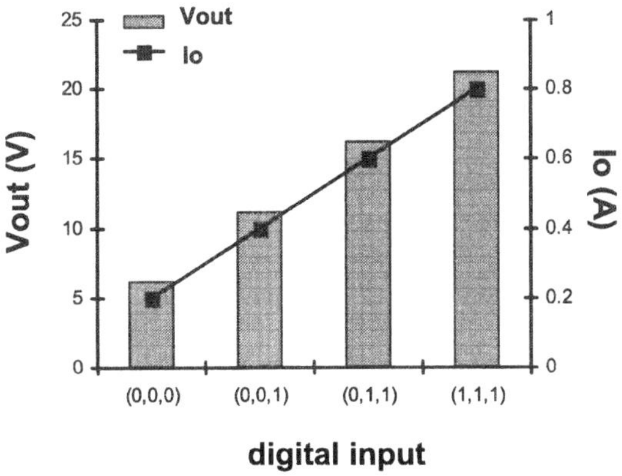

Figure 20. Simulated ideal response of V_{out} and I_o to the four unique digital input sequences.

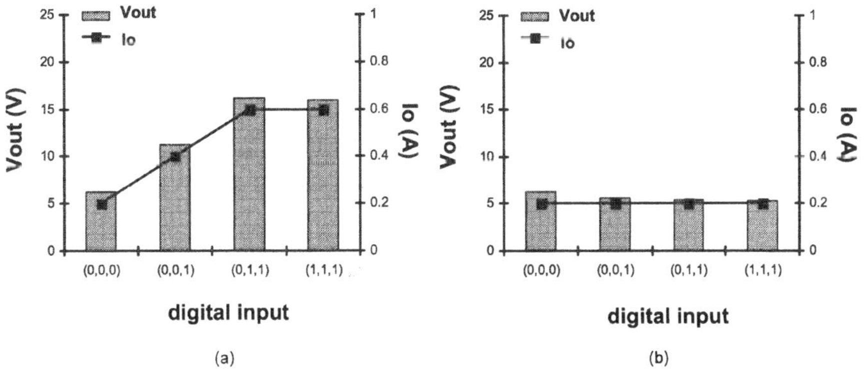

Figure 21. a) Simulated response of V_{out} and I_o to digital inputs when $I_o(max)$ is limited to 0.6 mA. b) Simulated response of V_{out} and I_o to digital inputs when $I_o(max)$ is limited to 0.2 mA.

B. Output Voltage Drift in LM117 Voltage Regulator

Radiation experiments on a LM117 voltage regulator indicated there was significant non-monotonic drift in the output voltage (V_{out}) in response to Co-60 irradiations at a standard laboratory dose rate of 150 rad(Si)/s. The results of these tests were originally presented in a 1996 study [22]. Some of the conclusions were discussed in section 2*C*. Through the use of analysis techniques similar to those used in the analysis on the LM111 TID response above (section 3*B*), the irregular output response of the irradiated LM117 regulator was effectively reproduced by circuit simulation, as shown in Figure 22. SPICE simulation enabled researchers to identify three competing circuit-level TID mechanisms for the LM117 radiation response. All three mechanisms were associated with degradation in the four transistors of the circuit's bandgap reference. Figure 23a shows the ideal topology the LM117's bandgap voltage reference, which is a version of the Brokaw cell [46]. This subcircuit is composed of two lateral pnp BJTs (Q3 and Q4) which form a current mirror to drive equal currents (I_o) through the npn transistors (Q1 and Q2). I_o ($\approx 25\ \mu A$) is

$$I_O \cong \frac{\Delta V_{be}}{R1}, \tag{8}$$

where ΔV_{be} is the base-emitter voltage differential between Q1 and Q2. The constant, supply-independent, reference voltage ($V_{ref} \approx 1.25$V) is the sum of voltage drops across the base-emitter junction of Q1 ($V_{be1} \approx 0.65$V) and the resistor R2 ($2I_o$R2 ≈ 0.6V). The relationship between the output voltage and V_{ref} is expressed in Equation 2.

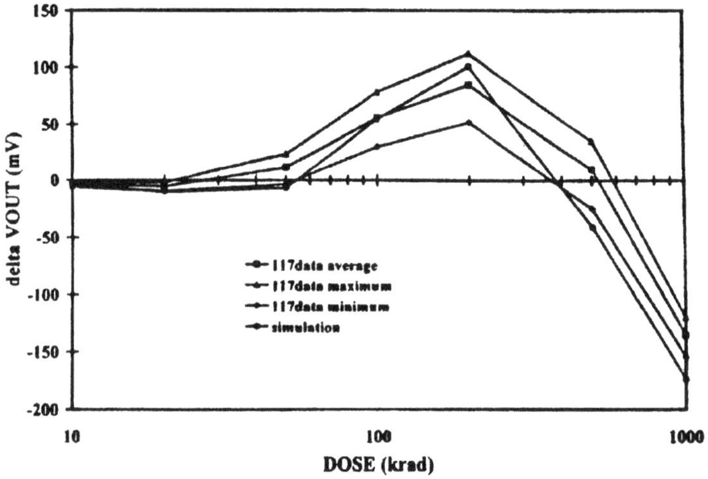

Figure 22. Increased input bias current and offset voltage in biased localized irradiations of the output stage of LM101 operational amplifiers. Data show both bias current degrades and offset voltage after 100 krad(SiO_2) [22].

TID degradation mechanisms for V_{ref} (and V_{out}) were identified as: 1) collector-to-emitter leakage in the two npn devices, 2) gain degradation in the two npn devices, and 3) gain degradation in the two pnp transistors [22]. A brief summary of these mechanisms is as follows:

1) Collector-to-emitter leakage (increased V_{out}): The increased V_{ref} (and V_{out}) is caused by a radiation-induced channel that develops in the p-type base of Q2. The channel establishes a resistive path (R_{leak}) between the collector-to-emitter, as shown in Figure 23b. The leakage path increases ΔV_{be} between Q1 and Q2, thereby increasing I_o, V_{ref}, and V_{out} [22].

2) Gain degradation in the pnp transistors (decreased V_{out}): Radiation-induced excess base current in transistors Q3 and Q4 reduces the collector currents of Q4 and Q1 (I_o - ΔI_o). This reduces both V_{be1} and the voltage drop across R2, causing the V_{ref}, and V_{out} to decrease with TID exposure [22].

3) Gain degradation in the npn transistors (decreased V_{out}): Simulations indicate that I_o remains unchanged with radiation-induced gain degradation in transistors Q1 and Q2. However, gain degradation does cause a reduction in V_{be1}. Moreover, because of the differences in emitter geometries, the emitter current of Q1 drops relative to Q2. The combined decrease in V_{be1} and voltage drop across R1 (reduced emitter current of Q1) causes V_{ref} and V_{out} to decrease [22].

The combined effects of these three circuit-level mechanisms cause the highly nonlinear response in the output characteristics of the LM117 voltage regulator. Through detailed hand analysis, supported by SPICE simulations, each mechanism and its effect on the circuit response can be characterized independently. This type of analysis can provide insight into potential low-cost hardening techniques for future linear integrated circuit designs.

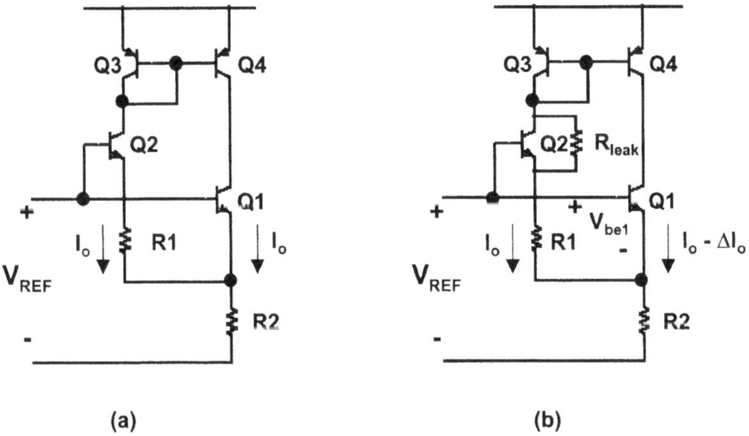

Figure 23. a) Ideal topology of Brokaw cell band-gap regulator [46] and b) the effects of radiation-induced gain degradation on band-gap regulator, i.e., collector-to-emitter leakage path and changes to V_{BE1} and ΔI_O [22].

5. Conclusions

This paper presents a comprehensive overview of TID effects in bipolar linear integrated circuits. However, unlike many studies in this field that focus primarily on basic TID mechanisms at either the process or transistor level, this text discusses the effects of TID from the circuit perspective, particularly the effects of transistor gain degradation on specific circuit parameters and the effects of circuit parameter degradation on select linear bipolar circuit applications.

Of course, it would be impossible to discuss and analyze on how TID exposure affects *all* linear bipolar circuits and their applications. Readers interested in obtaining more data on this topic should consult recently compiled compendiums of total dose effects on bipolar linear circuits found in references [43, 47]. The data and analysis shown in this text is provided to give a perspective on how systems are threatened by ionizing radiation if they utilize linear bipolar circuits to perform critical subsystem functions such as signal amplification, voltage regulation, and mixed signal processing. It should be noted that the author is by no means recommending bipolar linear parts be avoided when designing systems that operate in radiation environments; nevertheless, system designers must be aware of the hazards, especially those hazards that may be difficult to measure effectively, such as the enhanced sensitivity of circuit parameters to radiation exposures at low-dose-rates.

Acknowledgements

The author would to thank previous authors, R.L. Pease, S. L. Kosier, A. H. Johnston, and S. S. McClure, for graciously granting permission to present their data in this manuscript. This paper would not be possible without their scientific discoveries and insights. I would also like to acknowledge the critical scientific contributions of E. Enlow, R.D. Schrimpf, D. M. Schmidt, D.M Fleetwood, M. Shaneyfelt, C. Cirba, T. Turflinger, J. Titus, P. Cole, J. Krieg, P. Fouillat, S. C. Witczak, T. Carriere, and R. Ecoffet to this field of study. I thank D. Espinosa for his editorial comments. I would like to give special recognition to L. Cohn for his continued interest and enthusiastic support of this research.

References

1. D. S. Peck, R. R. Blair, W. L. Brown, and F. M. Smits, "Surface Effects of Radiation on Transistors," *Bell System Technical Journal*, vol. 42, pp. 95-129, 1963.
2. T. P. Ma and P. V. Dressendorfer, *Ionizing Radiation Effects in MOS Devices and Circuits*. New York: Wiley, 1989.
3. P. Adell, R. D. Schrimpf, H. J. Barnaby, R. Marec, P. Calvel, C. Barillot, O. Mion, and C. Chatry, "Analysis of Single Event Transients in Analog Circuits," *IEEE Trans. Nucl. Sci.*, vol. 47, 2000.
4. H. J. Barnaby, R. D. Schrimpf, A. L. Sternberg, V. Berthe, C. R. Cirba, and R. L. Pease, "Proton radiation response mechanisms in bipolar analog circuits," *IEEE Trans. Nucl. Sci.*, vol. 48, pp. 2074-2080, 2001.
5. B. G. Rax, A. H. Johnston, and C. I. Lee, "Proton Damage Effects in Linear Integrated Circuits," *IEEE Trans. Nucl. Sci.*, vol. 45, pp. 2632-2637, 1998.
6. B. G. Rax, A. H. Johnston, and T. Miyahira, "Displacement damage in bipolar linear integrated circuits," *IEEE Trans. Nucl. Sci.*, vol. 46, pp. 1660-1665, 1999.
7. R. L. Pease, "Long-term ionization (total dose) effects in bipolar devices and circuits," *IEEE Trans. Nucl Sic.*, vol. Special Issue, 2002.
8. H. J. Barnaby, R. D. Schrimpf, D. M. Fleetwood, and S. L. Kosier, "The Effects of Emitter Tied Field Plates on Lateral PNP Ionizing Radiation Response," *IEEE BCTM Technical Proceedings*, pp. 35-38, 1998.
9. H. J. Barnaby, C. Cirba, R. D. Schrimpf, S. L. Kosier, P. Fouillat, and X. Montagner, "Minimizing gain degradation in lateral pnp bipolar junction transistors using gate control," *IEEE Trans. Nucl. Sci.*, vol. 46, pp. 1652-1659, 1999.

10. H. J. Barnaby, C. Cirba, R. D. Schrimpf, S. L. Kosier, P. Fouillat, and X. Mantagner, "Modeling BJT radiation response with non-uniform distributions of interface traps," *IEEE Trans. Nucl. Sci.*, vol. 47, pp. 514-518, 1999.
11. P. Cazenave, P. Fouillat, X. Montagner, H. Barnaby, R. D. Schrimpf, L. Bonora, J. P. David, M. C. Calvet, and P. Calvel, "Total Dose Effects on Gate Controlled Lateral PNP Bipolar Junction Transistors," *IEEE Trans. Nucl. Sci.*, vol. 45, pp. 2577-2583, 1998.
12. E. W. Enlow, R. L. Pease, W. E. Combs, R. D. Schrimpf, and R. N. Nowlin, "Response of advanced bipolar processes to ionizing radiation," *IEEE Trans. Nucl. Sci.*, vol. 38, pp. 1342-1351, 1991.
13. D. M. Fleetwood, S. L. Kosier, R. N. Nowlin, R. D. Schrimpf, R. A. R. Jr., M. Delaus, P. S. Winokur, A. Wei, W. E. Combs, and R. L. Pease, "Physical mechansims contibutins to enhanced bipolar gain degradation at low dose rates," *IEEE Trans. Nucl. Sci.*, vol. 41, pp. 1871-1883, 1994.
14. D. M. Fleetwood, L. C. Riewe, J. R. Schwank, S. C. Witczak, and R. D. Schrimpf, "Radiation effects at low electric field in thermal, simox, and bipolar-base oxides," *IEEE Trans. Nucl. Sci.*, vol. 43, pp. 2537-2546, 1996.
15. S. L. Kosier, R. D. Schrimpf, D. M. Fleetwood, M. DeLaus, R. L. Pease, W. E. Combs, A. Wei, and F. Chai, "Charge Separation for Bipolar Transistors," *IEEE Trans. Nucl. Sci.*, vol. 40, pp. 1276-1285, 1993.
16. S. L. Kosier, A. Wei, R. D. Schrimpf, D. M. Fleetwood, and M. Delaus, "Physically Based Comparison of Hot-Carrier Induced and Ionizing Radiation Degradation in BJTs," *IEEE Trans. Electron Dev.*, vol. 42, pp. 436-444, 1995.
17. R. N. Nowlin, E. W. Enlow, R. D. Schrimpf, and W. E. Combs, "Trends in the Total dose Response of Modern Bipolar Transistors," *IEEE Trans. Nucl. Sci*, vol. 39, pp. 2026-2035, 1992.
18. D. M. Schmidt, A. Wu, R. D. Schrimpf, D. M. Fleetwood, and R. L. Pease, "Modeling ionizing radiation induced gain degradation of the lateral pnp bipolar junction transistor," *IEEE Trans. Nucl. Sci.*, vol. 43, pp. 3032-3039, 1996.
19. M. R. Shaneyfelt, J. R. Schwank, D. M. Fleetwood, P. S. Winokur, K. L. Hughes, and G. L. Hash, "Interface trap buildup rates in wet and dry oxides," *IEEE Trans. Nucl. Sci.*, vol. 39, pp. 2224, 1992.
20. S. C. Witczak, R. D. Schrimpf, H. J. Barnaby, R. C. Lacoc, D. C. Mayer, K. F. Galloway, R. L. Pease, and D. M. Fleetwood, "Moderated degradation enhancement in lateral pnp transistors due to measurement bias," *IEEE Trans. Nucl. Sci.*, vol. 45, pp. 2644-2648, 1998.
21. A. Wu, R. D. Schrimpf, H. J. Barnaby, D. M. Fleetwood, R. L. Pease, and S. L. Kosier, "Radiation-induced gain degradation in lateral pnp BJTs with lightly and heavily doped emitters," *IEEE Trans. Nucl. Sci.*, vol. 44, pp. 1914-1921, 1997.
22. H. J. Barnaby, H. J. Tausch, R. Turfler, P. Cole, P. Baker, and R. L. Pease, "Analysis of Bipolar Linear Circuit Response Mechanisms for High and Low Dose Rate Total Dose Irradiations," *IEEE Trans. Nucl. Sci.*, vol. 43, pp. 3040-3048, 1996.
23. H. J. Barnaby, R. D. Schrimpf, R. L. Pease, P. Cole, T. Turflinger, J. Krieg, J. Titus, D. Emily, M. Gehlhausen, S. C. Witczak, M. C. Maher, and D. VanNort, "Identification of degradation mechanisms in a bipolar linear voltage comparator through correlation of transistor and circuit response," *IEEE Trans. Nucl. Sci.*, vol. 46, pp. 1666-1673, 1999.
24. A. H. Johnston and R. E. Plaag, "Models for total dose degradation of linear

integrated circuits," *IEEE Trans. Nucl. Sci.*, vol. 34, pp. 1474-1480, 1987.

25. A. H. Johnston, G. M. Swift, and B. G. Rax, "Total dose effects in conventional bipolar transistors and linear integrated circuits," *IEEE Trans. Nucl. Sci.*, vol. 41, pp. 2427-2436, 1994.
26. A. H. Johnston, B. G. Rax, and C. I. Lee, "Enhanced damage in linear bipolar integrated circuits at low dose rate," *IEEE Trans. Nucl. Sci.*, vol. 42, pp. 1650-1659, 1995.
27. S. McClure, R. L. Pease, W. Will, and G. Perry, "Dependence of total dose response of bipolar linear microcircuits on applied dose rate," *IEEE Trans. Nucl. Sci.*, vol. 41, pp. 2544-2549, 1994.
28. R. L. Pease, L. M. Cohn, D. M. Fleetwood, M. A. Gehlhausen, T. L. Turflinger, D. B. Brown, and A. H. Johnston, "A proposed hardness assurance test methodology for bipolar linear circuits and devices in a space ionizing radiation environment," *IEEE Trans. Nucl. Sci.*, vol. 44, pp. 1981-1988, 1997.
29. R. L. Pease, M. Gehlhausen, J. Krieg, J. Titus, T. Turflinger, D. Emily, and L. Cohn, "Evaluation of proposed hardness assurance method for bipolar linear circuits with enhanced low dose rate sensitivity (ELDRS)," *IEEE Trans. Nucl. Sci.*, vol. 45, pp. 2665-2672, 1998.
30. J. L. Titus, D. Emily, J. F. Krieg, T. Turflinger, R. L. Pease, and A. Campbell, "Enhanced low dose rate sensitivity (ELDRS) of linear circuits in a space environment," *IEEE Trans. Nucl. Sci.*, vol. 46, pp. 1608-1615, 1999.
31. J. Beaucour, T. Carriere, A. Gach, and P. Poirot, "Total dose effects on negative voltage regulator," *IEEE Trans. Nucl. Sci.*, vol. 41, pp. 2420-2426, 1994.
32. J. Krieg, T. Turflinger, J. Titus, P. Cole, P. Baker, M. Gehlhausen, D. Emily, L. Yang, R. L. Pease, H. J. Barnaby, R. D. Schrimpf, and M. C. Maher, "Hardness assurance implications of bimodal total dose response in a bipolar linear voltage comparator," *IEEE Trans. Nucl. Sci.*, vol. 46, pp. 1627-1632, 1999.
33. S. N. Rashkeev, D. M. Fleetwood, R. D. Schrimpf, and S. T. Pantelides, "Defect generation by hydrogen at the Si-SiO2 interface," *Physical Review Letters*, vol. 87, 15 October 2001.
34. H. J. Barnaby, C. R. Cirba, R. D. Schrimpf, K. F. Galloway, M. Pagey, and R. Milanowski, "A Two Dimensional Engineering Model for Radiation-Induced Interface Trap Formation," *GOMAC 2000 Digest of Papers*, pp. 613 - 616, 2000.
35. M. R. Shaneyfelt, J. R. Schwank, S. C. Witczak, R. L. Pease, and D. M. Fleetwood, "Thermal stress effects on enhanced low-dose rate sensitivity of linear bipolar circuits," *IEEE Trans. Nucl. Sci*, vol. 47, 2000.
36. M. R. Shaneyfelt, R. L. Pease, J. R. Schwank, M. C. Maher, G. L. Hash, D. M. Fleetwood , P. E. Dodd, C. A. Reber, S. C. Witczak, L. C. Riewe, H. P. Hjalmarson, J. C. Banks, B. L. Doyle, and J. A. Knapp, "Impact of passivation layers on enhanced-low-dose-rate sensitivity and pre-irradiation elevated-temperature-stress in bipolar linear ICs," *IEEE Trans. Nucl. Sci*, vol. 49, pp. 3171-3179, 2002.
37. H. J. Barnaby, R. J. Milanowski, R. D. Schrimpf, L. W. Massengill, and M. Pagey, "Modeling Ionizing Radiation Effects in Lateral PNP BJT's with Non-uniform Trapped Charge Distributions," *GOMAC Digest*, pp. 585-588, 1998.
38. *National Semiconductor Corporation Data Book*, 1988 Edition.
39. A. S. Sedra and K. C. Smith, *Microelectronic Circuits*, fourth ed. New York: Oxford University Press, 1998.
40. T. Carriere, R. Ecoffet, and P. Poirot, "Evaluation of accelerated total dose testing of

linear bipolar circuits," *IEEE Trans. Nucl. Sci.*, vol. 47, pp. 2350-2357, 2000.

41. S. S. McClure, J. L. Gorelick, R. L. Pease, B. G. Rax, and R. L. Ladbury, "Total dose performance of radiation hardened voltage regulators and references," *Radiation Effect Data Workshop*, vol. 2001 IEEE, pp. 1-5, 2001.
42. R. L. Pease, S. McClure, J. Gorelick, and S. C. Witczak, "Enhance low-dose-rate sensitivity of a low-dropout voltage regulator," *IEEE Trans. Nucl. Sci.*, vol. 45, pp. 2571-2576, 1998.
43. R. L. Pease, W. E. Combs, A. Johnston, T. Carriere, and S. McClure, "A compendium of recent total dose data on bipolar linear microcircuits," *1996 IEEE Radiation Effects Data Workshop Record*, vol. No. 96TH8199, pp. 28-37, 1996.
44. G. Massobrio and P. Antognetti, *Semiconductor Device Modeling with SPICE*, second ed. New York: McGraw Hill, 1993.
45. J. A. Lipman, W. C. Bruncke, D. L. Crosthwait, K. F. Galloway, and R. L. Pease, "Use of scanning electron microscope for screening bipolar surface effects," *IEEE Trans. Nucl. Sci.*, vol. 21, pp. 383-386, 1974.
46. R. A. Pease, "The design of a band-gap reference circuit: trials and tribulations," *1990 IEEE BCTM Meeting*, 1990.
47. R. L. Pease, S. McClure, A. H. Johnston, J. Gorelick, T. L. Turfinger, M. Gehlhausen, J. Krieg, T. Carriere, and M. Shaneyfelt, "An updated data compendium of enhanced low dose rate sensitive (ELDRS) bipolar linear circuits," *2001 IEEE Radiation Effects Data Workshop*, pp. 127-133, 2001.

International Journal of High Speed Electronics and Systems
Vol. 14, No. 2 (2004) 543–561

World Scientific
www.worldscientific.com

HARDNESS ASSURANCE FOR COMMERCIAL MICROELECTRONICS

RONALD L. PEASE
RLP Research Inc., 8 Songbird Lane
Los Lunas, New Mexico 87031, USA

An approach to hardness assurance for commercial microelectronics is presented based on hardness assurance guidelines developed by the US Department of Defense in the late 1970s and early 1980s. Modifications are made to accommodate commercial variability in radiation response and frequent changes in design and process.

Keywords: radiation hardness assurance, radiation testing, total dose, displacement damage, single event effects, dose rate response, latchup, neutrons, heavy ions, protons

1. Introduction

Many electronic systems must operate either in the presence of a radiation environment or with the possible threat of a radiation environment. Examples of systems operating in the presence of radiation include spacecraft, high altitude avionics, nuclear power plant instrumentation, high-energy particle accelerator detectors, and nuclear waste management robotics. Many military systems must be prepared for the threat of nuclear weapon induced radiation, even though they may not otherwise be exposed to radiation. Earth satellites, including both commercial and military, are exposed to the natural space radiation environment and face the threat of radiation from high altitude or space based nuclear weapon detonations. For all of these systems some form of radiation hardening will normally be performed in the system design phase. This hardening may take the form of system level techniques such as shielding or using un-powered spares for ionizing radiation, the use of terminal protection devices for electromagnetic pulse mitigation, error detection and correction or triple modular redundancy for single event upset, and power supply re-cycling for dose rate induced latchup protection. However, one of the primary approaches to radiation design hardening is electronic piece-part selection. In the past the US military invested a significant amount of funding to develop microelectronic parts that were able to withstand high levels of radiation. Hence the piece-part selection for many hardened systems included a high number of radiation-hardened parts. While investment in radiation hardened microelectronic technology continues, although at a much lower level than in the past, many of the radiation hard parts that were developed in the past are no longer available. Many former suppliers of radiation hardened microelectronic parts have ceased production of hard parts because the demand is low and much greater profits can be made in the commercial markets. In addition the state-of-the-art of radiation-hardened parts is usually several generations behind the commercial state-of-the-art and many systems demand a better, faster, cheaper approach. This has lead, by necessity, to the widespread use of commercial microelectronics in systems that have radiation requirements.

Hardness assurance includes those activities required to guarantee that the hardened system design will remain hard in *production* and that fielded systems will be able to withstand the radiation environments that they were designed to meet. In a general sense hardness assurance covers all activities at the system level, sub-system level, board level and piece-part level to maintain all elements of the hardened design. However, since the topic of this paper is hardness assurance of commercial microelectronics, only piece-part hardness assurance will be considered herein.

In the mid 1970s the US Defense Nuclear Agency (now known as the Defense Threat Reduction Agency, DTRA) began a program to develop all of the documentation necessary to support piece-part hardness assurance, HA. The group responsible for carrying out this effort is the Space Parts Working Group (SPWG) Hardness Assurance Committee (HAC). This group was comprised of representatives from the Department of Defense, DoD, and the National Aeronautical and Space Agency, NASA, and their contractors. Electrical test methods and dosimetry standards were developed through the American Society for Testing and Materials, ASTM, and most radiation environment test methods and hardness assurance guidelines were developed as military standards and handbooks. The first 10 years of this effort is described in a summary review paper.[1] The following definition was adopted by the working group: "Piece-part hardness assurance consists of the procedures, controls, and tests used to insure that a purchased piece-part has a response to nuclear-induced stresses that is within known and acceptable limits."[1] If we eliminate the word "controls" and substitute the word radiation for "nuclear" the definition applies to commercial microelectronics today. While the focus of the DNA funded piece-part HA program was on mature hardened technologies, the methodology developed is generic and applies to any microelectronic piece-parts, whether commercial, radiation tolerant, or radiation hard, as we will attempt to demonstrate.

2. Piece-part HA Methodology Overview

The general approach to piece-part HA is illustrated in the flow diagram of Fig. 1 taken from reference 2. The starting point for the piece-part HA program is part categorization shown in the center of the figure.

Each microelectronic part in the system is categorized in one of four categories based on its application, radiation requirements, radiation response data and assigned probability of survival and confidence level. The four categories are as follows:

(i) Not acceptable- corrective action required,
(ii) Hardness Critical Category 1 (HCC-1)- radiation lot acceptance testing required on every lot,
(iii) Hardness Critical Category 2 (HCC-2)- radiation sample testing required on a periodic basis,
(iv) Hardness Non-Critical (HNC)- no further radiation testing required.

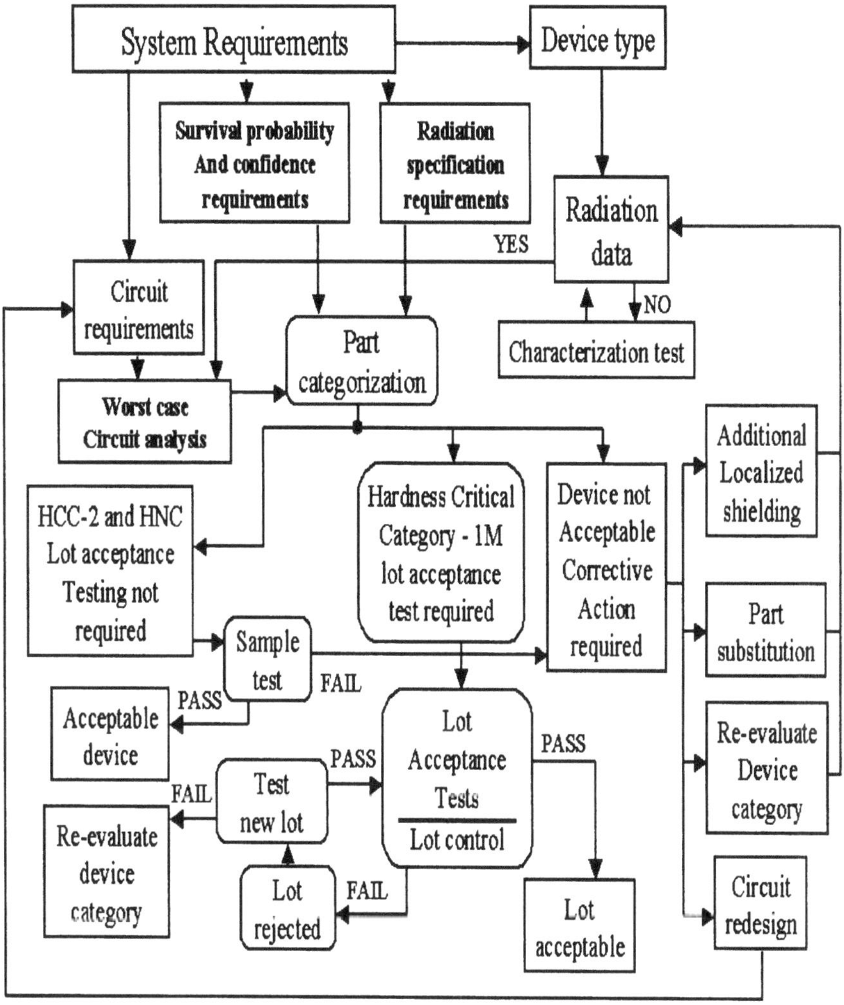

Fig. 1. Piece-part hardness assurance flow diagram from reference 2.

The method used to determine the category for each part in each application is based on the concept of radiation design margin. The radiation design margin, RDM, is defined as the mean of the radiation failure level, R_{mf}, divided by the radiation specification, R_{spec}, for the part.

The first issue to address is the determination of R_{mf}. There are three types of failures in microelectronic parts, parametric, functional and catastrophic. Catastrophic failure (permanent damage) can occur for some devices in the form of, for example, heavy ion or proton induced gate oxide rupture or latchup induced burnout. For these cases the failure definition is obvious. For parametric or functional failure, the failure definition depends on the application of the part, its criticality to the system mission and the radiation environment. For some radiation environments, such as long-term ionizing radiation effects (total dose) and displacement damage, electrical parameters usually degrade monotonically with increasing radiation dose or fluence. Failure is defined as an unacceptable change in a parameter critical for system operation. Functional failure can also occur in microcircuits from total dose or displacement damage when the internal parameter degradation prevents proper circuit operation. This is usually more of an issue for digital circuits rather than linear or mixed-signal circuits. For other environments, such as transient ionization (dose rate) or single event effects (SEE) from protons or heavy ions, there is usually functional failure (upset or latchup) but very little parametric degradation. Again, failure must be defined by the application of the part for each radiation environment based on an evaluation by the system designer to identify critical electrical parameters and allowed changes as well as the criticality of the part. A failure definition is required for each part in each application for each environment.

Once a failure definition is established a mean failure level can be determined based on one of several methods, including electrical screens, model based predictions or sample radiation testing. *In most cases the preferred method will be sample radiation testing.*

The most desirable method of determining the radiation failure level of a part would be based on a pre-irradiation electrical parameter value. However there are only a few cases where a reliable correlation exists between the radiation failure level and the value of a pre-irradiation electrical parameter, and those are for discrete devices and not microcircuits. For example a semi-empirical relation exists between the unity gain cutoff frequency of a bipolar transistor and the neutron induced degradation of current gain.[3] Another example is the correlation between pre-irradiation 1/f noise and channel resistance and the radiation induced increase in oxide trapped charge and interface traps, respectively, for total dose effects in MOSFETs.[4] However, for the vast majority of part types and radiation environments, no correlation has been found between pre-irradiation parameter values and radiation failure levels.

The prediction of radiation failure levels, based on radiation response models, is possible in some cases but it requires detailed information about the process flow and the circuit design and layout. For commercial microelectronic parts this information is seldom available, hence this approach is not considered viable if the RDM is to be known within, say, a factor of two for RDMs below 10. However, prediction of failure levels may be acceptable in cases where the RDMs are quite large and prediction is based on the knowledge of the process technology, the failure mechanisms and an historical set of data. Examples are displacement damage in CMOS circuits for non-ionization energy loss (NIEL) equivalent neutron fluence below 10^{13} n(1 MeV eq.)/cm^2, latchup in silicon-on-insulator (SOI) parts, and single event transients in bipolar linear circuits from protons or neutrons below 20 MeV. In these and other cases, prediction can be used to put part types in the HNC category.

The most reliable method of determining the mean failure level is based on sample radiation data. There are a number of radiation effects databases that have been compiled and are available to the public, including the DTRA supported Electronics Radiation Response Information Center (ERRIC) database (http://erric.dasiac.com), the NASA Jet Propulsion Lab (JPL) database (http://radnet.jpl.nasa.gov) and the NASA Goddard Space Flight Center (GFSC) database (http://radhome.gsfc.nasa.gov). Most of the data in the ERRIC database are on neutron and total dose effects and most of the data in the NASA databases are on total dose and SEE. Other sources include system contractor and/or government test reports and the IEEE Radiation Effects Data Workshop Record that includes workshop papers from the annual IEEE Nuclear and Space Radiation Effects Conference held in July. While published data exist on many commercial part types one has to be very careful using the data for the following reasons:

(i) Many of the data are on older date code parts and may not be applicable to parts manufactured today,
(ii) Sample sizes are often quite small,
(iii) The test parameters may not be the ones that are critical for the application being evaluated,
(iv) The test conditions (bias and load) may not be worst case for the intended application.

The use of database data for commercial parts is especially risky since process, circuit design and layout may change frequently (and without notice), which may affect the radiation response. If data are not available or are insufficient for one of the above reasons, then sample radiation tests must be run to determine the mean failure level.

3. Radiation Testing to Determine R_{mf}

To provide the radiation test data for the purpose of categorization the data should be taken to best simulate operating conditions and radiation environments that will be encountered in the system application. This can often be a challenge since the part may have many applications and may see extremes of operating biases, frequencies and temperatures. For this reason testing is usually done for a worst-case application, which is often different for different environments and sometimes even for different parameters for the same environment. For example, if a part is being tested for heavy ion SEE, the worst case bias for upset may be at the minimum supply voltage whereas for latchup it is the maximum supply voltage and operating temperature. The full range of applications and operating conditions should be considered when designing the radiation test.

The radiation environments encountered by the system can vary widely in particle type, energy, flux and fluence. For example in an earth orbit a satellite may encounter trapped protons and electrons in the Van Allen belts, heavy ions and high-energy photons from cosmic radiation and high fluxes of protons from solar flares. Instead of trying to accurately simulate each of these radiation environments, radiation testing is performed to measure the response of the microelectronic parts to different types of *effects*. These effects are associated with a characteristic of the environment. For total dose the effect is proportional to the ionization energy loss in the material measured in rads, where 1 rad is defined as 100 ergs/gm material. For commercial microelectronics the material of concern is usually silicon dioxide, so the measurement is rad(SiO_2). For transient ionization the effects are proportional to the dose rate measured in rad(Si)/s. For displacement damage from protons and electrons the effects on the part are proportional to the non-ionization energy loss or NIEL.[5] For SEE from heavy ions the important factor is the linear energy transfer or LET, measured in MeV-cm^2/mg or pC/μm. However, for proton induced SEE the important factor is the proton energy. Using these concepts, based on years of research in the radiation effects community, the testing can be performed using radiation sources that are not identical to the actual environment but can produce the desired *effect*. For example, total dose effects are usually simulated in the laboratory using Co-60 or Cs-137 photons rather than electrons or protons (the main source of ionization damage in space). While electrons and protons of the appropriate energies are available from various types of accelerators in the laboratory, it is much cheaper and easier to use Co-60 sources. ASTM standards address the measurement of the absorbed dose from Co-60 irradiation in rad(SiO_2).[6,7] Transport codes are available to calculate the absorbed dose in SiO_2 from the space environment.[8] While there is commonality for ionizing radiation between the space and laboratory environment through the measure of absorbed dose, a problem arises with respect to the flux or dose rate. As shown in Fig. 2,[9] in space the *average* dose rate can vary between about 10^{-3} rad/s to less than 10^{-4} rad/s whereas typical laboratory dose rate are orders of magnitude higher. The instantaneous dose rate in space varies from about 1 rad/s to near zero. The default dose rate for the US military test standard is 50-300 rad/s whereas in the European standard there are two ranges of dose rate, 2-10 and 0.1 to 0.01 rad/s. There are two issues that must be addressed because of differences in the response of microcircuits at different dose rates. One is labeled "time dependent effect" and the other "true dose rate effect". Time dependent effects refer to those processes that occur following irradiation, e.g. annealing or compensation of trapped charge and buildup of interface traps. True dose rate effects occur in thick oxides at low electric field and are thought to be a result of space charge in the oxide that builds up at high dose rate but not at low dose rate.[10,11] True dose rate effects are a problem for many bipolar linear circuits.[12] Time dependent effects are treated with a post irradiation elevated temperature anneal (incorporated in the test standard for CMOS parts used in space)[13,14] and true dose rate effects are treated either with a sufficiently low dose rate irradiation (undesirable for high specification doses) or with an elevated temperature irradiation.[15,16] In March 2003 the US military standard test method for total dose testing was revised to include low dose rate or elevated temperature irradiation tests for bipolar linear and mixed-signal parts.[14]

While Co-60 does a good job of simulating the ionization effects of electrons and protons, it does not simulate displacement damage effects. Another radiation source is required to simulate the displacement damage, such as neutrons from a nuclear reactor. However, if the ionization and displacement damage effects of protons are to be

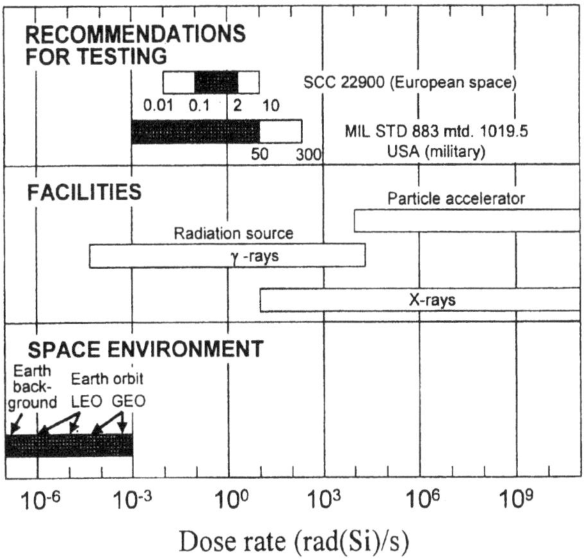

Fig. 2. Typical dose rates in space compared to laboratory facilities and test standards.[9] (© 2002 IEEE)

simulated in the laboratory using a combination of Co-60 and neutrons then the irradiations have to be performed consecutively on the same samples since the effects are interactive and not simply additive.[17] When neutrons are used as an irradiation source for displacement damage and compared to other forms of irradiation through the common factor NIEL, the neutron fluence must be expressed in the form of 1 MeV equivalent silicon damage fluence. This is done by integrating the differential neutron fluence as a function of neutron energy with the silicon cross section for displacement damage as a function of neutron energy. Using this notation the space particle environment (electrons, protons and heavy ions) can be expressed as a 1 MeV eq neutron flux or fluence through the calculation of NIEL. Displacement damage from space radiation affects primarily silicon solar cells, bipolar linear devices and circuits and opto-electronics.

Laboratory tests to simulate single event effects in space are performed on the basis of LET for heavy ions and proton energy for protons. Although LET is the primary factor relating SEE in space to the laboratory it is not the only factor. In space the heavy ions are very energetic, having energies that range up to several GeV per nucleon. At these energies the particles can penetrate the skin of the spacecraft, the microcircuit package and the entire chip. In the laboratory the accelerators used for heavy ion testing are either in the range of 1-10 MeV/nucleon or 10-100 MeV/ nucleon. For the high LET particles, even with the package lids removed the ions may only penetrate the top few microns of the chip. While this can be a problem for some circuits, e.g. bipolar linear circuit and power MOSFETs, the lower energy ions are adequate for simulating the effects of the higher energy space particles for many circuit types.[18]

Proton energies in space can range up to a few hundred MeV. These energies are easily simulated with laboratory accelerators. Attempts have been made to establish correlation between proton induced SEE and heavy ion induced SEE so that data taken in one environment can be used to imply microcircuit response in the other environment.[19] However, while this may work for some circuit types it is not universal.[20]

For the transient ionization (dose rate) environment, encountered primarily for a nuclear weapon event, the response (upset, latchup and/or burnout) is both a function of the dose rate and the pulse width. Hence the mean failure level is usually determined for both a very short pulse width, say, 10-20 ns, to cover the short pulse high dose rate regime and a wide pulse, say a few μs, to cover the equilibrium, wide pulse regime. The short pulse response can be measured using a flash x-ray machine or an electron linear accelerator (LINAC) while the wide pulse response usually requires a LINAC.

While we have discussed the issues with performing laboratory radiation tests to simulate the response to actual radiation environments, we have not discussed the requirements for part selection and sampling. This is a very important aspect of the categorization process that is often ignored. The test samples must be representative of the parts that will actually be used in the fielded systems. This is especially important if the parts are to be considered HNC since these parts will not be tested any further and system contractors are motivated to get as many parts in this category as possible, because of the cost savings. Because of the wide range of radiation responses that can be expected from the same part type from different manufacturers, each vendor's part must be considered a different part type. Also because the radiation response may change when the design, layout or process is modified, and there are no controls on these factors for a commercial part, samples must be drawn from multiple lots over a period of time. If this is not possible then lot sample testing must be performed and the part must be categorized as HCC-1. Sample sizes per lot should be a minimum of 5-10. This will be discussed in more detail in a later section.

4. Determination of R_{spec}

Since we are interested in establishing a value of RDM for each part we also must determine the radiation environment that each part will see during the mission. The radiation environment can vary widely within a system depending on how much material there is between the microcircuit chip and the radiation source. In the case of a spacecraft parts that are buried in the center of an electronic sub-system may get considerably less dose than a solar cell that is placed outside the spacecraft. Although the radiation environment should be determined for each application of each part type, a conservative approach is usually taken initially by assuming the radiation environment is the free field environment. The only problem with this assumption is that for some cases the radiation environment at the part may actually exceed the free field environment because of dose enhancement.[21] This problem is usually ignored since there are a number of built-in safety factors in the failure definition, the "worst case" radiation test conditions and the categorization criteria. If the part is found to be unacceptable because of the value of RDM based on a free field calculation of the environments then the environment is transported through the spacecraft material to the part location to get a better estimate of the environment. As discussed in the previous section the radiation environment must be represented using the same factors that are used in the testing. For the space environment the total dose is represented by rads (dose) and rad/s (dose rate), the heavy ion environment by the LET fluence, and the displacement damage by the 1 MeV eq. neutron fluence. The proton environment is usually expressed as a proton flux or fluence vs. proton energy. Codes are available both for estimating various components of the space radiation environment for given missions and for transporting the environments through various densities and thickness of materials.[8]

5. Categorization

As mentioned previously there are four HA categories for microelectronic parts, unacceptable, HCC-1, HCC-2 and HNC. There are two methods for determining what category a part is placed in based on RDM.

5.1 *Design margin breakpoint method*

The first method was developed for systems with moderate radiation requirements and has been applied by the military mostly for manned systems. It has been named the design margin breakpoint (DMBP) method and is based on historic radiation data for older parts with many part types lumped together for each radiation environment. This DMBP method is shown in Fig. 3. The breakpoints in this method are very conservative and for commercial microelectronic parts would probably be considered unrealistic.

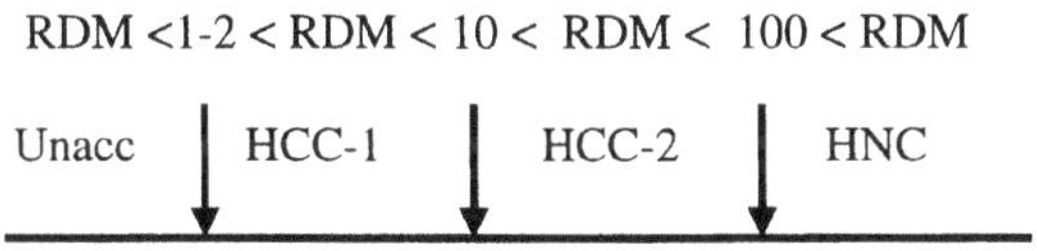

Fig. 3. HA categories using DMBP method.

For example using this method, if one wanted to avoid performing any further testing for total dose for a mission with a 20 krad specification (typical of many NASA missions), a part would have to demonstrate an average failure dose of 2 Mrad. Even periodic testing would be required for parts with an average failure dose of 200 krad. The breakpoint between unacceptable and HCC-1 is given as 1-2. If RDM equals 1 then half of the population would fail at the specification level. This is clearly unacceptable. However, if RDM equals 2 then a large % of the population may pass the specification level depending on the statistical distribution. While the DMBP method may be useful for certain military systems using radiation tolerant or hard parts, it is not recommended for systems using a large percentage of commercial parts.

5.2 *Part categorization criteria method*

The second method for categorization is based on the statistical distribution of the radiation failure levels and is called the Part Categorization Criteria (PCC) method. The breakpoints for this method are shown in Fig. 4.

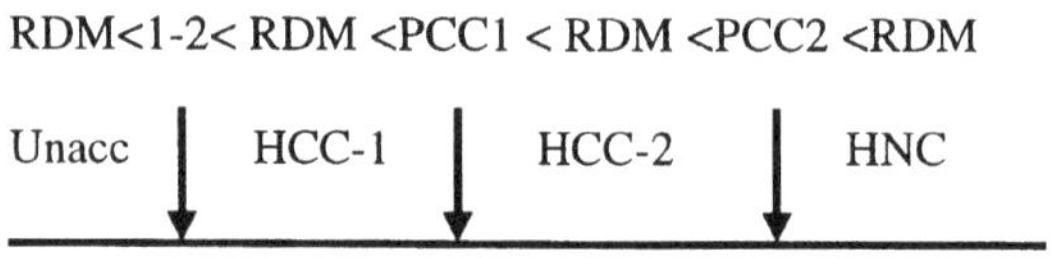

Fig. 4. HA categories using PCC method.

The application of this method requires knowledge of the distribution of radiation failure levels. This issue has been addressed for many part types in many environments and there is no general consensus about the nature of the distribution. Several parametric distributions, such as normal, lognormal and Weibull have been investigated as well as a non-parametric distribution. While it is not universal, the lognormal distribution is usually assumed. The lognormal distribution is favored because the values are always positive, it is easy to work with and it fits reasonably well for many part types in many environments. Assuming a lognormal distribution the geometric mean, $R_{mf} = \exp\{ave[\ln(R_{fail})]\}$, and standard deviation, $s[\ln(R_{fail})]$, are calculated for each set of test samples from the natural logarithms of the radiation failure levels, R_{fail}. The PCC is then calculated from the following relation:

$$PCC = \exp\{K_{TL} * s[\ln(R_{fail})]\}.$$

Assuming that PCC1 is just equal to PCC in the above equation, if RDM < PCC1 then the part is category HCC-1. K_{TL} is the one-sided tolerance limit factor and is a function of the sample size, n, the probability of survival, P_S, and the confidence level, C. Plots of K_{TL} vs. sample size for various values of P_S are shown in Fig. 5 for a confidence level of 0.90 (90%), which is the usual specified value of C.

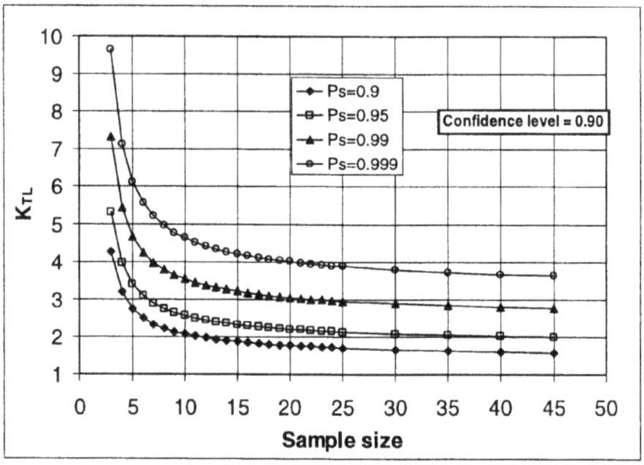

Fig. 5. K_{TL} vs. sample size for various values of P_S for C = 0.90.

Many space systems require a high level of survival probability since a part failure can mean the failure of the mission and repair and/or replacement is usually not an option. If we require, for example, a P_S of 0.999 with a C of 0.9, then for a sample size of 15 (5 each from three lots) K_{TL} is about 4.2. This means that for the part to avoid lot sample testing the average failure level minus 4.2 standard deviations must be greater than the specification radiation level.

The decision as to whether periodic sample testing or no further testing is to be performed is based on what we have called PCC2. If a part has a sufficient RDM then, in principle, lot acceptance is not required. Under the categorization system developed for mature parts, the assumption is that in those cases where many lots of a part type will be purchased over a long period of time, some parts should be periodically checked to assure that the RDM and standard deviation of the failure levels is maintained. It is unlikely that this approach will be workable for commercial part types since most commercial parts would not be considered mature in the sense of long-term stability without changes in process or design. Therefore for commercial parts the HCC-2 category can be eliminated and a safety margin can be used to establish the breakpoint between HCC-1 and HNC. The system program office should do the assignment of the breakpoint. However a reasonable number would be 2PCC1 as the breakpoint between HCC-1 and HNC. Another approach might be to increase the P_S requirement. For example, if the system required a P_S of 0.999, then the breakpoint between HCC-1 and HNC could be set using a PCC calculated for a P_S of 0.9999.

The breakpoint between HCC-1 and unacceptable was addressed for the DMBP method. With the PCC method this breakpoint is the same. The actual value of this breakpoint will be set by the system HA engineers and will usually be determined by how much risk they are willing to take in rejecting lots. For those parts that are found to be unacceptable there are several options, as shown in Fig. 1. One may add shielding to reduce the radiation environment at the part level or, if free field environments were used, perform a calculation of the actual environment at the part location. This may lower the value of R_{spec}. One may perform additional radiation testing if the original sample sizes were too low. One may also re-evaluate the application of the part to see if the failure definition is realistic or too conservative or if a design change can relax the failure definition. If none of these approaches result in the part being re-categorized as acceptable, then the part must be replaced with one that is acceptable. In summary, for commercial parts the PCC method should be used. The PCC-2 category should be eliminated and the breakpoint between HCC-1 and HNC should be set somewhat higher than PCC1.

6. Lot Acceptance Testing

For all of those parts that are category HCC-1 radiation lot sample testing must be performed on each lot. The first issue that must be addressed is the definition of a lot. As a minimum for commercial parts the lot is defined as a date code lot, i.e. all parts assembled in the same week. While this may include parts from more than one process lot, if the test samples are chosen randomly, and the sample size is sufficient, then the probability is good that parts from all the process lots will be represented in the test sample. It is much better to define a lot as a process or wafer lot where possible. For example, if parts are purchased in die form, because of special packaging requirements, they can usually be purchased from a single process or wafer lot. The lot sample test approach can take two forms, a test by attributes or a variables data test.

6.1 *Test by attributes*

The test by attributes is also known as a go/no-go test. In this type of test the test samples are randomly selected from the lot and subjected to the radiation test at the specification level. Based on the failure criteria (parametric or functional) an accept/reject (go/no-go) decision is made for each sample. This test requires a very large sample size to establish a high P_S. As an example for a sample size of 11 and an accept number of zero (no failures), the lot tolerance percent defective, LTPD, would be 20, corresponding to a P_S of only 0.80, for a confidence level of 0.9. To achieve a P_S of 0.999 (LTPD of 0.1), as in our earlier example, a sample of 2303 with no failures would be required. To overcome this problem of high sample size, which is clearly unacceptable, overtesting may be used.[22] An overtest is a test at a higher level than the specification level where the overtest factor is the ratio of the test level (R_{test}) to the specification level. Procedures have been developed for calculating an overtest factor based on sample size, accept number, P_S and C. However, to apply this approach one must make an estimate of the *maximum* standard deviation of the radiation failure levels for the part type and environment. For a lognormal distribution the maximum standard deviation is calculated from the logarithms of the radiation failure levels, $\sigma[\ln(\max)]$. For purposes of illustration let us assume that $\sigma[\ln(\max)]$ is 0.5. Table 1 is a list of the overtest factors that would be required for $P_S = 0.999$ and $C = 0.9$ as a function of sample size for zero failures.

Table 1. Overtest factors vs. sample size

Sample size (n)	Overtest factor (R_{test}/R_{spec})
5	3.97
10	3.11
11	3.02
15	2.75
20	2.53

For our example of a test to the specification level, we see that an overtest of 11 parts at roughly three times the specification level will give us the same P_S and C as a test of 2303 samples at the specification level (assuming $\sigma[\ln(\max)] = 0.5$).

6.2 *Test by variables data*

The variables data test method is the same approach that we discussed for categorization using the PCC method. One usually assumes a lognormal distribution of radiation failure levels and calculates a geometric mean and standard deviation. If $\exp\{\text{ave}[\ln(R_{fail})]-K_{TL}*s[\ln(R_{fail})]\}$ is greater than the specification level then the lot passes, otherwise it fails.

There are a number of cases where this seemingly straightforward approach does not work well. For example the test approach most often used to determine a failure level in environments such as total dose and displacement damage is called the step-stress approach. In the step-stress approach the parts are taken to discrete radiation levels and are measured to determine parametric change or functional failure. If the parts pass at one level and fail functionally at the next higher level some means of interpolating the failure level must be established. If some parts fail at the first level and others pass at the highest level then one must extrapolate to a failure level, which entails a considerable risk of error. Methods have been developed to handle these situations[23,24] but additional tests should be run to completely bound the failure levels for all samples. Another approach that may be used in the case of parametric failures is to base the accept/reject criterion on parametric changes at a fixed radiation level, either at the specification level or higher. This approach is valid only if the variation of the parameter with the radiation environment is linear. If the response is sub-linear then the results will be conservative. However, if the response is super-linear, as is the case with abrupt failures, then the approach is non-conservative and should not be used. Another class of exceptions is where the maximum level that can be reached by the laboratory radiation source is insufficient to cause failure in any of the test samples. In this case the overtest method should be used and a value of $\sigma[\ln(\max)]$ must be assumed.

6.3 *An example*

An example is provided to illustrate the procedure in performing a lot acceptance test based on the variables data method. This example is taken from real data from a semiconductor manufacturer. The part type is a commercial LM117 adjustable voltage regulator, which is used in many space systems. The critical parameter that determines the failure criterion is the reference voltage, Vref, with a nominal value of 1.25V. The failure definition will be taken as Vref dropping below the minimum specification value of 1.2V. The worst-case bias for this part is for dc operation with a maximum input voltage. A sample of 18 parts from the same wafer lot (6 each from three different wafers) is irradiated in a step-stress test and the value of Vref is measured at 3, 10, 30 and 50 krad. Let us assume the specification dose for this part is 7 krad and a P_S of 0.99 is required at a confidence level of 90%. Figure 6 is a plot of Vref vs. dose for the 18 samples with the scale expanded to show the behavior near the failure level of 1.2V (1200 mV).

The failure doses are interpolated graphically from the lines crossing 1200 mV. A plot of the inverse cumulative probability distribution of the ln(failure dose) is given in Fig. 7 for the 18 samples. This function (inverse cumulative distribution) has a mean of zero and a standard deviation of 1, as shown in the plot.

If the distribution is lognormal then the data will fall on a straight line when plotted as ln(failure dose). We can see from Fig. 7 that the assumption of lognormal is reasonable. For a P_S of 0.99 and confidence of 0.9 the K_{TL} is 3.1. The geometric mean failure dose is found to be 9.3 (10.9 krad) and the standard deviation 0.18. The value of $\exp\{\text{ave}[\ln(R_{fail})]-K_{TL}*s[\ln(R_{fail})]\}$ is 6.15 krad so the lot fails the 7 krad requirement.

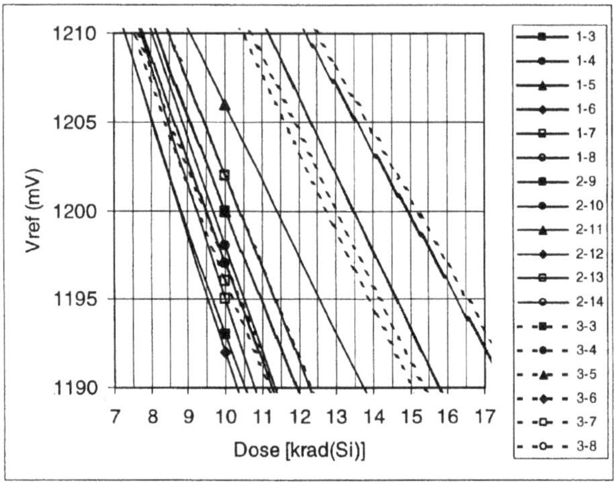

Fig. 6. Vref vs. dose for LM117 adjustable voltage regulator.

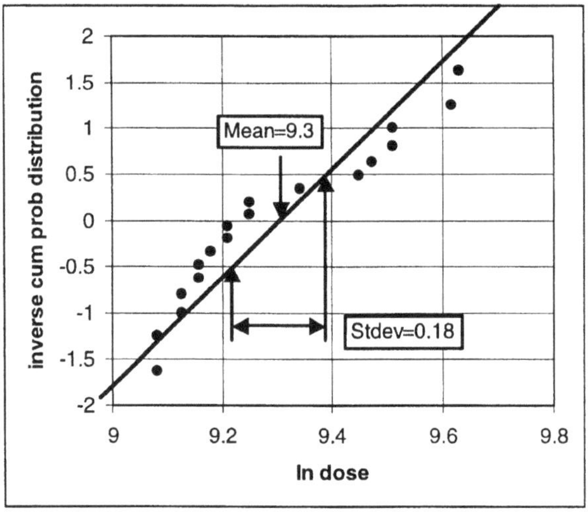

Fig. 7. Inverse normal cumulative distribution vs. ln failure dose for 18 samples of an LM117 voltage regulator.

6.4 *Radiation screens*

Radiation screens (100% radiation tests) have been used as a substitute for lot acceptance testing in some cases. For a radiation screen to be acceptable it must be repeatable and non-destructive since the irradiated parts will be used in the flight hardware. An attempt to develop a screen for total dose was only partially successful.[25] Two approaches were tried: 1) irradiate to a low dose to engender a measurable parameter change and extrapolate to a parametric failure level, and 2) irradiate to the specification dose, determine if the part passes or fails, and if it passes, anneal out the degradation at elevated temperature. Since total dose degradation is often very non-linear the extrapolation technique did not work for many parts. Also because the anneal temperature must be restricted for reliability reasons, the irradiate-and-anneal method was not successful. One case where radiation screens have been successful is for dose rate induced latchup. All parts are given a dose rate pulse of radiation to a specified dose and if they latch-up they are eliminated from the lot. This approach has been used on several military systems. Of course in those cases where radiation screens are employed, the sample lot acceptance test does not apply since an accept/reject decision is made on every sample on the basis of the screen test.

7. Modifications to the HA Methodology to Accommodate Commercial Parts

While the original US military developed piece-part hardness assurance methodology was designed to address mature radiation tolerant/radiation hard parts, the HA approach is generic and does not require extensive modification to apply to commercial parts. The concept of categorization based on RDM should apply to any electronic piece-part and any type of system with radiation requirements. However some of the details require modification in dealing with commercial parts, which do not have controls over processes, circuit designs and layouts that may affect radiation response. For this reason data in databanks are unlikely to meet the criteria for being representative of the lot(s) that will be used in fielded systems. If it can be determined that the databank data are from lots that are manufactured with the same process, circuit design and layout as those that will be purchased for system use then they may be used if they meet the additional criteria of using appropriate test conditions, electrical measurements and sample sizes. If not then characterization data will have to be taken on samples that are representative of lots that will be used in fielded systems. For those cases where there is a lifetime buy of the parts, including tests samples, system hardware and replacement parts (if required) all from the same lot, then the characterization test becomes, by default, the lot acceptance test. In this case there are only two categories, acceptable or not acceptable.

The other modification for commercial parts would be in the method for categorization. The DMBP method is not recommended even with a change in the breakpoints, since there is insufficient historical radiation data on commercial parts to support generic breakpoints. The PCC method should be used with the elimination of HCC-2 and a re-definition of the breakpoint between HCC-1 and HNC. The category of HCC-2 has always been a gray area with somewhat imprecise definitions for the breakpoints between HCC-1 and HCC-2 as well as HCC-2 and HNC. It makes some sense when a product will be on the market for many years and a long life system will be using the part in relatively high volume for many lots and many years. However, such a case will probably never occur for commercial parts. Even the CMOS digital small scale ICs and the bipolar linear circuits that have been available for many years undergo die shrinks and process modifications that change radiation response. Therefore, for commercial microelectronics only three categories are required with a modification to the breakpoint between HCC-1 and HNC as suggested above.

While it may seem difficult to get commercial parts into the HNC category, there are many microcircuit technologies in many radiation environments that can be placed in the HNC category based on an understanding of mechanisms, modeling/analysis or historical data. Some examples were given in section II. There are many other cases where it is reasonable to place certain parts in the HNC category based on the radiation level, the failure definition or the criticality of their use in the system.

Many commercial parts, e.g. linear and mixed-signal circuits, large complex application specific integrated circuits (ASICS) and processors, display multiple failure mechanisms and complex radiation responses such as true dose rate response, sensitivity to pre-irradiation elevated temperature stress (PETS), and time dependent effects for the "total dose" environment. For the high energy particle and heavy ion environments there are numerous possible effects including single event upset (SEU), single event latchup (SEL), single event transients (SET), single event functional interrupt (SEFI), single event gate rupture (SEGR) and single event burnout (SEB), any combination of which may occur in a single microcircuit. These multiple failure mechanisms must be addressed by the radiation test method and do not affect the overall HA methodology. For example high energy protons cause both total dose degradation and displacement damage in many commercial bipolar linear circuits. To test for degradation from both of these effects the same set of test samples would need to be exposed to both an ionizing radiation source, e.g. Co-60, and a displacement damage source, e.g. neutrons, unless the tests were performed with a proton source.[26] Another example is rebound in CMOS circuits where the total dose tests must be performed on the same samples in two steps, an initial irradiation to look for oxide trapped charge induced failures, followed by an additional irradiation and an elevated temperature anneal to look for interface trap related failures.[14] In the case where multiple effects may occur from the same radiation exposure, e.g. heavy ion induced SEU and SEL, and the total fluence must be restricted because of ionization induced degradation, two set of test samples may be required. One set would be tested at room temperature with V_{CC} at the minimum value and the other set tested at an elevated temperature with V_{CC} at the maximum.

Several forms of high speed microcircuits, such as emitter coupled logic (ECL), GaAs and other 3-5 technologies and SiGe heterojunction bipolar transistor (HBT) technologies, are very radiation tolerant for certain radiation environments and effects (displacement damage and often total dose). In those cases where it can be demonstrated that the parts are HNC at the specification level, no radiation lot sample testing is required. However, when radiation lot sample testing is required, the high speed part types are treated as any other part type in terms of the HA methodology. In the case of SiGe HBTs, the microcircuits are often fabricated in a BiCMOS technology and it is the CMOS portion of the circuit that dominates the radiation response.[27]

8. Conclusion

From the 1960s through the 1980s military systems had a significant share of the microelectronics market and had influence with the semiconductor manufacturers. During this time many radiation hardened process technologies and families of parts were developed and made available to hardened systems. Hardness assurance test methods and guideline documents were developed and a general HA methodology was established with funding from DTRA. In the 1990s, with the commercial microelectronics industry experiencing exponential growth and the military portion of the market rapidly decreasing, many of the producers of radiation-hardened parts left the hardened parts market. Many commercial parts are now being used in systems that have radiation requirements. Because of this, a piece-part hardness assurance approach is required to deal with these parts. We have presented the hardness assurance approach developed for the US military and have argued that, with minor modifications, it is applicable for commercial microelectronics.

9. Acknowledgements

This work was funded by the Defense Threat Reduction Agency under Contract N00164-97-D-0013 between NAVSEA Crane and Mission Research Corporation. The author would like to thank Mr. Lew Cohn for his support.

References

1. E. A. Wolicki, I. Arimura, A. J. Carlan, H. A. Eisen and J. J. Halpin, "Radiation hardness assurance for electronic parts: accomplishments and plans," *IEEE Trans. Nucl. Sci.*, vol. 32, no. 6, pp. 4230-4236, Dec. 1985.

2. Military Handbook, "Ionizing dose and neutron hardness assurance guidelines for microcircuits and semiconductor devices," MIL-HDBK-814, February 1994.

3. G. C. Messenger and M. S. Ash, *The Effects of Radiation on Electronic Systems*. New York, NY: Van Nostrand-Reinhold, 1986.

4. J. H. Scofield and D. M. Fleetwood, "Physical basis for nondestructive tests of MOS radiation hardness," *IEEE Trans. Nucl. Sci.*, vol. 38, no. 6, pp. 1567-1577, Dec. 1991.

5. G. P. Summers, C. J. Dale, E. A. Burke, E. A. Wolicki, P. Marshall, M. A. Gehlhausen and M. A. Xapsos, "Correlation of particle-induced displacement damage in silicon," *IEEE Trans. Nucl. Sci.*, vol. 34, no. 6, pp. 1134-1139, Dec. 1987.

6. ASTM standard E666, Practice for calculating absorbed dose from gamma or x-ray radiation.

7. ASTM standard E668, Practice for Application of Thermoluminescence Dosimetry Systems for Determining Absorbed Dose in Radiation-Hardness Testing of Electronics.

8. J. Barth, "Modeling space radiation environments," IEEE NSREC Short Course Notes, July 1997.

9. C. Poivey, "Radiation hardness assurance for space systems," IEEE NSREC Short Course Notebook, July 2002.

10. D. M. Fleetwood, S. L. Kosier, R. N. Nowlin, R. D. Schrimpf, R. A. Reber, Jr., M. DeLaus, P. S. Winokur, A. Wei, W. E. Combs and R. L. Pease, "Physical Mechanisms Contributing to Enhanced Bipolar Gain Degradation at Low Dose Rates," *IEEE Trans. Nuc. Sci.* vol. 41, no. 6, pp. 1871-1883, Dec. 1994.

11. D. M. Fleetwood, L. C. Riewe, J. R. Schwank, S. C. Witczak and R. D. Schrimpf, "Radiation Effects at Low Electric Fields in Thermal, SIMOX, and Bipolar-Base Oxides," *IEEE Trans. Nuc. Sci.* vol.43, no. 6, pp. 2537-2546, Dec. 1996.

12. R. L. Pease, W. E. Combs, A. Johnston, T. Carriere, C. Poivey, A. Gach, and S. McClure, "A Compendium of Recent Total Dose Data on Bipolar Linear Microcircuits," IEEE Rad. Effects Data Workshop Record, 28-37, 1996.

13. D. M. Fleetwood, P. S. Winokur, L. C. Riewe and R. L. Pease, "An improved standard total dose test for CMOS space electronics," *IEEE Trans. Nucl. Sci.*, vol. 36, no. 6, pp. 1963-1970, Dec. 1989.

14. MIL-STD-883, Test Method 1019, "Ionizing radiation (total dose) test procedure."

15. R. L. Pease, L. M. Cohn, D. M. Fleetwood, M. A. Gehlhausen, T. L. Turflinger, D. B. Brown and A. H. Johnston, "A Proposed Hardness Assurance Test Methodology for Bipolar Linear Circuits and Devices in a Space Ionizing Radiation Environment," *IEEE Trans. Nucl. Sci.* vol. 44, no. 6, pp. 1981-1988, Dec. 1997.

16. American Society for Testing and Materials (ASTM) F1892 "Standard guide for ionizing radiation (total dose) effects testing of semiconductor devices."

17. H. J. Barnaby, S. K. Smith, R. D. Schrimpf, D. M. Fleetwood, and R. L. Pease, "Analytical Model for Proton Radiation Effects in Bipolar Devices," *IEEE Trans. Nucl. Sci.* vol. 49, no. 6, pp., Dec. 2002.

18. P. E. Dodd, O. Musseau, M. R. Shaneyfelt, F. W. Sexton, C. D'hose, G. L. Hash, M. Martinex, R. A. Loemker, J. L. Leray and P. S. Winokur, "Impact of ion energy on single-event upset," *IEEE Trans. Nuc. Sci.* vol.45, no. 6, pp. 2483-2491, Dec. 1998.

19. E. L. Petersen, "The relationship of proton and heavy ion upset thresholds," *IEEE Trans. Nuc. Sci.* vol.39, no. 6, pp. 1600-1604, Dec. 1992.

20. R. Koga, P. Yu, K. Crawford, S. Crain and V. Tran, "Comparison of heavy ion and proton induced single event effects (SEE) sensitivities," *IEEE Trans. Nucl. Sci.*, vol. 49, no. 6, pp. 3135-3141, Dec. 2002.

21. D. M. Long, D. G. Millward and J. Wallace, "Dose enhancement effects in semiconductor devices," *IEEE Trans. Nucl. Sci.*, vol. 29, no. 6, pp. 1980-1984, Dec. 1982.

22. A. I. Namenson, "Hardness assurance and overtesting," *IEEE Trans. Nucl. Sci.*, vol. 29, no. 6, pp. 1821-1826, Dec. 1982.

23. A. I. Namenson and I. Arimura, "Estimating electronic parameter end points for devices which suffer abrupt functional failure during radiation testing," *IEEE Trans. Nucl. Sci.*, vol. 32, no. 6, pp. 4250-4253, Dec. 1985.

24. A. I. Namenson, "Statistical analysis of step stress measurements in hardness assurance," *IEEE Trans. Nucl. Sci.*, vol. 31, no. 6, pp. 1398-1401, Dec. 1984.

25. A. G. Stanley and W. E. Price, "Irradiate-anneal screening of total dose effects in semiconductor devices," *IEEE Trans. Nuc. Sci.* vol. 23, no. 6, 2035-2040, Dec. 1976.

26. H. J. Barnaby, S. K. Smith, R. D. Schrimpf, D. M. Fleetwood and R. L. Pease, "Analytical model for proton radiation effects in bipolar devices," *IEEE Trans. Nucl. Sci.*, vol. 49, no. 6, pp. 2643-2649, Dec. 2002.

27. J. A. Babcock, J. D. Cressler, L. S. Vempati, S. D. Clark, R. C. Jaeger and D. L. Harame, "Ionizing radiation tolerance of high-performance SiGe HBTs grown by UHV/CVD," *IEEE Trans. Nuc. Sci.* vol.42, no. 6, pp. 1558-1566, Dec. 1995.

International Journal of High Speed Electronics and Systems
Vol. 14, No. 2 (2004) 563–574

World Scientific
www.worldscientific.com

IONIZING RADIATION EFFECTS ON ULTRA-THIN OXIDE MOS STRUCTURES

A. CESTER AND A. PACCAGNELLA

Department of Information Engineering,
University of Padova, via Gradenigo 6/b, 35131 Padova, Italy

We have briefly reviewed the most important degradation mechanisms affecting ultra-thin gate oxides after exposure to ionizing irradiation. The increase of the gate leakage current seems the most crucial issue for device lifetime, especially for non-volatile memory and dynamic logic. The build-up of positive charge in the oxide and the subsequent threshold voltage shift, which was the major concern for thicker oxide, are no longer appreciable in today's devices due to the reduced oxide thickness permitting a fast recombination of trapped holes with electrons from interfaces. Among the leakage currents affecting thin oxides we have considered here the Radiation Induced Leakage Current (RILC) and the Radiation Soft Breakdown (RSB). RILC is observed after irradiation with a low Linear Energy Transfer (LET) radiation source and comes from a trap-assisted tunneling of electrons mediated by the neutral traps produced by irradiation. RILC depends on the applied bias during irradiation and the maximum is measured when devices are biased in flat band. Contrarily to RILC, RSB is observed after irradiation with high LET ions and derives from the formation of several conductive paths across the oxide corresponding to the ion hits. RSB conduction is explained by the theory of the Quantum Point Contact as also proposed for the electrically induced Soft breakdown. Finally, we present some preliminary results, which indicate that although the direct effects of irradiation (in terms of gate leakage current increase) are small for oxide thinner than 3nm, it is possible that these devices may experience an accelerated wear-out and/or breakdown after subsequent electrical stress relative to a fresh (not irradiated) device.

Keywords: MOS devices; radiation effect; CMOS device reliability; Soft Breakdown; oxide wear-out.

1. Introduction

It is well known that exposure to ionizing radiation degrades the electrical properties of solid-state electronics. The effects of γ-rays, x-rays, electron, proton, heavy ion exposure on MOS devices characteristics have been the topic of several works, books, and review articles over the past two decades.[1-4] In principle, the main effect of an ionizing particle is the generation of electron-hole pairs along the particle track, in a number and density depending on the particle type and energy and on the material characteristics. The energy lost in ionization processes is usually measured by the Linear Energy Transfer[5] (LET) parameter with dimension $MeV \cdot g^{-1} \cdot cm^2$. In irradiated MOS structures, carriers generated in the silicon substrate may produce transient phenomena possibly leading to soft error but also to catastrophic consequences such as Latch Up, mainly due to charge collection at reverse biased junctions: these phenomena will not be considered in this work. Instead we'll examine the radiation effects on the gate oxide layers, where the majority of the radiation generated carriers are promptly recombined shortly after thermalization. Some are left behind: while electrons can be rapidly swept out by the oxide field owing to their high mobility, slow residual holes may be trapped in pre-existing defects or in radiation induced defects inside the oxide. The resulting positive charge trapped in the gate oxide has been the main reliability concern in irradiated MOS along with the radiation induced Si/SiO_2 interface defects, related as well to radiation induced hole trapping and defect generation. As a consequence substantial threshold voltage (V_T) shift and transconductance (g_m) reduction have been measured as the typical results of exposing MOSFETs to ionizing radiation. CMOS hardening technologies were developed ad hoc to prevent or limit the occurrence of these degradation phenomena, permitting the use of hardened MOS IC's in radiation harsh environments.

In parallel, Moore's law itself was playing an underground role in enhancing the radiation tolerance of CMOS components, a role appeared only in recent years. In fact, shrinking the device dimensions following Moore's law can be accomplished only by the parallel thinning of the gate oxide to preserve the transistor current driving capability[6,7]. How the gate oxide thinning could enhance the device radiation tolerance was foreseen in the pioneer work by Saks and Ancona.[8] In fact, being the electron tunneling distance around 3 nm in SiO_2, when the oxide thickness reaches 6 nm or less the radiation induced oxide positive charge is easily recombined or neutralized by electrons tunneling from the gate and/or Si substrate. This oxide thickness was approximately reached at the CMOS technological node of 0.25μm, and the first results reporting the excellent radiation tolerance of quarter-micron CMOS components soon appeared.[9,10] Positive charge accumulation in the thick field oxide is not removed for both LOCOS and shallow trench isolation and it may produce the onset of the parasitic nMOSFETs between source and drain. However, this problem can be successfully circumvented by using enclosed (gate-all-around) transistor layouts and proper guard-rings.[9] For these devices nominal radiation tolerance over 10 Mrad(Si) were measured, i.e., exceeding by far the typical requests even for long space missions. Following the CMOS technological evolution the gate oxide thickness has now reached 2 nm or so, becoming even less sensitive to positive charge trapping problems.

If radiation induced charge trapping problems have been overcome owing to the gate oxide transparency to electrons, this characteristic has become the weak point from a reliability viewpoint. Ionizing radiation (as well as electrical stresses) can in fact produce defects acting not as trapping centers but as the agents of leakage paths across the gate oxide driving an excess gate current. Gate leakage adversely affects the overall circuit power consumption, which has been often taken as the key parameter for reliability predictions. Different leakage currents developed depending on the oxide thickness and radiation LET. In this work, we have reviewed the main issues of Radiation Induced Leakage Current (RILC) and Radiation Soft-Breakdown (RSB), as well as the long term wear out of irradiated oxides. In section 2 we have analyzed the RILC conduction mechanism, its dependence on the applied bias during irradiation and the growth kinetics with the total dose. In section 3 we summarize the main aspects of RSB: the conduction mechanism, the dependence on applied bias during irradiation, the gate current noise, and the temperature dependence. Finally, in section 4 we present the radiation induced wear-out and breakdown of gate oxides, which derives from the combined effect of irradiation plus a successive electrical stress.

2. Radiation Induced Leakage Current

Radiation Induced Leakage Current was firstly observed in oxides with thickness in the range of 4-8 nm.[11] The gate current density-oxide field characteristics (J_g-E_{ox}) of a 6-nm oxide are shown in Fig. 1a before and after irradiation with 8 MeV electrons, whose LET is 1.8 $MeV \cdot g^{-1} \cdot cm^2$. The main effect of irradiation is the increase of the low-field gate current, observed between E_{ox}=3MV/cm and E_{ox}=6MV/cm in Fig. 1a. The oxide trapped charge in the irradiated MOS capacitors is negligible, as deduced from the overlap of the high-field characteristics (over 6 MV/cm) of irradiated and unirradiated devices, corresponding to the Fowler-Nordheim electron tunneling regime.[12,13] RILC can be evaluated as the excess current J_e arisen after stress:

$$J_e = J_g - J_0 \tag{1}$$

where J_g and J_0 are the gate current density after and before the stress, respectively.

RILC shares several characteristics with the SILC (Stress Induced Leakage Current) produced by electrical stresses at high oxide fields.[14-18] As for SILC, RILC has been attributed to a trap-assisted tunneling of electrons across the oxide, mediated by neutral traps generated by irradiation.[14,15] An analytical model of RILC has been developed for ultra-thin oxides submitted to ionizing radiation, based on the solution of the Schrödinger equation for a simplified oxide band structure.[19,20] Here RILC occurs through a two-step process (see Fig. 1b): first, an electron tunnels into the oxide defect from the cathode conduction band edge. Then, the electron tunnels out the trap after having lost approximately 1.5eV[19,20], in agreement with previous findings for SILC.[14,15] Consequently, the final process results to be an inelastic trap assisted tunneling. The mathematical support of the model is quite heavy and even the final analytical expression of the tunneling current is pretty complex. They show that the RILC is definitely field dependent and that the same trap parameters (energy position, cross section) can be used independently of the total dose, indicating that similar defects are generated at the beginning of, during, and at the end of irradiation . Noticeably, only defects close to the center of the oxide thickness can contribute to the excess gate leakage, owing to the decrease of the cumulative tunneling probability for defects near to the interfaces.[20]

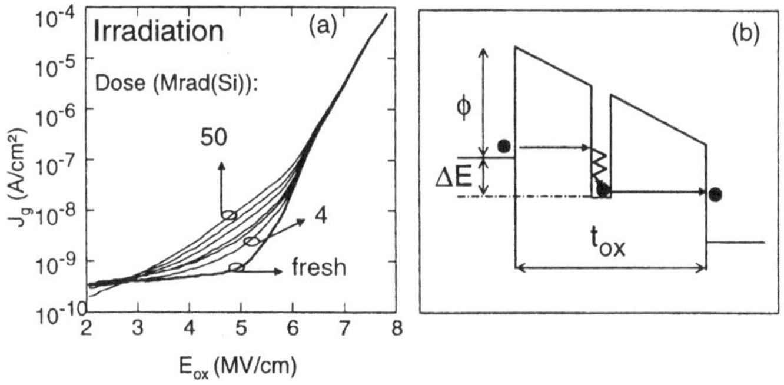

Fig. 1. J_g-E_{ox} a) curves measured before (fresh) and after irradiation for various doses ranging from 4 to 50 Mrad(Si); b) Simplified band structure for the inelastic trap assisted tunneling mediating RILC. [From: M. Ceschia, et al. IEEE – Trans. Nucl. Sci. (**45**) 1998 - © 1998 IEEE]

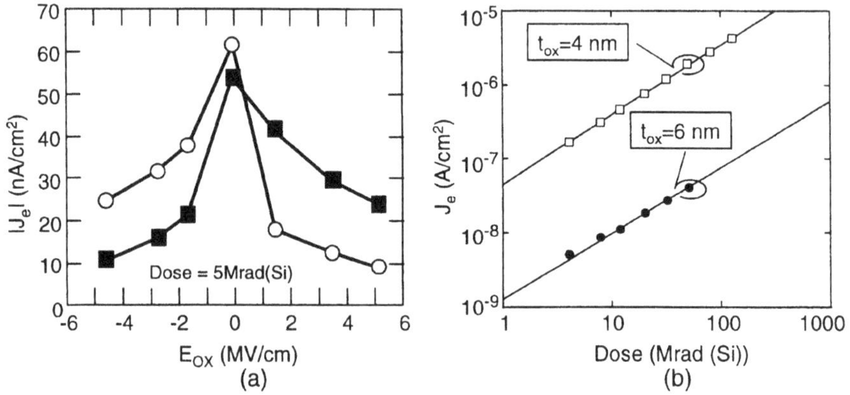

Fig. 2. a) Excess current read at $|E_{OX}|$ = 5MV/cm for different gate voltages applied during irradiation for t_{ox}=4nm (O=positive RILC; ■=negative RILC); b) RILC kinetics (negative excess current) for t_{ox}=4nm and 6nm. [From: M. Ceschia, et al. IEEE – Trans. Nucl. Sci. (**45**) 1998 - © 1998 IEEE]

Even though traps involved in RILC conduction are neutral, RILC depends on the oxide field applied during irradiation, as illustrated in Fig. 2a for t_{ox}=4nm.[19] RILC is read at $|E_{ox}|$=5 MV/cm. Positive RILC is higher than the negative one on capacitors negatively biased during irradiation. The opposite holds true on devices positively biased when irradiated. Such difference disappears and RILC is the maximum when the oxide field is close to zero during irradiation. These results show that the trap distribution is controlled by the oxide field during irradiation. This fact cannot be explained if neutral traps were directly generated by the impinging 8-MeV electrons, as in case of knock-on displacement damage of the oxide lattice, for instance. Neutral defects should be homogeneously generated in the oxide layer, and any effect due to the oxide field should be negligible. Instead, charged precursors must be involved in a field sensitive generation process of RILC defects. Neutral defects may result from the neutralization of an E′ centers after capturing a hole.[21,22] In other words, a neutral electron trap could result from a hole capture at a weak Si-Si covalent bond, which may relax into a $Si:^{-}\ Si^{+}$ neutral amphoteric defect after the hole compensation. Even though holes appear as the first candidate for generating neutral defects, the possible role of hydrogen in principle cannot be neglected.[23] In fact, ionizing radiation can easily break the weak bonds of bonded H atoms, which can migrate under an electric field and be trapped in different sites, possibly generating again neutral defects in the oxide and/or leaving electron traps in the original site.

Remarkably, RILC can be generated even at zero applied oxide field (it is maximum at E_{ox}=0) indicating that high oxide fields are not necessary to produce the neutral defects, but only instrumental to inject energetic electrons (and holes) during high field stresses producing SILC.

RILC intensity grows with the defect density and consequently with the irradiation dose. Fig. 2b shows the RILC growth kinetics as a function of the total dose for 4-nm and 6-nm oxide irradiated with 8 MeV electrons. RILC is measured at E_{ox}=6 MV/cm. For a given radiation dose, the excess current is much higher in the 4-nm oxides, owing to the higher tunneling probability across the barrier, which is thinner than in the 6-nm oxide. RILC data can be well fitted by using the following relation:[19]

$$J_e = K_R \cdot \text{Dose}^{\beta} \qquad (2)$$

where $\beta \cong 0.9$ for t_{ox}=6 nm and $\beta \cong 0.94$ for t_{ox}=4 nm, while K_R is a constant that depends on the read-out gate voltage and oxide thickness. Approximately, RILC grows linearly with the radiation dose, indicating that the oxide defect density follows a linear growth rate as well. This characteristic is strikingly different from that observed for SILC produced by electrical stresses: in fact, SILC tends to saturate at high stress fields, indicating that the defect generation mechanisms slow down when a high defect density is reached in the oxide layer.[18] Instead, radiation induced defects increase at a rate independent of the density of defects that are already present in the oxide, and the resulting leakage currents are higher than those achievable by electrical stresses. In other words, it is possible to generate more oxide defects by ionizing radiation than by electrical stress, where maximum density is limited by the phenomena of dielectric breakdown. However, radiation induced breakdown is not generally observed for low LET radiation sources (i. e., low energy electrons or gamma rays) even at large total doses, but is observed for high LET heavy ions as discussed in the following section.

3. Radiation Soft Breakdown

Recently it was shown that Radiation Soft Breakdown (RSB) can be produced in oxides thinner than 4-nm after exposure to high LET ions (>40MeV·mg^{-1}·cm^2).[24-28] RSB derives from the residual damage of dense ion tracks across the oxide and appears similar to the Soft Breakdown produced by electrical stresses[29-31]. RSB has been observed even after small ion fluences where the corresponding conductive paths are likely distributed on the device surface reproducing the ion hit distribution. The dependence of the excess gate current on the LET of the incident particles is summarized in Fig. 3 for the total dose of 0.6 Mrad(Si) for Ag ions (Fig. 3a) and 80 Mrad(Si) for Si ions (Fig. 3b) in 4-nm oxides. While ions with LET<20 MeV·mg^{-1}·cm^2 can produce only RILC, ions with LET>40 MeV·mg^{-1}·cm^2 (approximately) can induce RSB. From the analysis of the results for electrons and Si ions, RILC seems to be not sensitive to LET variation for a given radiation dose. Instead, RSB occurs only for high LET values and increases with LET. In fact, only ion tracks generating high density of e-h pairs can produce high density of oxide defects, resulting in conductive paths across the oxide through neighboring defects.

In contrast with RILC, which is associated to a tunneling process across a single trap, RSB conduction can be successfully explained on the basis of the Quantum Point Contact (QPC) model.[28] The QPC model was first introduced for electrically induced oxide Hard Breakdown and SB.[32,33] The basic idea beyond the QPC picture is that the conduction is strongly localized and electrons flow through a single conductive path. In the case of radiation, each ion crossing the oxide has a non-zero probability of generating a RSB leakage path.

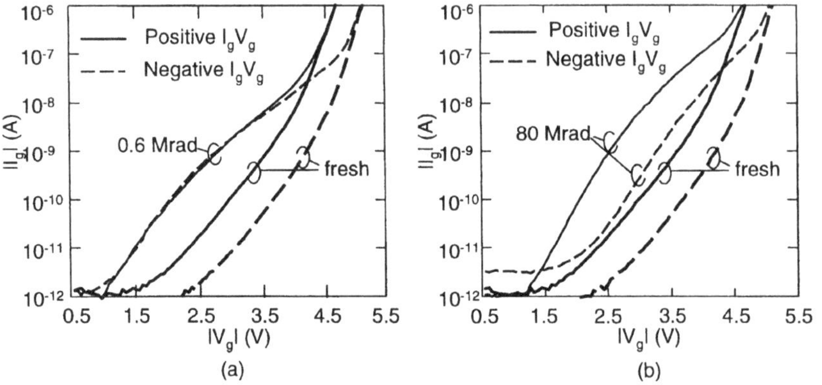

Fig. 3. a) Positive and negative I_g-V_g before and after: a) 257 MeV Ag irradiation (LET = 52.2 MeV·mg^{-1}·cm^2); b) 158 MeV Si irradiation (LET= 2.5 MeV·mg^{-1}·cm^2) [From: M. Ceschia, et al. IEEE – Trans. Nucl. Sci. (**47**) 2000 - © 2000 IEEE]

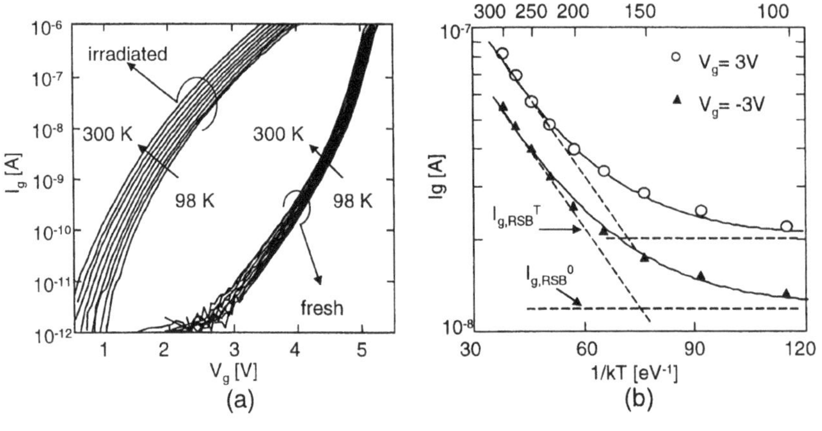

Fig. 4. a) Ig-Vg curves before and after irradiation measured at temperatures ranging between 98K and 300K in a MOS capacitor with t_{ox} = 4 nm. [From: A. Cester, et al. Apl. Phys Lett. (**79**) 2001 - © 2001 American Institute of Physics]
b) RSB current measured at Vg = ±3V as a function of temperature in the same device. The experimental data are well fitted by solid lines representing Eq.(3). Dashed curves correspond to the temperature dependent and independent current components. [From: A. Cester, et al. IEEE – Trans. Nucl. Sci. (**48**) 2001 - © 2001 IEEE]

Based on the QPC model, the lateral dimension of the breakdown paths is so small that the momentum in the direction perpendicular to propagation is quantized. If the path is narrow enough, the ground sub-band E_0 is above the Fermi level at the cathode, and the conduction can only take place by tunneling across an effective 1-dimensional potential barrier, whose height is just E_0. This barrier is not material-related, but its height cannot be larger than that of the Si/SiO_2 system because electrons would be otherwise no longer laterally confined.[33] This model leads to the exponential relation for the gate current across a single SB spot:

$$I_g = A \cdot \exp[B \cdot (V_g - V_0)] \tag{3}$$

The parameters A and B have been related to the height and thickness of the local barrier in the breakdown path. The offset voltage V_0 is due to some potential drop in the substrate. In case of RSB, where several spots can be activated after irradiation, Eq. 3 should be modified accordingly as:

$$I_g = N \cdot A \cdot \exp[B \cdot (V_g - V_0)] \tag{4}$$

N being the number of active spots.[34] Noticeably, RSB is dependent on the gate voltage, as well illustrated in Fig.3a, while RILC is controlled by the oxide field, and an offset appears between positive and negative curves as in Fig.3b.

RSB depends on the measurement temperature. For instance, Fig. 4 shows nine I_g-V_g measurements taken at different temperatures after irradiation with $32 \cdot 10^9$ Ag ions/cm^2 (LET = 52.2 MeV·mg^{-1}·cm^2) on a 4-nm oxide. The corresponding I_g-V_g characteristics before irradiation are plotted for comparison. Fig. 4b shows the RSB current measured at a constant gate voltage Vg=±3V in the same device of Fig. 4a. In fresh devices the gate leakage current derives from electron tunneling, which is a direct tunneling across a trapezoidal potential barrier for V_g<4V and a Fowler-Nordheim tunneling across a triangular barrier for V_g>4V. Being controlled by the direct or FN tunneling, the fresh I_g-V_g characteristics are almost independent on temperature and the different curves almost overlap in Fig.4a. The temperature effect is larger on the RSB current, which decreases at low temperature but it is still orders of magnitude larger than in non-irradiated devices. From these results the gate current can be (at least empirically) modeled as the superposition of two contributions, corresponding to two the dashed lines in Fig.4b:[30]

$$I_{gRSB}(T) = I_{gRSB,0} + I_0 \cdot \exp(-E_a/kT) \tag{5}$$

$I_{gRSB,0}$ is the temperature-independent component, while the second term is the temperature-dependent component following the usual exponential law for thermal activation. The activation energy E_a has been estimated around $E_a = 0.057 \pm 0.005$ eV, independent on the measurement gate voltage.[30] Such low E_a value reflects the weak temperature dependence of tunneling-based phenomena. Hence, the increase of the device temperature during normal operation will be accompanied by a limited increase of the RSB current. For instance, a RSB current of 1 µA at room temperature (300 K) reaches only 1.5 µA at 375K, in 4-nm oxides.

One of the main features of the RSB conduction is the noisy behavior, which is never observed in case of low LET irradiation (that is, in case of RILC). In Fig. 5 we have plotted the gate current measured at V_g = -2.7V in a 4-nm oxide after irradiation with 257 MeV I ions (LET = 61.6 MeV·mg^{-1}·cm^2). The RSB current approximately behaves as a multi-level Random Telegraph Noise (RTN).[30] Such fluctuations correspond to the activation/inactivation of conductive paths inside the radiation induced oxide damaged areas.

By zooming the gate current a "small" Random Telegraph Noise (RTN) appears superimposed to the main "large" fluctuations (see the two zooms in Fig. 5), pointing out to the complex nature of the radiation induced weak spots. Mixed "large" and "small" RTN fluctuations have been obtained for different oxide thicknesses, radiation doses, measurement polarities, and sampling frequencies. Noticeably, "small" fluctuations occur much more frequently than "large" ones, and both are usually attributed to an electron trapping/detrapping process.

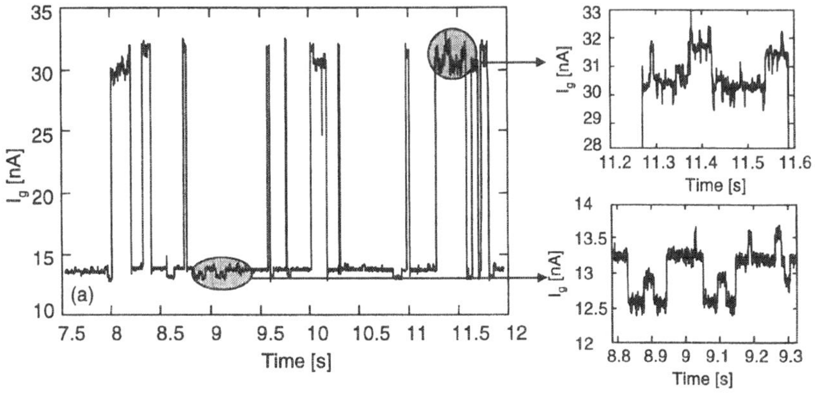

Fig. 5. Gate current measured at Vg = -2.7 V as a function of measurement time in a 4-nm oxide after $7 \cdot 10^6$ I ions/cm2. Sampling frequency was 4 kHz. [From: A. Cester, et al. IEEE – Trans. Nucl. Sci. (**48**) 2001 - © 2001 IEEE]

The correlation between switching frequency and fluctuation amplitude is not clear at present, but this result may provide further understanding of the microscopic nature of the RSB spot. In fact, it is reasonable to assume that fast fluctuations can be attributed to electron trapping/detrapping in shallow traps, hence characterized by fast trapping/detrapping kinetics. On the contrary, deep traps could be responsible for the slow fluctuations due to their longer capture and emission time. The current fluctuations are either "slow and large" or "fast and small": this suggests that shallow electron traps should modify the conductance of the RSB spot much less than slow deep traps. In a simple view, this indicates that deep traps could reside in a central region of the RSB spot, which can be effectively clogged by a trapped electron resulting in larger RSB current variations. Instead, shallow traps should be located peripherally with respect to the RSB spot, where their impact on the driven current is relatively modest.

RILC and RSB may negatively impact the device reliability. For instance they surely hamper the data retention in memory cell such as Flash EPROM or DRAM and can be the origin of read disturbs of non-volatile memories[35]. Nevertheless, RILC and RSB have been demonstrated not to be an enhanced failure mode of significance for Very Large Scale Integration (VLSI) static logic.[36] Those observations were based on the assumption that the device operation integrity is basically preserved owing to the small gate leakage. In addition the impact of RILC and RSB is further decreased due to the enhanced tunneling current featured by the thinner oxide (<3nm) employed in today and future technologies. Moreover, the RSB probability appears extremely low in space missions. These conclusions are basically true for RSB itself, nevertheless recent studies have shown that RSB is an indication that other problems are about to occur, as we will discuss in the next section. Furthermore, the RSB impact on the performance of MOSFETs with tiny gate area, proper of the new decananometer CMOS generation, is still an open issue worth to be investigated. In fact, the gate area involved in RSB may become a significant fraction of the whole gate area, thus affecting the overall MOSFET electrical characteristics, such as drain current and transconductance.

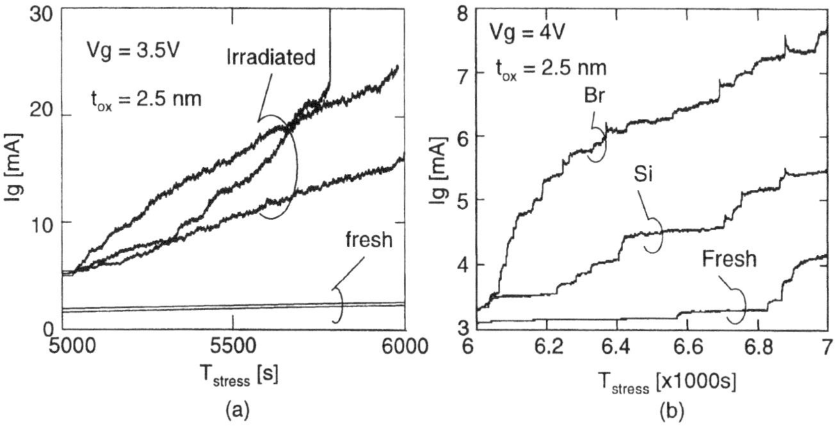

Fig. 6. Gate current evolution during Constant Voltage Stress for: a) three fresh and three irradiated devices with $6.5 \cdot 10^7$ I ions/cm^2; b) gate current evolution for a fresh and two irradiated devices ($6.5 \cdot 10^7$ Si and Br ions/cm^2).

4. Radiation Induced Wear-Out and Breakdown

Recently, it has been reported that radiation is not the only concern when considering long term device reliability, as gate oxides are subjected to high electric fields during normal circuit operations. Oxide wear-out due to the applied oxide fields is usually investigated on large area MOS capacitors through accelerated electrical stresses at fields higher than those of normal device operation. The cumulative effect of ionizing radiation and accelerated electrical stresses has been considered a few times in literature[11,37-39], mainly for oxides irradiated with low LET radiation (γ-rays, electrons, X-rays). It was found that radiation damage does not significantly impact the oxide reliability parameters, such as time-to-breakdown and time-zero-breakdown, at least for relatively high oxide thickness (>12nm). In case of RILC, a modification due to irradiation was reported for the growth kinetics of the excess current during electrical stresses after γ-irradiation.[11]

Some fundamental differences appear under high-LET-ion irradiation. In this case an anomalous and accelerated wear-out appears when Constant Voltage Stress (CVS) is applied to an irradiated device. The effect of irradiation on the wear-out of the oxide insulating properties is illustrated in Fig. 6, where we present the gate current evolution during a CVS at V_g = 3.5 V in a MOS capacitor with gate area of 10^{-2}cm^2. Here we show the comparison between three fresh and irradiated devices with $6.5 \cdot 10^7$ I ions/cm^2.

For irradiated devices the current increase is fast and no clear SB event can be identified, while a progressive degradation of the oxide insulating characteristics appears. Instead, no appreciable degradation can be seen in the three fresh devices. Similar behavior has been observed for different ions (Si, Br, I, and Au) and the degradation rate has been found increasing with increasing LET.

The progressive degradation can be attributed either to the enlargement of a single weak spot driving most of current, or to the sequential opening of several small weak spots, which appears to be more likely. In case of a single spot we should expect some SB-like signatures, such as discrete current jumps and random telegraph signal noise, which are not observed indeed. Heavy ion irradiation can instead enhance the defect production during electrical stresses in several weak points corresponding to the ion hits (65000 ions hit the oxide surface), so that electrons injected during CVS require less energy to generate defects and produce some SB leakage paths.

These results have important reliability implications. From the point of view of device application in radiation environments we may pose the fundamental question of the RSB impact on the CMOS circuit performance. Could RSB be a concern for space applications? In principle it shouldn't, at low doses and for thin gate oxides. Anyway, it is not clear if even the low level damage will act (or less) as the seed for further oxide degradation due to electrical stress during the device operating life. On one side this opens the question of the RSB impact on long term failure predictions of irradiated oxides.

5. Conclusions

In this work we have reviewed some of the basic degradation mechanisms affecting ultra-thin gate oxides. Positive charge trapping is no longer a real concern for device lifetime, since the reduced oxide thickness enhances the probability that holes, generated by ionizing radiation and then captured in the oxide, are easily recombined by electrons injected from the interfaces. More serious problem comes from the neutral trap generation inside the oxide, which contribute to increase the gate leakage current. Different leakage currents have been observed after irradiation mainly dependent on the Linear Energy Transfer (LET) of radiation sources. Among them:

- Radiation Induced Leakage Current (RILC) is generally observed after irradiation with γ-rays, electrons, X-rays and ions with low LET mainly and derives from a trap assisted tunneling through the defects generated by irradiation.
- Radiation Soft Breakdown, deriving from the formation of several localized conductive path where the ions hit the oxide surface. On the contrary of RILC, RSB is characterized by a multi-level Random Telegraph Noise.

The non-zero gate current surely hampers the long-term reliability of non-volatile memories and can affect circuits such as dynamic logic. Even though effect of the gate leakage currents, which add up to the tunneling current of a fresh device, can be pretty small and might not be a real concern for the majority of logic circuits, heavy ion irradiation can generate weak and rich-defect regions corresponding to the ion hits. Such regions may act as a seed for further degradation during a subsequent electrical stress, leading to an accelerated wear-out and/or breakdown, and seriously hampering the device lifetime.

Acknowledgements

We would like to thank Andrea Scarpa, Andrea Candelori, Marco Cerchia, Salvatore Cimino, Leonardo Bandiera, Luca Larcher, Giorgio Cellere, and Gabriella Ghidini.

References

1. T. P. Ma and P. V. Dressendorfer, "Ionizing radiation effects in MOS devices and circuits", John Wiley & Sons, 1989.
2. A. Holmes-Siedle and L. Adams, "Handbook of radiation effects" – 2nd Edition, Oxford University Press, 2002.
3. T. R. Oldham, "Ionizing radiation effects in MOS oxides", World Scientific, 2000.
4. IEEE – Trans. Nuclear Science - Special issue on Single-Event Effects and space radiation environment, April 1996.
5. J. F. Ziegler, and J. P. Biersack, "Stopping and Range of Ions in Matter", Version 2000.37, Jan. 1, 2000.
6. B. Davari, R. H. Denard, and G. Shahidi, "CMOS Scaling for High Performance and Low Power – The Next Ten Years", Proc. of IEEE, Vol. 83, pp. 595, 1995.
7. D. J. Frank, R. H. Dennard, E. Nowak, P. M. Solomon, Y. Taur, and H.-S. P. Wong, "Device Scaling Limits of Si MOSFETs and Their Application Dependencies", Proc. of IEEE, Vol. 89, pp. 259, 2001.
8. N. S. Saks and M. G. Ancona, "Generation of Interface States by Ionizing radiation at 80K Measured by charge pumping and subthreshold Slope Techniques", IEEE – Trans. Nucl. Sci., Vol. 34, No. 6, pp. 1172, 1987.
9. A. Giraldo, A. Paccagnella, C. Dachs, F. Faccio, E. Heijne, P. Jarron, K. Kloukinas, A. Marchioro, "Total dose behaviour of commercial submicron VLSI technologies at low dose rate", Proc. 3rd Workshop on Electronics for LHC Experiments, London, CERN Technical note CERN/LHCC97-60, p. 139-143, 1997.
10. R. C. Lacoe, D. C. Mayer, J. V. Osborn, and G. Yabiku, "Total dose hardness of three commercial CMOS microelectronics foundries", 4th RADECS, pp. 265. 1997.
11. A. Scarpa, A. Paccagnella, F. Montera, G. Ghibaudo, G. Pananakakis, G. Ghidini, and P. G. Fuochi, "Ionizing Radiation Induced Leakage Current on Ultra-Thin Gate Oxides", IEEE Trans. Nucl. Sci., vol. 44, p. 1818, 1997.
12. D. J. DiMaria, "Determination of insulator bulk trapped charge densities and centroids from photocurrent-voltage characteristics of MOS structures", J. Appl. Phys., Vol. 47, pp. 4073, 1976.
13. M. Lenzlinger, E .H. Snow, "Fowler-Nordheim tunnelling into thermally grown SiO_2", J. Appl. Phys., Vol. 40, pp. 278, 1969.
14. E. Rosenbaum and L. F. Register, "Mechanism of stress induced leakage current in MOS capacitors", IEEE Trans. on Electron Device, Vol. 44, pp.317, 1997.
15. S. Takagi, N. Yasuda and A. Toriumi, "Experimental Evidence of Inelastic Tunneling in Stress-Induced Leakage Current", IEEE Trans. on Electron Devices, Vol. 46, pp. 335, 1999.
16. E. F. Runnion, S. M. Gladstone, R. S. Scott, D. J. Dumin, L. Lie, and J. C. Mitros, "Thickness Dependence of Stress-Induced Leakage Current in Silicon Dioxide", IEEE Trans. on Electron Devices, Vol. 44, pp. 993, 1997.
17. K. Sakakibara, N. Ajika, M. Hatanaka, H. Miyoshi, and A. Yasuoka, "Identification of Stress Induced Leakage Current Components and Corresponding Trap Models in SiO_2 Films", IEEE Trans. on Electron Devices, Vol. 44, pp. 986, 1997.
18. A. Scarpa, P. Riess, G. Ghibaudo, A. Paccagnella, G. Pananakakis, M. Ceschia, and G. Ghidini, "Electrically and radiation induced leakage current in thin oxides", Microelectronics Reliability, Vol. 40, pp. 57, 2000.
19. M. Ceschia, A. Paccagnella, A. Cester, A. Scarpa, and G. Ghidini, "Radiation Induced Leakage Current and Stress Induced Leakage Current in Ultra-Thin Gate Oxides", IEEE Trans. on Nucl. Sci., vol. 45, No. 6, p.2375, Dec. 1998.
20. L. Larcher, A. Paccagnella, M. Ceschia, and G. Ghidini "A Model of Radiation Induced Leakage Current (RILC) in Ultra-Thin Gate Oxides", IEEE Trans. on Nucl. Sci., vol. 45, p.1553, 1999.
21. A. J. Lelis, H. E. Boesch Jr., T. R. Oldham, and F. B. McLean, "Reversibility of trapped hole annealing", IEEE Trans. Nucl. Sci., vol. 35, pp. 1186 -1191, 1988.
22. M. Walters and A. Reisman, "Radiation-Induced Neutral Electron Trap Generation in Electrically

Biased Insulated Gate Field Effect Transistor Gate Insulators", J. Electrochem. Soc., vol. 138, pp. 2756, 1991.
23. D. J. DiMaria and E. Cartier, "Mechanism for stress-induced leakage currents in thin silicon dioxide films", J. Appl. Phys, vol. 78, pp. 3883, 1995.
24. M. Ceschia, A. Paccagnella, S. Sandrin, G Ghidini, J. Wyss, M. Lavalle, and O. Flament "Low Field Leakage Current and Soft Breakdown in Ultra-Thin Gate Oxides After Heavy Ions, Electrons or X-ray Irradiation", IEEE Trans. on Nucl. Sci., Vol. 47, p.566, 2000.
25. M. Ceschia, A. Paccagnella, A. Cester, G Ghidini, and J. Wyss, "From Radiation Induced Leakage Current to Soft-Breakdown in irradiated MOS devices with ultra-thin gate oxide" Mat. Res. Soc. Symp. Proc., Vol. 592, pp. 201, 2000.
26. M. Ceschia, A. Paccagnella, M. Turrini, A. Candelori, G Ghidini, and J. Wyss, "Heavy ion irradiation of thin gate oxides", IEEE Trans. on Nucl. Sci., vol. 47, p.2648, 2000.
27. A. Cester, L. Bandiera, M. Ceschia, G Ghidini, and A. Paccagnella, "Noise Characteristics of Radiation-Induced Soft Breakdown Current in Ultrathin oxides", IEEE – Trans. Nucl. Sci., Vol. 48, pp. 2093, 2001.
28. J. F. Conley, Jr., J. S. Suehle, A. H. Johnston, B. Wang, T. Miyahara, E. M. Vogel, and, J. B. Bernstein, "Heavy ion induced Soft Breakdown of thin gate oxides", IEEE Trans. on Nucl. Sci., vol. 48, p.1913, 2001.
29. E. Miranda, J. Suñe, R. Rodriguez, M. Nafría, X. Aymerich, L. Fonseca, and F. Campabadal, "Soft Breakdown conduction in ultrathin (3-5nm) gate dielectrics", IEEE – Trans. Electron Devices, vol. 47, p. 82, 2000
30. A. Cester, L. Bandiera, G Ghidini, I. Bloom, and A. Paccagnella, "Soft Breakdown Current Noise in Ultra-thin Gate Oxides", Solid-St. Electron., vol. 46, p. 1019, 2002.
31. S. Bruyere, E. Vincent, and G Ghibaudo, "Quasi-Breakdown in ultra-thin SiO_2 films: occurrence characterization and reliability assessment methodology", Proc. 38th IEEE – IRPS, 2000, p. 48-54.
32. E. Miranda, J. Suñe, M. Naifría, and X. Aymerich, "Modeling of Soft Breakdown spots in silicon dioxide films as point contacts", Appl. Phys. Lett., Vol. 74, p. 959, 1999.
33. A. Cester, L. Bandiera, J. Suñe, A. Paccagnella, L. Boschiero, and G Ghidini, "A Novel Approach to Quantum Point Contact for post Soft Breakdown conduction", Proc. of IEEE – IEDM, 2001.
34. A. Cester, A. Paccagnella, J. Suñe, and E. Miranda, "Post-radiation-induced soft breakdown conduction properties as a function of temperature", Appl. Phys. Lett., Vol. 70, pp. 1336, 2001.
35. G Cellere, A. Chimenton, L. Larcher, A. Modelli, A. Paccagnella, P. Pellati, J. Wyss, "Radiation effects on floating-gate memory cells", IEEE – Trans. Nucl. Sci., Vol. 48, pp. 2222, 2001
36. L. W. Massengill, B. K. Choi, D. M. Fleetwood, R. D. Schrimpf, K. F. Galloway, M. R. Shaneyfelt, T. L. Meisenheimer, P. E. Dodd, J. R. Schwank, Y. M. Lee, R. S. Johnson, and G Lucovsky, "Heavy-ion-induced breakdown in ultra-thin gate oxides and high-k dielectrics", IEEE Trans. Nucl. Sci., vol. 48, pp. 1904, 2001.
37. A. Paccagnella, A. Candelori, A. Milani, E. Formigoni, G Ghidini, F. Pellizzer, D. Drera, P. G Fuochi, and M. Lavale, "Breakdown properties of irradiated MOS capacitors", IEEE Trans. Nucl. Sci., vol. 43, p. 2609, 1996.
38. S. R. Anderson, R. D. Schrimpf, K. F. Galloway, and J. L. Titus, "Exploration of heavy ion irradiation effects on gate oxides reliability", Microelectron. Reilab., vol. 35, p. 603, 1995.
39. T. Brozek, R. Wisniewski, R. B. Beck, and A. Jakubowski, "Effect of radiation on breakdown of electrically pre-degraded oxides in MOS structures", Microelectron. Eng., vol. 28, p. 349, 1995.

International Journal of High Speed Electronics and Systems
Vol. 14, No. 2 (2004) 575–580

HYDROGEN AT THE Si/SiO_2 INTERFACE: FROM ATOMIC-SCALE CALCULATIONS TO ENGINEERING MODELS

S. N. RASHKEEV,[1] D. M. FLEETWOOD,[2,1] R. D. SCHRIMPF,[2]
AND S. T. PANTELIDES[1,3]

[1]*Department of Physics and Astronomy, Vanderbilt University, Nashville, TN 37235, U.S.A.*

[2]*Department of Electrical Engineering and Computer Science, Vanderbilt University, Nashville, TN 37235, U.S.A.*

[3]*Solid State Division, Oak Ridge National Laboratory, Oak Ridge, TN 37831, U.S.A.*

Two contrasting behaviors have been observed for H in Si/SiO_2 structures: a) Radiation experiments established that protons released in SiO_2 migrate to the Si/SiO_2 interface where they induce new defects; b) For oxides exposed first to high-temperature annealing and then to molecular hydrogen, mobile positive charge believed to be protons can be cycled to and from the interface by reversing the oxide electric field. First-principles density functional calculations identify the atomic-scale mechanisms for the two types of behavior and conditions that are necessary for each. Using the results of the atomic-scale calculations we develop a model for enhanced interface-trap formation at low dose rates due to space charge effects in the base oxides of bipolar devices. We find that the hole trapping in the oxide cannot be responsible for all the Enhanced Low-Dose-Rate Sensitivity (ELDRS) effects in SiO_2, and the contribution of protons is also essential. The dynamics of interface-trap formation are defined by the relation between the proton mobility (transport time of the protons across the oxide) and the time required for positive-charge buildup near the interface due to trapped holes. The analytically estimated and numerically calculated interface-trap densities are found to be in very good agreement with available experimental data.

Keywords: bipolar transistors; radiation effects; interface phenomena; space technology.

1. Introduction

The transport and reactions of hydrogen-related species play critical roles in determining the ionizing radiation response and long-term reliability of metal-oxide-semiconductor (MOS) and bipolar microelectronic technologies.[1] As a consequence of device processing, hydrogen is present in abundant quantities in bound forms in the oxide, the polysilicon gates and metal interconnects. When a MOS device is irradiated, hydrogen is released and transports to the Si/SiO_2 interface where it can induce new defects.[2-4] When significant densities of oxygen vacancies are present near the interface, generated protons do not react with the interface, and they can be moved back into the oxide by reversing the electric field.[5] Here we describe the results of our recent first-principles density-functional calculations that allowed us to identify atomic-scale mechanisms that are responsible for this behavior.[6-9] Using these results we develop a model for enhanced interface-trap formation at low dose rates in bipolar devices.[10] This result illustrates how the atomic-scale calculations may be used in developing engineering models.

2. Interaction between hydrogen and the Si/SiO_2 interface (atomic-scale calculations)

Hydrogen is released and migrates through the oxide as a proton because H^0 is not a stable charge state. Our calculations also show that neutral hydrogen is not stable in the interfacial region as well.[7,8] This means that a proton arriving at the interface interacts with the hydrogen passivated dangling bond (Si-H) directly, without being neutralized by an electron from the Si

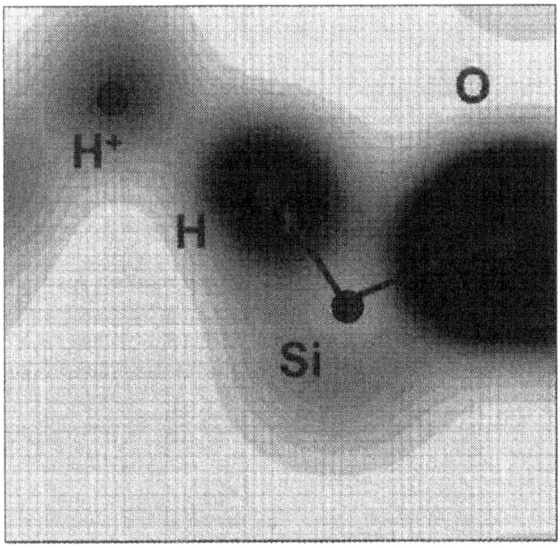

Fig. 1. Electronic density via density functional theory when a proton approaches a Si-H bond at or near Si/SiO_2 interface (after Refs. 7 and 8).

side of the interface as was suggested in some works (Fig. 1). The Si-H bond is highly polar, with about 1.6 electrons located near the H atom, and this negatively charged part of the dipole interacts with the proton directly, forming an H_2 molecule and leaving the positively charged defect (dangling bond without an electron) located at the Si atom. The energy level of this defect is located within the Si band gap, i.e., its charge state can be manipulated by the potential at the interface.

The behavior of H^+ is different at interfaces with different local coordination. At an abrupt interface without suboxide bonds (a Si-Si bond at the SiO_2 side of the interface), the proton does not feel any energy barrier higher than 1 eV (Fig. 2a). It can penetrate into Si or return back into the oxide. However, it probably would prefer to migrate laterally at the interface because the migration energy for such a motion is extremely low (0.3-0.5 eV) before it reacts with some hydrogen passivated dangling bond in a way described above and creates a new defect. A suboxide bond can trap a proton in a position with a very asymmetric barrier (Fig. 2b). The proton would prefer to return back in the oxide, rather than to penetrate into Si.

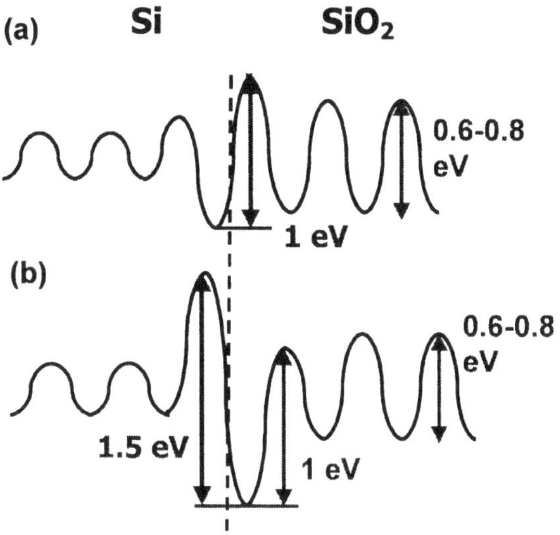

Fig. 2. A schematic for the proton potential energy across the Si/SiO_2 interface (shown by the dashed line): a) for an abrupt intrinsic interface; b) for an interface with a Si-Si suboxide bond.

Therefore, in the samples with a high density of suboxide bonds the flood of protons gets captured by the suboxide bonds, enabling one to observe cycling between the two interfaces when the electric field in the oxide is periodically reversed.[5]

3. Enhanced interface-trap formation for low dose rates (physical model)

The radiation-induced gain degradation of many types of bipolar transistors (BJTs) is greater at low dose rates than at high rates. The enhanced low dose rate sensitivity (ELDRS) is an important problem for space systems, and much work has been done to develop hardness-assurance methods that take ELDRS into account.[11] The dose-rate dependence of interface-trap formation can be modeled as a space-charge effect in the oxide.[12,13] It has been suggested that a key element is the difference in transport rates for holes and protons in SiO_2.[4] The fact that at least two different types of radiation-induced trapped positive charge can be observed in high quality gate oxides has been shown experimentally by Freitag et al.[14,15]

Recently, we developed a physical model for the ELDRS phenomenon using differential rate equations for electrons, holes, and protons plus Poisson's equation.[10] Fortunately, in some important cases analytical solutions describing the dynamics of the system may be obtained. The information about the physical mechanisms responsible for radiation-induced interface defect formation was taken from first-principles calculations.[6–9] The proton transport and interfacial reactions described above have been incorporated into a multiple-trapping-detrapping (MTD) code (ECORCE) similar to the one used to study the contribution of oxide trapped charge to ELDRS described by Graves et al.[16] For simplicity (and for convenience of comparison with analytic models), we describe the transport of holes and protons using drift-diffusion equations; a refined model that incorporates dispersive proton motion gives only quantitative changes of the final results and does not provide any new effects.

Interface-trap formation (related to proton reactions at the Si-SiO_2 interface) at high dose rates is reduced due to hole-related positive charge buildup in the Si-SiO_2 interfacial region. The mobility of holes is much higher than the mobility of protons, and the trapped holes can form an electrostatic barrier that prevents the protons from reaching the interface and generating interface traps. At low dose rates, the charge buildup process is reduced, relative to charge transport processes, and both holes and protons can reach the interface and participate in oxide- and interface-trap formation. Our analytical and numerical calculations are in reasonable agreement with available experimental data (Fig. 3).

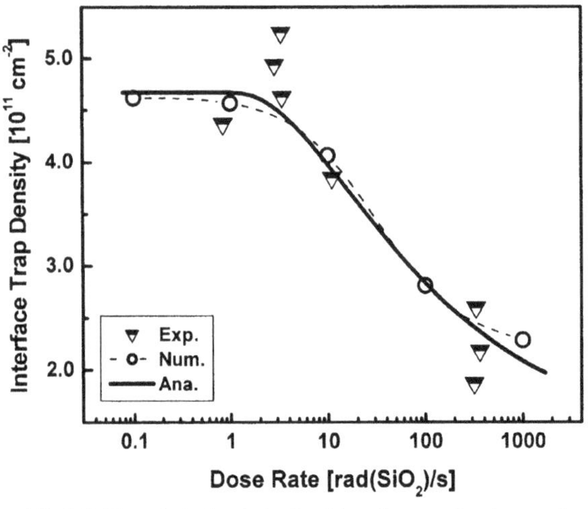

Fig. 3. Experimental (Ref. [12]), analytical and simulated interface trap density as a function of dose rate for an Analog Devices' RF25 capacitor irradiated to a total dose of 200 krad(SiO_2). The hole and proton mobility used in the simulator were $\mu_P = 1\times10^{-5}$ cm^2/Vs and $\mu_{H^+} = 1\times10^{-11}$ cm^2/Vs. Interface trap build-up in the flatband to midgap energy range of the silicon band-gap arises from the contribution of 10% of the protons reaching the interface. The applied electric field is zero in numerical simulations and 0.01 MV/cm in analytical calculations

They also explain the dependence of the positive charge buildup on the mobility of the carriers and on the external electric field. Figure 4 shows the results of simulations for a range corresponding to practical electric fields observed by the transistors built with the RF25 process, e.g. a typical V_{BE} = 0.6 V operating bias would apply a field of 0.1 MV/cm over the base oxide. When the electric field becomes larger, the ELDRS effects become less pronounced because at larger fields it is harder to reach the field reversal regime. This means that all the generated positive charges will reach the interfacial region, where they get a chance to react and create an oxide- or interface- trap. In this case, the concentration of radiation-induced traps is independent of the dose rate.

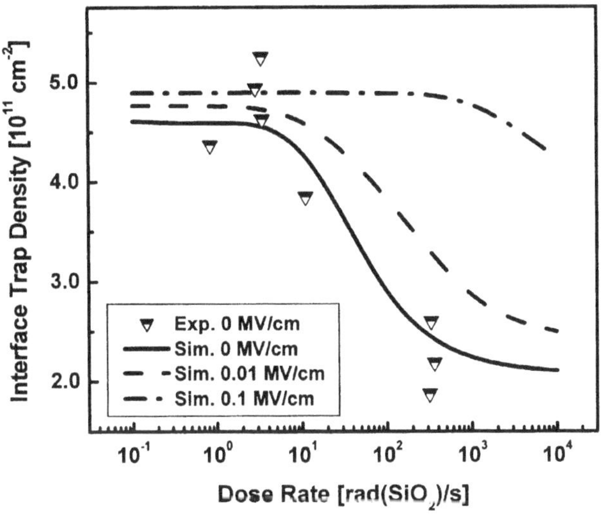

Fig. 4. Experimental (Ref. [12]) interface trap density as a function of dose rate for an Analog Devices' RF25 capacitor irradiated to a total dose of 200 krad(SiO_2). Simulations have been done for different values of the applied electric field. The hole and proton mobility used in the simulator were $\mu_P = 1\times10^{-5}$ cm^2/Vs and $\mu_{H^+} = 1\times10^{-11}$ cm^2/Vs.

4. Conclusion

Using results of first-principles atomic-scale calculations we have developed an analytical model of enhanced low dose rate effects in bipolar base oxides and compared its results with one-dimensional simulations. Our results strongly reinforce the idea that space-charge limited transport of both holes and protons is critical to an understanding of ELDRS effects in bipolar-base oxides. Also, they illustrate the usefulness of first-principles calculations in developing engineering models.

Acknowledgements

The work was supported in part by AFOSR Grant F-49620-99-1-0289, the US Navy, the Defense Threat Reduction Agency under contract No. DTRA01-00-C-0010, and by the William A. and Nancy F. McMinn Endowment at Vanderbilt University. Research at

ORNL was sponsored by the Division of Materials Sciences, US Department of Energy, under contract DE-AC05-00OR22725 managed by UT-Batelle, L.L.C.

References

1. D. M. Fleetwood, "The role of hydrogen in MOS and bipolar radiation response and long-term reliability," *Microelectron. Reliab.*, vol. 42, p. 523, 2002.
2. A. G. Revesz, "Chemical and structural aspects of the irradiation behavior of SiO_2 films on silicon," *IEEE Trans. Nucl. Sci.*, vol. 24, p. 2102, 1977.
3. F. B. McLean, "A framework for understanding radiation-induced interface states in SiO_2 MOS structures," *IEEE Trans. Nucl. Sci.*, vol. 27, p. 1651, 1980.
4. D. B. Brown and N. S. Saks, "Time-dependence of radiation-induced interface trap formation in metal-oxide-semiconductor devices as a function of oxide thickness and applied field," *J. Appl. Phys.*, vol. 70, p. 3734, 1991.
5. K. Vanheusden, W. L. Warren, R. A. B. Devine, D. M. Fleetwood, J. R. Schwank, M. R. Shaneyfelt, P. S. Winokur, and Z. J. Lemnios, "Nonvolatile memory device based on mobile protons in SiO_2 thin films," *Nature (London)*, vol. 386, no. 6625, p. 587, 1997.
6. S. T. Pantelides, S. N. Rashkeev, R. Buczko, D. M. Fleetwood, and R. D. Schrimpf, "Reactions of hydrogen with Si-SiO_2 interfaces," *IEEE Trans. Nucl. Sci.*, vol. 47, p. 2262, 2000.
7. S. N. Rashkeev, D. M. Fleetwood, R. D. Schrimpf, and S. T. Pantelides, "Defect generation by hydrogen at the Si-SiO_2 interface," *Phys. Rev. Lett.*, vol. 87, p. 165506, 2001.
8. S. N. Rashkeev, D. M. Fleetwood, R. D. Schrimpf, and S. T. Pantelides, "Proton-induced defect generation at the Si-SiO_2 interface," *IEEE Trans. Nucl. Sci.*, vol. 48, p. 2086, 2001.
9. S. N. Rashkeev, D. M. Fleetwood, R. D. Schrimpf, and S. T. Pantelides, "Dual behavior of H^+ at Si-SiO_2 interface: mobility vs trapping," *Appl. Phys. Lett.*, vol. 81, p. 1839, 2002.
10. S. N. Rashkeev, C. R. Cirba, D. M. Fleetwood, R. D. Schrimpf, S. C. Witczak, A. Michez, and S. T. Pantelides, "Physical model for enhanced interface-trap formation at low dose rates," *IEEE Trans. Nucl. Sci.*, vol. 49, p. 2650, 2002.
11. R. L. Pease, L. M. Cohn, D. M. Fleetwood, M. A. Gehlhausen, T. L. Turflinger, D. B. Brown, and A. H. Johnston, "A Proposed Hardness Assurance Test Methodology for Bipolar Linear Circuits and Devices in a Space Ionizing Radiation Environment," *IEEE Trans. Nucl. Sci.*, vol. 44, p. 1981, 1997.
12. D. M. Fleetwood, L. C. Riewe, J. R. Schwank, S. C. Witczak, and R. D. Schrimpf, "Radiation Effects at Low Electric Fields in Thermal, SIMOX, and Bipolar-Base Oxides," *IEEE Trans. Nucl. Sci.*, vol. 43, p. 2537, 1996.
13. S. C. Witczak, R. C. Lacoe, D. C. Mayer, D. M. Fleetwood, R. D. Schrimpf, and K. F. Galloway, "Space Charge Limited Degradation of Bipolar Oxides at Low Electric Fields," *IEEE Trans. Nucl. Sci.*, vol. 45, p. 2339, 1998
14. R. K. Freitag, D. B. Brown, and C. M. Dozier, "Experimental Evidence of Two Species of Radiation Induced Trapped Positive Charge," *IEEE Trans. Nucl. Sci.*, vol. 40, no. 6, p. 1316, 1993.
15. R. K. Freitag, D. B. Brown, and C. M. Dozier, "Evidence for Two Types of Radiation-Induced Trapped Positive Charge," *IEEE Trans. Nucl. Sci.*, vol. 41, p. 1828, 1994.
16. R.J. Graves, C.R. Cirba, R.D. Schrimpf, R.J. Milanowski, A. Michez, D.M. Fleetwood, S.C. Witczak, and F. Saigne, "Modeling Low-Dose-Rate Effects in Irradiated Bipolar-Base Oxides," *IEEE Trans. Nucl. Sci.*, vol. 45, p. 2352, 1998.

International Journal of High Speed Electronics and Systems
Vol. 14, No. 2 (2004) 581–603

World Scientific
www.worldscientific.com

SWITCHING OXIDE TRAPS

TIMOTHY R. OLDHAM

NASA Goddard Space Flight Center/QSS Group, Inc.
Code 561.4, Greenbelt, MD 20771, USA

Positive oxide trapped charge is one of the main factors determining the radiation response of a CMOS device. The most widely accepted model for oxide-trapped charge is the dipole model, originally proposed by Lelis et al. The annealing of radiation-induced positive trapped charge proceeds (usually) via the tunneling of electrons, which form metastable dipoles, compensating the trapped positive charge without removing it. Under appropriate bias, these compensating electrons can tunnel back to the Si substrate, restoring the trapped positive charge. The experimental work leading to the development of this model is summarized. By now there is a large body of experimental and theoretical work by others, confirming and extending the original model. In particular, the relevance of the model to some electron trapping studies has been shown, and its application to the larger topic of oxide reliability is discussed.

Keywords: radiation effects, oxide trap charge, oxide reliability

1. Introduction

We consider radiation-induced charge trapping in SiO_2 dielectric layers, primarily from the point of view of CMOS devices. However, SiO_2 insulators are used in many other ways, and the same defects occur in other contexts. The key studies, which determined the nature of the oxide charge traps, were done primarily on gate oxides in CMOS devices, because that was the main radiation problem in CMOS at one time. There are two major reviews of radiation-induced oxide charge trapping already in the literature, which discuss the subject in far greater detail than is possible here. The first of these was by McLean et al.[1] in 1989, and the second, ten years later, was intended as an update, because of additional, new work that had been reported.[2]

Basically, the picture that has emerged is that ionizing radiation creates electron-hole pairs in the oxide, and the electrons have much higher mobility than the holes. Therefore, the electrons are swept out of the oxide very rapidly by any field that is present, leaving behind any holes that escape the initial recombination process.[1] These holes then undergo a polaron hopping transport toward the Si/SiO2 interface (under positive bias).

Near the interface, some fraction of them fall into deep, relatively stable, long-lived hole traps. See reference 1 and its bibliography for more details on the charge generation, recombination and transport processes. The nature and annealing behavior of these hole traps is the main focus of this paper, and of reference 2, Chapter 2. One of the main points in this discussion will be that some of these traps form dipoles, which exchange charge reversibly with the Si substrate. This process introduces a source of instability into the transistor threshold voltages, which could be important in complex high-speed circuits. These circuits depend critically on precise threshold control, and often on precise threshold matching for their performance. Any source of variation or instability, therefore, has the potential to degrade performance, or even to cause outright functional failure.

2. The Interfacial Transition Layer

The presence of hole traps in the oxide is a fundamental consequence of the oxidation process by which the oxide is grown. There is a nonstoichiometric transition layer in the oxide near the Si/SiO_2 interface, due to the incomplete oxidation of the Si. An early discussion of this layer was provided by Deal et al.,[3] in a 1967 review article, which was based on work done even earlier. Basically, Si right at the interface cannot be oxidized completely because of the lattice mismatch, so there is always some excess Si. As oxide grows, the interface moves, so the excess Si, in effect, moves into the oxide, where it is eventually consumed. But new excess Si is always being formed at the interface, so a steady state concentration profile of excess Si is eventually established in the interfacial region, with stoichiometric SiO_2 in the bulk oxide. Deal et al. concluded that there were positively charged defects associated with the excess Si, with energy levels apparently outside the Si bandgap. Etchback experiments showed these defects to be within 20 nm of the interface (and the total oxide thickness was much greater in those days). Of course, in present-day samples, the total oxide thickness is typically much less than 20 nm, so this picture has to be modified somewhat. But the process-induced positive charge described by Deal et al. is very consistent with our present understanding of radiation-induced trapped positive charge, except that it was not induced by radiation. Deal et al. also noted that low temperature oxidation increased the concentration of excess Si at the interface, and high temperature annealing tended to reduce it. We will discuss process dependences later, but we note that radiation hardened oxides often use low temperature oxidation and low temperature annealing, both conditions which increase the excess Si concentration.

Later, Grunthaner et al. performed a series of XPS (x-ray photoemission spectroscopy) studies to determine the physical and chemical properties of this interface layer, which they summarized in a review article.[4] They also summarized dozens of experiments by others which also supported the idea that there is a unique, nonstoichiometric transition layer with excess Si (or oxygen deficiency, depending on one's point of view) with a high degree of local strain (see bibliography of reference 4).

3. Early Spectroscopic Studies

Spectroscopic studies on quartz and bulk glasses have identified a large number of paramagnetic defects, which could be studied by electron spin resonance measurements. Of these, the one that correlates with radiation-induced oxide positive charge trapping, is the so-called E_γ' center. This correlation was first shown by Lenahan and Dressendorfer,[5] and later confirmed by others. The E′ center was first detected by Weeks[6] in α-quartz, and associated with a single oxygen vacancy by others. At first, the idea that the center was due to a single oxygen vacancy was controversial, until Feigl et al.[7] showed that it was correct. The basic idea is illustrated in Fig 1. Two Si atoms are joined by a weak, strained Si-Si bond, where an oxygen atom should be, and each is back-bonded to three oxygen atoms. When a positive charge is captured, the Si-Si bond is broken, and the lattice relaxes. The new idea introduced by Feigl et al. was that the lattice relaxation is asymmetrical. The positively charged Si relaxes back into the plane of its oxygens, or even past it,[8] so that it puckers in the opposite direction. The neutral Si, with an unpaired electron in one orbital, relaxes toward the vacancy. Then Lenahan and Dressendorfer showed that oxide trapped charge, measured as ΔV_{MG}, the voltage shift at midgap, was removed at the same rate as the E′ signal in a series of isochronal annealing measurements, shown in Fig. 2. The correlation between these two measurements is strong evidence that the E′ signal and positive oxide charge are due to the same defect, and the correlation has been observed repeatedly on different samples by now.

4. Annealing Studies

Although oxide trapped charge is sometimes referred to as "fixed" charge, this description is correct only in a relative sense. That is, radiation-induced oxide trapped charge is stable on the time scale of many experiments, but it undergoes a long-term annealing process, which has a complex dependence on time, temperature, and applied field. The annealing process can proceed by either of two mechanisms, tunneling or thermal excitation. The tunneling process can be analyzed in terms of a tunneling front, as illustrated in Fig. 3.[9, 10] The probability of tunneling to a particular trap site is an exponential function of the barrier height and the depth of the trap in the oxide, there will be position in the oxide at any given time, where the holes to the right will have been neutralized, and the holes to the left will remain. This position, called the tunneling front, is actually about 0.2 nm wide, and moves into the oxide with a velocity of about 0.2 nm

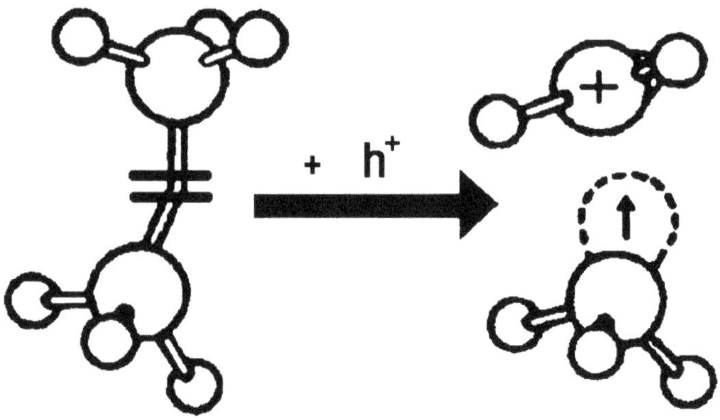

Figure 1. Oxygen vacancy acting as a hole trap.

per decade of time, according to model predictions. This ln t dependence provides an analytical basis for the ln t annealing behavior observed empirically in many MOS oxides.

Thermally activated annealing occurs when a carrier is thermally excited from the SiO_2 valence band to the trap level, neutralizing the trap. Both the tunneling front model, and the thermal emission model were combined in an elegant fashion by McWhorter et al.,[11] which is illustrated in Fig. 4. Basically, they show both a tunneling front and a thermal activation front, where the position of both fronts varies as ln t. Clearly, both mechanisms can lead to ln t annealing behavior, but they require different trap energy distributions to do so. For a trap with a single, discrete energy level, the tunneling process will predict a straight-line ln t annealing response, if the distribution of traps is uniform with increasing depth into the oxide. This result is often observed in unhardened, commercial oxides. If the spatial distribution of traps is peaked at the interface, and drops off with increasing depth in the oxide, the slope of the annealing curve will decrease with increasing time, which is often observed in hardened gate oxides. To predict ln t annealing by the thermal process, one has to assume a uniform distribution of trap states across the bandgap (as McWhorter has done in Fig. 4), rather than a discrete energy level. The distribution of trap states as a function of energy level in the bandgap has been measured for many oxides, and there are well-defined peaks rather than a uniform distribution. But the peaks have a finite width, so the assumption of a uniform distribution may be justified in a limited range, but not in general. The response of all oxides is temperature independent at room temperature and over a reasonable range of higher temperatures. If one raises the temperature enough, the

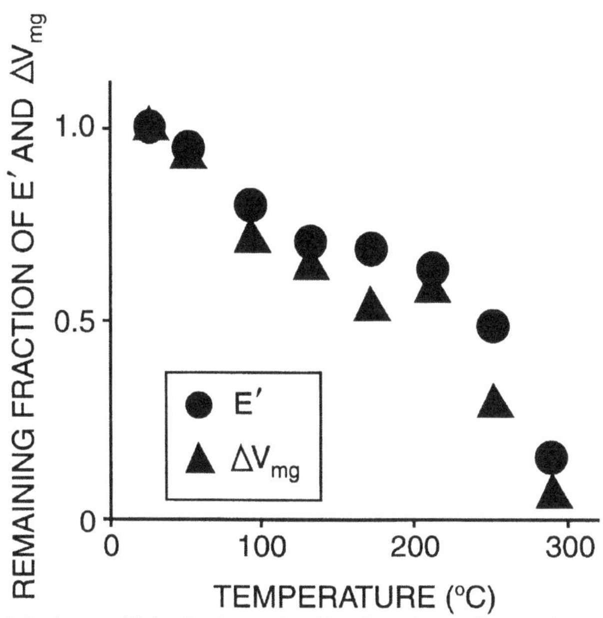

Figure 2. Correlation between E' signal and trapped positive charge in annealing experiments, after ref. [5], copyright American Inst. of Physics, reproduced by permission.

thermally activated process will become important sooner or later in all oxides. The temperature at which thermal activation becomes important varies form one oxide to another. In some oxides thermal activation is important even before 100 °C. In others, the response remains independent of temperature well above 100 °C. But by 200 °C, all oxides will undergo significant thermal annealing in a short time.

5. Negative Bias Annealing and the Oxygen Vacancy Dipole Model

Although the annealing process for trapped holes had been studied, and understood to a degree, the literature contained little discussion of what happened at the atomic level during annealing. Many researchers seem to have assumed, without saying so specifically, that an electron tunneled to a trapped hole, neutralized it, and reformed the broken Si-Si bond, so that the damage to the oxide was completely removed. A critical result, shown in Fig. 5,[12] forced many new studies, and eventually led to many new insights. Schwank et al. annealed an irradiated sample at +10V and 100 °C for about a week, until all the trapped positive charge (as determined by ΔV_{MG}) had annealed. Then they reversed the bias, to negative 10V, and continued annealing at 100 °C. Within about a day, nearly half of the "annealed" positive charge was restored. This result indicated

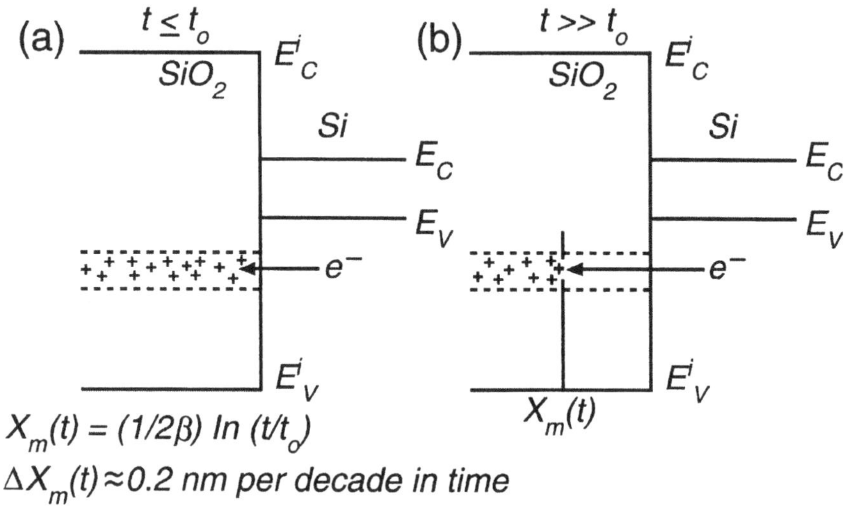

Figure 3. Tunneling front analysis; at any time there is a position where trapped positive charge to the right has been neutralized by electron tunneling, and trapped charge to the left remains.[2]

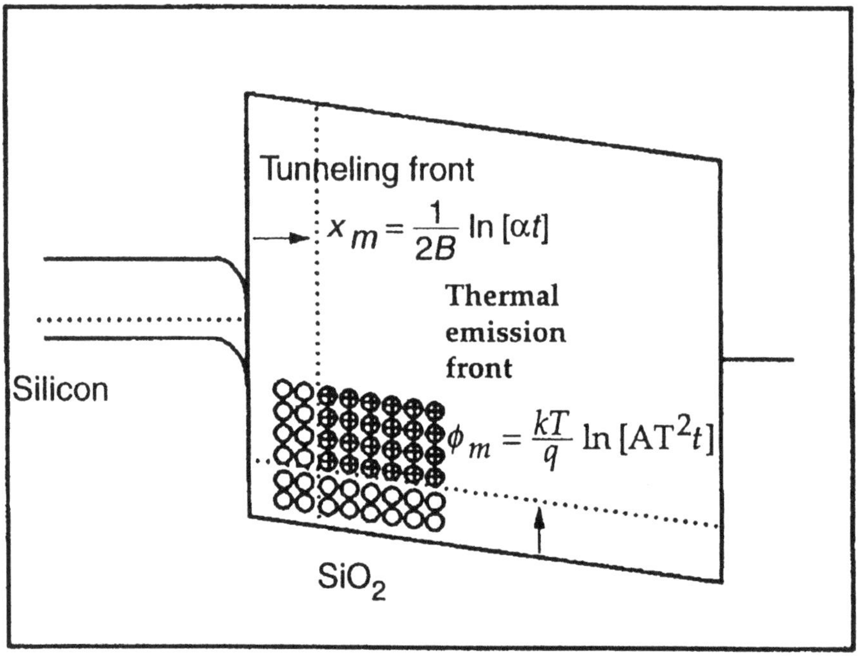

Figure 4. Annealing by the tunneling front and the thermal activation front, combined,[11] copyright IEEE, reproduced by permission.

that the annealing process involved some kind of compensation process, because the defects were neutralized without being removed (and unirradiated controls showed no such instability). Eventually, Lelis et al.[13, 14] carried out a careful study of this effect.

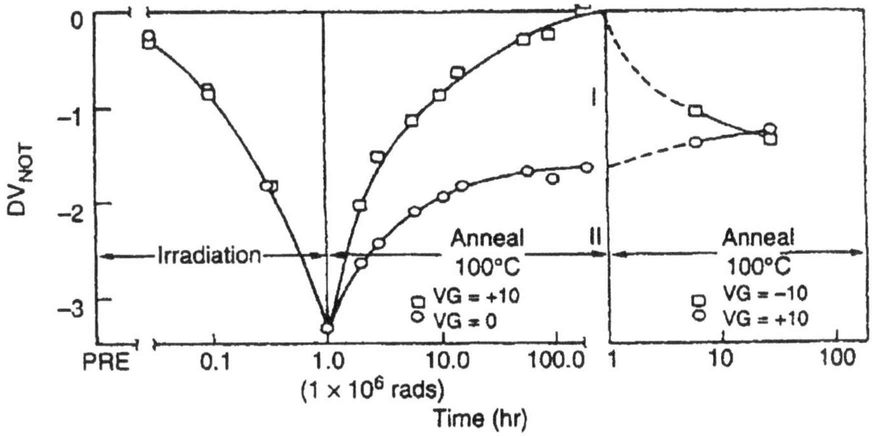

Figure 5. Trapped hole annealing results, showing that annealed holes are not really removed,[12] copyright IEEE, reproduced by permission.

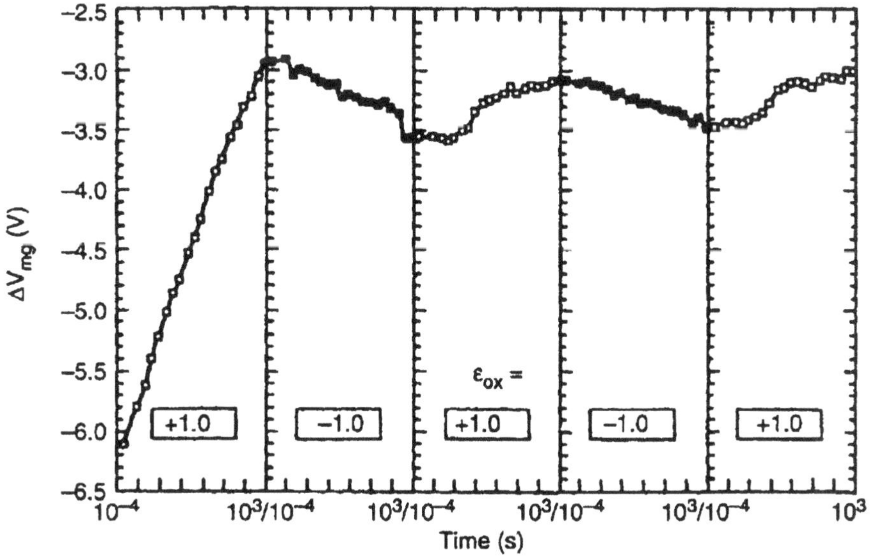

Figure 6. Alternate positive and negative bias annealing, showing reversible annealing over multiple cycles.[14]

One of their key results is shown in Fig. 6, which was obtained by irradiating a hardened oxide with a short Linac pulse, and annealing with the applied field alternately +/- 1MV/cm. The negative bias reverse annealing amounts to about 0.5V, which is significant, but it is clearly only a fraction of the total annealing. They interpreted these results as indicating two different processes – a permanent "true" annealing, and a reversible compensation process, where trapped charge was neutralized without being removed. The charge transfer seemed to be more or less constant over several cycles, and roughly logarithmic with time. These observations were taken to mean that charge was tunneling in and out of a hole trap defect in a more or less reversible way. The model proposed by Lelis et al. to explain these results is illustrated in Fig. 7. They proposed that the electron tunneling into the hole trap goes to the neutral side of the E′ center, creating a dipole structure, a positively charged Si adjacent to a negatively charged Si. In the interface region where the E′ centers are found, there is a high degree of local strain because the lattice mismatch between the Si substrate and the oxide is being accommodated. Therefore, one would expect a distribution of separation distances between the Si atoms in the E′ centers. Lelis et al. suggested that if the positive and negative Si atoms were close together, the coulomb attraction would be sufficient to pull them together, reforming the broken bond, resulting in true annealing. If the relaxation were greater, so that the Si atoms were far apart, then the dipole would be stable, or at least metastable, and it could exchange charge with the substrate in a reversible manner, as in Fig. 6. Other results,[15] not shown here, also suggested that if one cycled charge long enough, some of the dipoles would undergo true annealing – that is, even if the Si atoms were relatively far apart, random thermal vibration might eventually bring them close enough together to reform the broken Si-Si bond. Lelis et al. also noted that there was significant variation from one oxide to another in the amount of negative bias reverse annealing. The only unhardened, commercial oxide in their initial set of samples had almost no reversible annealing, and the hardened oxides all had a significant amount. This pattern held when other oxides were tested later. We will return to this point later, in Section 10.

The dipole hypothesis proposed by Lelis et al. was attractive because it explained complex annealing behavior, true annealing and reversible annealing, in terms of a single defect that was already well known. It was consistent with the existing spectroscopic data,[5] which showed only a single defect contributing to the hole trapping. The ESR technique detects unpaired electron spins, and the unpaired spins are eliminated in the dipole structure as the positive charge is compensated. In fact, the spectroscopic data was one thing that led Lelis et al. to propose dipoles instead of having the electron tunnel to the positive Si, without reforming the broken bond. Then the number of unpaired spins would increase, not decrease, as the positive charge was compensated. Almost immediately, independent experiments by other groups began to provide experimental

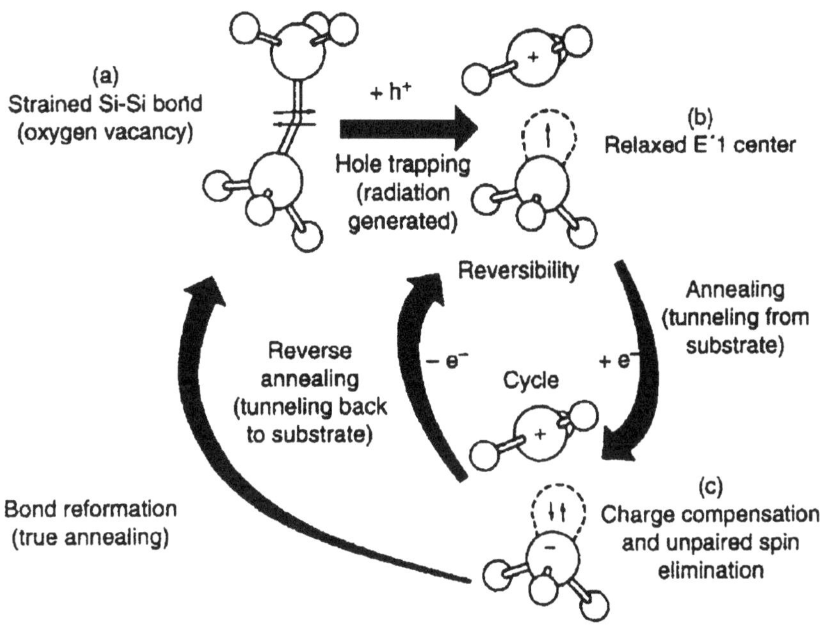

Figure 7. Model for hole trapping, permanent or "true" annealing, and reversible compensation processes.[14]

results that could be interpreted as supporting the dipole hypothesis, and we will discuss some of these in the next several sections. But we note that it was also controversial for several years. Three different groups presented papers which, at least implicitly, criticized it.[16 19] We cannot discuss these papers fully here, but reference 2 contains a relatively complete discussion. Freitag et al.[16, 17] and Edwards[18] concluded that the electrical switching behavior was due to some other (unknown) defect, not to E′ centers. And Warren et al.[19] concluded it was too early to say where the electron tunneling into the hole trap went (in 1994, six years and many confirming experiments later), that the topic needed further study. The experiment that settled these controversies was reported by Conley et al.[20, 21] in 1995, seven years after the first dipole model paper by Lelis et al. It is discussed in Section 9, below.

6. TSC Studies

There have been numerous TSC (thermally stimulated current) measurements on MOS devices, because the technique is a good way to determine the energy levels of trapped charges. Two of these studies are particularly relevant here. The first of these was done by Shanfield et al.[22] in 1988, about the same time as the first paper by Lelis et al., but not published in the open literature until later. Shanfield et al. used capacitor samples from the same source as some of the transistors used by Lelis et al., and they obtained results very consistent with those of Lelis et al. That is, they found significant compensation in the hardened oxide, but not in the unhardened oxide. The key point is that in a TSC measurement, one measures the current flow of charge freed by thermal excitation. If both electrons and holes are trapped, the current reflects the sum of these two components. In a C-V measurement, the voltage shift reflects the net trapped charge – that is, the difference between the trapped electrons and holes. If $Q_{TSC}>Q_{CV}$, then it is evidence for significant electron trapping. In the hardened oxide, where Lelis et al. had seen significant charge switching, the TSC results also indicated significant electron trapping. In the unhardened oxide, where Lelis et al. had seen no switching, and no evidence of compensation, Shanfield et al. saw no evidence for compensation in the TSC results, either. For this reason, the Shanfield et al. experiments were the first independent confirmation of the Lelis et al. model, in the sense that the model could easily explain the results, but they were difficult to explain with any other model.

Later, Fleetwood et al.[23] did another TSC measurement, intended to repeat and extend the Shanfield et al. work. They measured TSC as a function of applied bias, and confirmed that space charge effects had been significant in the work on unhardened oxides by Shanfield et al. They used samples from a different source than Shanfield et al. or Lelis et al., but also had a hardened gate oxide, and a very soft, thick field oxide. They found significant compensation in the hardened oxide ($Q_{TSC}>Q_{CV}$), but not in the soft oxide ($Q_{TSC}= Q_{CV}$, within experimental uncertainty). The Fleetwood et al. experiment was the third time a different group had reported the same basic result, compensation in hard oxides but not in soft oxides, and for the second source of samples. We will discuss this further in Section 10.

7. 1/f Noise Measurements

Many authors have used 1/f noise to characterize radiation damage in MOS devices, starting with Scofield et al.[24] in 1989. They concluded that 1/f noise correlated with oxide-trapped charge, not with interface trapped charge. This conclusion may have been surprising to some initially, but it has since been confirmed by others. Scofield et al. cited the first paper by Lelis et al.[13] to support the idea that there is an electron trap associated with the hole trap that can exchange charge with the substrate. Fleetwood et al.[25] have recently presented further analysis, DFT (Density Functional Theory) modeling, which suggests several possible mechanisms which may account for different experimental observations. They note that capture cross sections inferred from 1/f noise, hole trapping, and trapped hole neutralization are all about the same, and the energy levels for the defects are also about the same; all of which suggests that the defects actually are the same. They argue that the E_γ' and the E_δ' can both contribute to 1/f noise. They distinguish between the $E_{\gamma4}'$ and the $E_{\gamma5}'$, where the $E_{\gamma4}'$ has the neutral silicon atom puckered, and back-bonded to a fourth Si. The $E_{\gamma5}'$ is similar, except that a nearest neighbor Si is also close enough to interact, which shifts the energy levels so that the complex no longer forms a stable dipole. They suggest that the noise is mediated by lattice relaxation, as the structure shifts back and forth between these two configurations, it gains and loses the dipole electron. The E_δ', or dimer, may account for an unexplained observation, namely, that pre-irradiation 1/f noise correlated with post-irradiation hole trapping. If the oxygen vacancy precursor loses an electron to the substrate, it would create an E_δ' center initially, which could alternately gain and lose an electron, which would also be mediated by lattice relaxation. These ideas from the modeling have not been tested experimentally, but they provide plausible explanations for a number of observations.

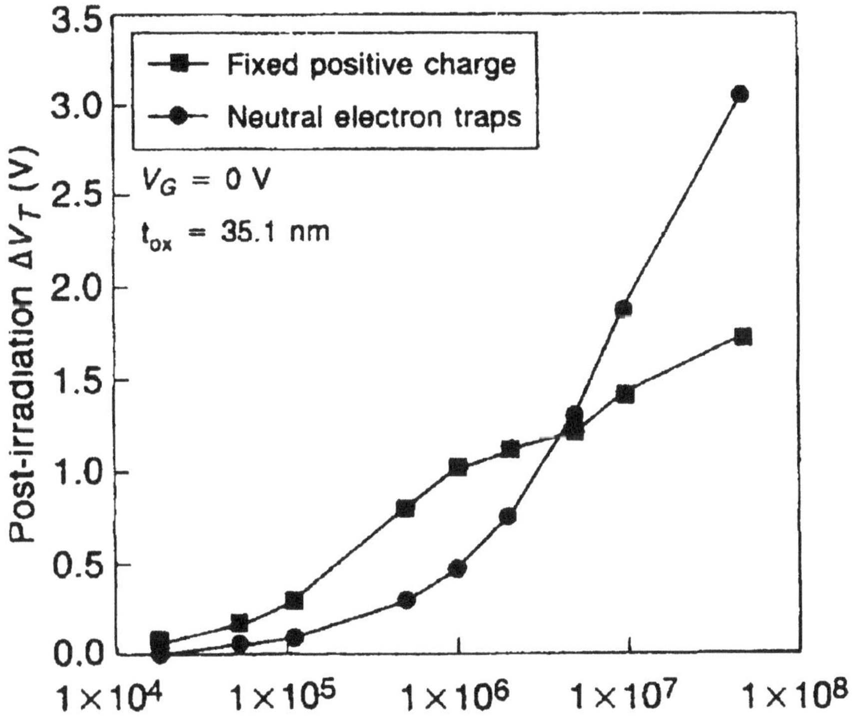

Figure 8. Positive trapped charge and neutral centers as a function of dose in rad (SiO_2),[32] copyright Electrochemical Society, reproduced by permission.

8. Neutral Electron Traps

Electron trapping in oxides exposed to intense radiation during processing, either by electron beam or x-ray lithography, or by plasma etching, ion implantation, reactive ion etch (RIE), or other processes, has been studied for more than 25 years. Early studies[26-28] showed that radiation in processing created large amounts of positive charge, which could be annealed fairly easily. But after the positive charge was removed, large numbers of neutral centers remained. These neutral centers could trap electrons during normal device operation, and they were very difficult to remove.[29, 30] An important early insight was provided by Aitken, who suggested the traps were dipolar in nature – the positive end of the dipole capturing electrons, and the negative end capturing holes. He also suggested that a distribution of separation distances between the ends of the dipole would lead to a wide range of field strengths, explaining why electron capture cross sections vary by orders of magnitude, even when measured by the same experimenter on the same sample. (The idea that electron traps and hole traps have a common origin was later supported by Aslam,[31] who showed that process steps that increase the density of one also increase the density of the other.) In 1980, when Aitken suggested a dipole, there was no dipolar structure that could be suggested, that seemed to be relevant. However, Lelis et al. did eventually propose one. An experimental study was eventually done by Walters et al.,[32] to test the idea that the dipole structure proposed by Lelis et al. was the neutral trapping center proposed by Aitken. One of their key results is shown in Fig. 8. He measured positive charge trapping, and electron trapping as a function of dose, using optically assisted electron injection. Positive charge is observed to buildup first, but the number of positive centers starts to saturate around 10^6 rads. In terms of the dipole model, this means that electron-hole pairs are being created in a region with a significant density of trapped positive charges, which recombine with the electrons to form dipoles. The relative saturation means that the positive charges are being consumed almost as fast as they are created, but then the number of neutral centers is growing rapidly. Then, as they continued to inject electrons, the amount of electron trapping increased very rapidly, becoming the dominant effect by 10^7 rads. Walters performed the experiment in Fig. 8 with no bias applied during irradiation, but he also irradiated with bias. And he found that positive bias, which increases hole trapping, also increased the formation of neutral centers. He also found that the spatial distribution of neutral centers was the same as the distribution of hole traps, within his experimental resolution. For these reasons, he concluded that the dipole structure proposed by Lelis et al. was the neutral electron trap. Under appropriate experimental conditions, the positive end of the dipole could capture a second electron, making the whole complex net negatively charged. These results were a significant confirmation of the dipole model, and also a significant extension of it. Certainly, the fact that another group had used the model to make a prediction, and that was tested and confirmed experimentally was important independent support for the dipole model. And by showing the connection of the dipole model to electron trapping, it was shown to be relevant to processing and reliability problems outside the traditional radiation effects area. Walters was the first to propose any specific structure for an electron trap, so his work was a breakthrough in that area.

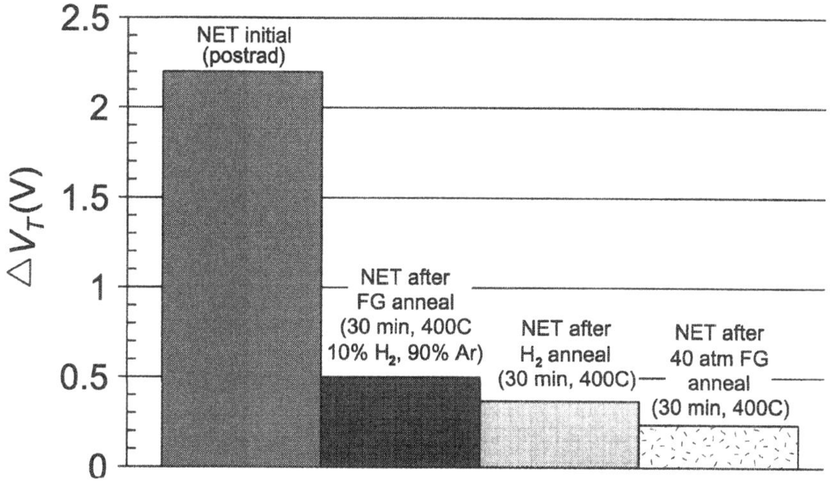

Figure 9. Annealing of neutral electron traps by exposure to hydrogen.[2]

Efforts to anneal neutral electron traps have mostly involved attempts to passivate the centers by annealing in hydrogen rich ambients. Results of some of these experiments are summarized in Fig. 9.[29, 30] For the oxides in these experiments (t_{ox} = 35 nm), the initial NET density was about 1.3 x 10^{12}/cm^2. The samples were then exposed to a series of annealing experiments, where each test involved more hydrogen than the one before, the number of unpassivated defects was reduced in a consistent manner. But even after extreme measures (annealing in 50 atm forming gas, which is equivalent to 5 atm pure hydrogen), more than 10^{11}/cm^2 NET remained. The simplest view of how these results fit together is illustrated in Fig. 10.[2] Parts a, b, and c are the same as Fig. 7; part d is the electron trap, as described by Walters; and part e is the neutral electron trap (dipole) passivated by reaction with hydrogen. The experimental difficulty in getting the hydrogen passivation reaction to go to completion suggests that the passivation and depassivation reactions are in equilibrium, and the depassivation reaction is strong enough that the defects are never fully passivated.

In a wafer that has been through a CMOS process, one would expect a large number of trapped holes (part b in Fig. 10) at some point, because radiation is so common in CMOS processing. (In fact, DeLaus[33] made the comment that the most severe radiation environment most parts ever see, is in the process line.) High temperature processing will neutralize these by exciting carriers from the valence band, leaving a high concentration of dipole states (part c). Many of these will be passivated, forming Si-H bonds (part e), because hydrogen-rich ambients are commonly used in processing. But the reaction is not efficient enough to passivate them all, so many NETs will remain, which can subsequently trap electrons (part d). The arrows between part c of Fig. 10, and parts d and e, run in both directions because other work by Thompson and Nishida,[34] and by Hsu et al.[35] Thompson and Nishida studied the detrapping of trapped electrons as a function of temperature and applied field, and showed that the traps are shallow, and relatively easy to empty. Hsu et al. also showed that in hot electron injection experiments there is a trap filling process, and a trap creation process. That is, there are pre-existing traps which can be filled, consistent with the transition from part c to part d in Fig. 10. And if the injected electrons are energetic enough, they can create "new" traps, by breaking a Si-H bond, for example (10(e) to 10(c) or 10(d)). The dipole model, together with follow-up experiments by others, provides a framework, then, which can be very useful for interpreting a variety of reliability physics experiments. We will discuss this further in Section 11.

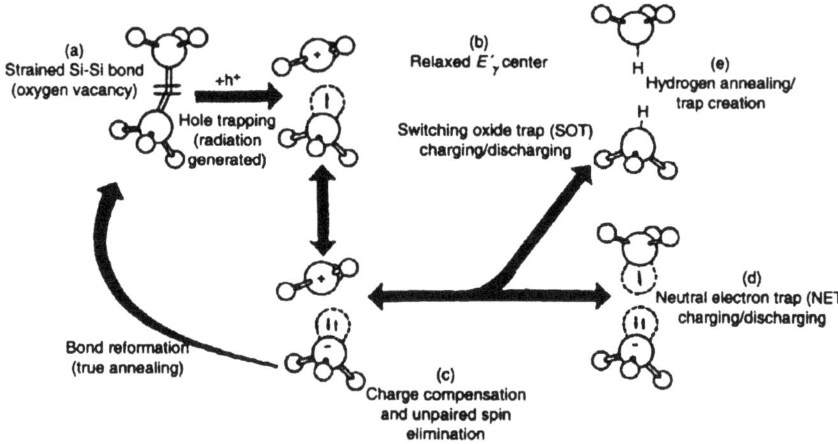

Figure 10. Schematic representation of hole traps, dipoles (compensated hole traps), NETs, and hydrogen annealed NETs.[2]

9. Recent Spectroscopic and Theoretical Studies

Two ESR spectroscopic studies and two theoretical modeling efforts bearing on the Lelis et al. dipole model have been published, several years after the model was originally published. The first of these was by Warren et al.,[19] which concluded that the "fixed" oxide charge and the switching oxide charge were probably both E_γ' centers. Of course, this conclusion is completely consistent with the dipole model, but we should note that Warren et al. did not accept this idea. They pointed out that Edwards[18] had argued that the dipole configuration was energetically unfavorable, because of the electron-electron repulsive energy. And that Fleetwood had studied the charge exchange process, without reaching a conclusion on exactly where electrons tunnel to, when they tunnel into a hole trap. For these reasons, they concluded it was premature to specify the exact location of the tunneling electron, and that the question should be studied further.

The second ESR study was by Conley et al.,[20, 21] who monitored the E_γ' signal during a series of alternating bias anneals. The samples were irradiated under positive bias, then annealed unbiased for a few days, then annealed again under alternating bias, negative bias first.

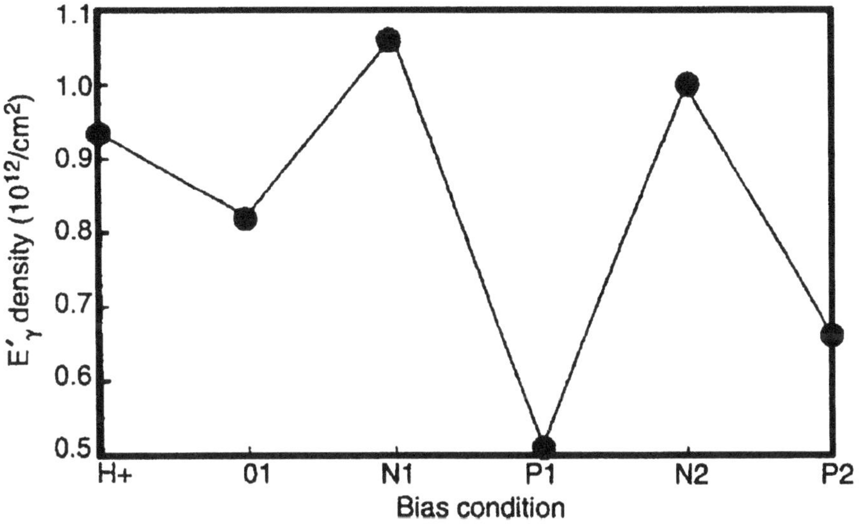

Figure 11. E' density during alternating positive and negative bias annealing.[20, 21]

Typical results are shown in Fig. 11. The E_γ' signal increases under negative bias, and decreases under positive bias, the same as positive oxide trapped charge in electrical measurements by Lelis et al., and others. This result is very strong evidence for the dipole model, because the unpaired spin detected by ESR and the positive charge detected electrically are increasing or decreasing together, even though they reside on different Si atoms. The dipole model is the only model that has been proposed, which is consistent with this result. (We note that Fleetwood et al.[36] has also proposed hemi-E′ centers as possible trap structures – in effect, half an E′ center. The trouble with these structures is that they have an unpaired spin in the neutral state, not in the charged state.

In Fig. 11, they would have valleys where the peaks are, and peaks where the valleys are. Fleetwood later agreed these centers are not consistent with the results in Fig. 11.[37]) From this result, it is clear that the dipole structure of Lelis et al. is the dominant switching oxide trap, or border trap. Of course, it is always possible that other structures will later be shown to produce small, second order effects.

Two recent theoretical studies have also supported the Lelis et al. model. The first of these was by Karna et al.,[38] who concluded, based on first principles quantum mechanical calculations, that the E′ center can take two forms. The first of these is the E_δ' center, which is a dimer structure where the hole is delocalized – spread over both Si atoms. This structure is symmetrical, with the Si atoms not relaxing far away from each other. When an electron tunnels to this structure, the broken bond reforms, and the trap is permanently annealed. The second form is the E_γ', where the Si atoms relax asymmetrically, and one of them ends up puckered away from the other, back-bonded to another oxygen atom. The electron-electron repulsion energy means this state has a higher energy than the ground state, but it forms metastable dipoles. This calculation was the first to provide a theoretical explanation for stable dipoles.

The second theoretical study was by Nicklaw et al.,[39] who used density functional theory (DFT), a different mathematical method, but reached very similar conclusions. The main difference is that, depending on the positions of nearby Si neighbors, the E_γ' can assume two configurations, and only one of them produces a metastable dipole.

Taken together, these four follow-up studies by different groups have provided strong additional support for the dipole model of Lelis et al.

10. Process Dependencies

As we have noted already, some soft, commercial oxides seem to have no significant charge switching, or compensation, while hardened oxides generally do. In fact, Conley et al.[21] reported that test chips from three high volume commercial process lines (Texas Instruments (TI), National Semiconductor, and Micron) had been evaluated, and none showed any significant radiation-induced switching oxide traps. This is despite the fact these are all relatively soft oxides, with a high density of other oxide traps. On the other hand, three hardened oxides ((TI), Sandia National Laboratories, and IBM/Loral/BAE) all did show switching behavior by a significant fraction of their oxide traps. The TI example is particularly instructive, since the hardened and unhardened oxides were from the same split lot, and differed in only one process step (the post-oxidation anneal, or POA). See reference 2 for a full discussion, but the main difference was that the hardened oxide received a POA at a lower temperature. The different annealing responses of these two oxides are shown in Fig. 12, where we have plotted $\Delta V_T/t_{ox}^2$ because the oxide thicknesses are different. If one compares the response at a hundred seconds, or at a thousand seconds, the hardened oxide has about an order of magnitude smaller shift. But at earlier times the difference between the oxides is smaller, and if one extrapolates back to the end of the radiation pulse (4 μs), it is possible there is no difference at all in the trapping of the two oxides. This conclusion depends on how one extrapolates, or whether one believes the extrapolation, but there is clearly less difference earlier than there is later, because the hard oxide recovers faster. These results suggest there is actually little or no difference in the number of oxide traps in the hard and soft oxides, but they are closer to the interface in the hard oxide, and so are neutralized by tunneling electrons more quickly. Recall that the hard oxide in Fig. 12 has significant negative bias reverse annealing, or charge switching, and the soft oxide has essentially none; and that these oxides differ in only one process step. So the process change that makes the hard oxide hard also introduces the radiation-induced negative bias instability, and increases the annealing rate.

We note that Deal[40] discussed a process-induced negative bias instability, which he attributed to excess Si near the interface. This was originally an important commercial reliability problem, but Deal showed that the instability could be eliminated by a high temperature (at the oxidation temperature) POA in an inert ambient, as was done on the unhardened oxide here. He inferred that the effect of the high temperature anneal was to remove the excess Si from the interface region. The reason high temperature anneals have been a fixture in commercial processes since Deal's early work is that he (and others later) showed it did remove excess Si from the interface region, eliminating an important instability. It is important to realize that excess Si in the interface region is an important commercial reliability problem, even in the absence of radiation. But in the results in Fig. 12, the hard oxide is hard because the instability has been built back in by eliminating the normal POA. It has been argued[41] that high temperature annealing steps increase the excess Si concentration at the interface by reducing the oxide. While this is certainly true if the temperature is high enough, it seems not to be a significant effect at the temperatures used in these oxides. Instead, the effect of the high temperature POA seems to be to cause the oxygen vacancies to diffuse into the oxide, away from the interface, making them less active. But they still contribute to radiation damage for a long time in the soft oxide, because they are so stable. For a more complete discussion, see Chapter 2 of reference 2, and its bibliography.

11. Reliability Physics and Electron Trap Creation

We have already introduced the subject of electron trap creation, citing Hsu et al.[35] Their work was consistent with earlier work by DiMaria,[42] who showed that when electrons are heated above about 2 eV by the field, Si-H bonds start to be broken. Although the average electron energy never reaches 9 eV, there is a high-energy tail to the distribution above 9 eV, which gives rise to impact ionization. The important application of these

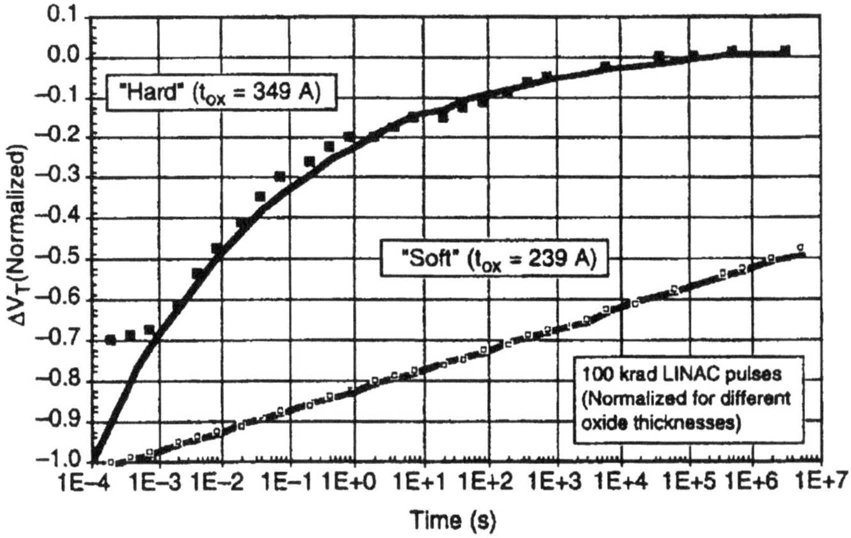

Figure 12. Normalized post-irradiation responses of "hard" and "soft" oxides, which differ only in one process step (POA).[21]

results is in breakdown studies. Generally, the oxide is considered to go through a wear-out phase, during which trapped charge builds up, followed by breakdown. At one time, there was some controversy over the details of the wear-out process, with some authors arguing that impact ionization led to positive charge buildup as the dominant mechanism.[43, 44] Others[45, 46] argued against impact ionization, and for a buildup of stable electron traps as the main cause of breakdown. More recently, it has become clear that both positive and negative charge buildups can occur, and both can cause breakdown. The work by DiMaria shows that both processes can occur at the same time. The basic idea is that charge builds up near the injecting electrode, usually the Si substrate. Then a very high field develops between the oxide trapped charge and the image charges at the interface. When the field becomes large enough, avalanche multiplication starts, which leads to local heating, which destroys the oxide. Clearly, both polarities of charge can develop large space charge fields, and both can be explained is different aspects of the same oxygen vacancy defect complex we have already described. The oxygen vacancy model presented here may not explain all the breakdown results in the literature, but it seems to provide a framework, which can account for many of the known results.

12. Oxide Traps as Interface Traps

A discussion of interface traps is beyond the scope of this paper, but there is one area where the dipole model helped to resolve a controversy about radiation-induced interface traps. There was a school of thought in the 1980s and early 1990s,which supported that idea that interface traps arose from a defect conversion process, that oxide traps somehow were converted to interface traps.[47-51] One weakness of this argument is that no one was ever able to propose a plausible mechanism for the conversion process. In a review paper, Oldham et al.[52] summarized other arguments against this idea, concluding that hole removal and interface trap formation had different time dependences, different field dependences, and different temperature dependences. Except for the fact that both processes increased with increasing dose, they appeared to be completely independent. Oldham et al. did, however, offer a new explanation of experimental results, which had been offered to support the defect conversion idea. They pointed out that no defect conversion was really necessary – oxide traps that remained oxide traps could account for the observations. The Lelis et al. dipole model, and the charge transfer mechanisms built into it, could account for the results. The fundamental question was how to define an interface trap. Most researchers took an interface trap to be a state in the Si band-gap in equilibrium with the substrate, meaning that it responded to small voltage changes, such as the 15 mV AC probe voltage in a high frequency CV measurement. Oxide traps (dipoles) do not respond to small voltage changes because they are not in equilibrium with the substrate, but they do respond to large voltage changes (e.g., from accumulation to inversion, or the reverse). For this reason, oxide traps were being interpreted as interface traps in some experiments.

A few years after Oldham et al. presented this discussion, Fleetwood[53] proposed that oxide traps with these electrical properties should be called "border traps." At that time, the first Lelis et al. paper proposing the dipole model had been in the literature four years, and was already well-known. Now ten more years have passed, and the dipole model is still the only experimentally confirmed model for border traps (or for switching oxide traps, as we have usually preferred to call them here).

13. Oxide Thickness and Scaling

In the history of the semiconductor industry, scaling to smaller feature sizes has been a pervasive theme – almost everything has depended on it. One consequence of scaling has been that the gate oxide has become thinner every year, and thinner oxides are less sensitive to radiation damage. From Q=CV, one would predict that ΔV_T is proportional to t_{ox}^2. McGarrity[54] estimated the hardening that could be achieved by thinning the oxide, without special processing. The most important deviation from the thickness squared dependence is that below about 10 nm, ΔV_T falls off even faster than t_{ox}^2 would predict.[55,56] Almost all the trapped positive charge is neutralized by tunneling electrons, because the tunneling distance is a large fraction of the total thickness, and tunneling from both electrodes is significant. Present day gate oxides are so thin, that gate oxide hardening is no longer a practical problem. The main radiation problem now is in the isolation structures. In oxide isolation structures, the same defects and the same physical mechanisms determine the response, but there are differences because all the mechanisms have complex time dependences, field dependences, and temperature dependences. For example, the fields are lower, which means the yield of charge from recombination is less, the transport is slower, thicker oxide also means slower transport and so on. Generally, the response of LOCOS (LOCal Oxidation of Silicon) field oxides varies widely,[57, 58] probably because the processing varies widely. STI (shallow trench isolation) structures are replacing LOCOS, but have not been studied as much.[59]

One area where the dipole model has been used effectively is in understanding the ELDRS (enhanced low dose rate sensitivity) of some bipolar isolation oxides. ELDRS is at least partly due to space charge effects, and to different degrees of compensation (dipole formation) at different dose rates.[60, 61]

One radiation effect in the gate oxide has been reported, which is a consequence of gate oxide thinning – RILC (radiation induced leakage current).[62] This leakage current is thought to be a form of trap-assisted tunneling, where an electron tunnels from the substrate to a trap in the oxide. Then it tunnels from the trap to the other electrode, contributing to gate oxide leakage. Of course, the oxygen vacancy dipole model is an obvious candidate for the trap state. The same effect is observed when electrical stresses, rather than radiation, generate a trapped hole or a dipole state. Then it is called SILC (stress-induced leakage current).[63]

14. Conclusions

The oxygen vacancy dipole model, proposed by Lelis et al., was developed in response to radiation experiments showing what was first called negative bias reverse annealing. But it has been confirmed in numerous other experiments, and extended by other groups. The later work by Walters, showing the connection with neutral centers and electron trapping is especially significant. The dipole model is the basis for a comprehensive model of oxide charge trapping (of both polarities), applicable to many oxide reliability problems. Although the model grew out of radiation effects studies, it is now of interest to a much wider community, for this reason.

References

1. F.B. McLean, H.E. Boesch, Jr., and T.R. Oldham, Electron-Hole Generation, Transport, and Trapping in SiO_2, Chapter 3 of Ionizing Radiation Effects in MOS Devices and Circuits, edited by T.P Ma and P.V. Dressendorfer, Wiley-Interscience, New York (1989).
2. T.R. Oldham, Ionizing Radiation Effects in MOS Oxides, World Scientific Publishing, Inc., Singapore (1999).
3. B.E. Deal, M. Sklar, A.S. Grove, and E.H. Snow, Characteristics of Surface State Charge (Q_{ss}) of Thermally Oxidized Silicon, J. Electrochem. Soc., 114, 266 (1967).
4. F.J. Grunthaner and P.J. Grunthaner, Chemical and Electronic Structure of the SiO_2/Si Interface, Materials Science Reports, North Holland, Amsterdam, 1, 65 (1986).
5. P.M. Lenahan and P.V. Dressendorfer, Hole Traps and Tri-Valent Silicon Centers in Metal/Oxide/Silicon Devices, J. Appl. Phys., 55, 3495 (1984).
6. R. A. Weeks, Paramagnetic Resonance of Defects in Irradiated Quartz, J. Appl. Phys., 27, 1376 (1956).
7. F.J. Feigl, W.B. Fowler, and K.L. Yip, Oxygen Vacancy Model for the E_1' Center in SiO_2, Solid State Comm., 14, 225 (1974).
8. K.C. Snyder and W.B. Fowler, Oxygen Vacancy in α-Quartz: A Possible Bi- and Metastable Defect, Phys. Rev., B48, 13238 (1993).
9. S. Manzini and A. Modelli, Tunneling Discharge of Trapped Holes in Silicon Dioxide, Insulating Films in Semiconductors, J.F. Verweij and D.R. Wolters, ed., Elsevier Science Publishers B. V. (North Holland), p. 112 (1983).
10. T.R. Oldham, A.J. Lelis, and F.B. McLean, Spatial Dependence of Trapped Holes Determined from Tunneling Analysis and Measured Annealing, IEEE Trans. Nucl. Sci., NS-33, 1203 (1986).
11. P.J. McWhorter, S.L. Miller, and W.M. Miller, Modeling the Anneal of Radiation-Induced Trapped Holes in a Varying Thermal Environment, IEEE Tran. Nucl. Sci., NS-37, 1682 (1990).
12. J.R. Schwank, P.S. Winokur, P.J. McWhorter, F.W. Sexton, P.V. Dressendorfer, and D.C. Turpin, Physical Mechanisms Contributing to Device "Rebound," IEEE Trans. Nucl. Sci., NS-31, 1434 (1984).
13. A.J. Lelis, H.E. Boesch, Jr., T.R. Oldham, and F.B. McLean, Reversibility of Trapped Hole Annealing, IEEE Trans. Nucl. Sci., NS-35, 1186 (1988).
14. A.J. Lelis, T.R. Oldham, H.E. Boesch, Jr., and F.B. McLean, The Nature of the Trapped Hole Annealing Process, IEEE Trans. Nucl. Sci., NS-36, 1808 (1989).
15. A.J. Lelis and T.R. Oldham, Time Dependence of Switching Oxide Traps, IEEE Trans. Nucl. Sci., NS-41, 1835 (1994).
16. R.K. Freitag, D.B. Brown, and C.M. Dozier, Experimental Evidence for Two Species of Radiation-Induced Trapped Positive Charge, IEEE Trans. Nucl. Sci., NS-40, 1316 (1993).
17. R.K. Freitag, D.B. Brown, and C.M. Dozier, Evidence for Two Types of Radiation-Induced Trapped Positive Charge, IEEE Trans. Nucl. Sci., NS-41, 1828 (1994).
18. A.H. Edwards and W.B. Fowler, Final Report – ONR Contract N00014-92-J-2001, 1995.
19. W.L. Warren, M.R. Shaneyfelt, D.M. Fleetwood, J.R. Schwank, P.S. Winokur, and R.A.B. Devine, Microscopic Nature of Border Traps in MOS Oxides, IEEE Trans. Nucl. Sci., NS-41, 1817 (1994).
20. J.F. Conley, P.M. Lenahan, A.J. Lelis, and T.R. Oldham, Electron Spin Resonance Evidence for the Structure of a Switching Oxide Trap: Long Term Structural Change at Silicon Dangling Bond Sites in SiO_2, Appl. Phys. Lett., 67, 2179 (1995).
21. J.F. Conley, P.M. Lenahan, A.J. Lelis, and T.R. Oldham, Electron Spin Resonance Evidence E_γ' Centers Can Behave as Switching Oxide Traps, IEEE Trans. Nucl. Sci., NS-42, 1744 (1995).

22. Z. Shanfield, G.A. Brown, A.G. Revesz, and H.L. Hughes, A New MOS Radiation-Induced Charge: Negative Fixed Interface Charge, IEEE Trans. Nucl. Sci., NS-39, 303 (1992).
23. D.M. Fleetwood, R.A. Reber, Jr., and P.S. Winokur, Effect of Bias on Thermally Stimulated Current (TSC) in Irradiated MOS Devices, IEEE Trans. Nucl. Sci., NS-38, 1066 (1991).
24. J.H. Scofield, T.P.Doerr, and D.M. Fleetwood, Correlation of Pre-Irradiation 1/f Noise and Post-Irradiation Oxide Trapped Charge in MOS Transistors, IEEE Trans. Nucl. Sci., NS-36, 1946 (1989).
25. D.M. Fleetwood, H.D. Xiong, Z.Y. Lu, C.J. Nicklaw, J.A. Felix, R.D. Schrimpf, and S.T. Pantelides, Unified Model of Hole Trapping, 1/f Noise, and Thermally Stimulated Current in MOS Devices, IEEE Trans. Nucl. Sci., NS-49, 2674 (2002).
26. J.M. Aitken, D.R. Young, and K. Pan, Electron Trapping in Electron-Beam Irradiated SiO_2, J. Appl. Phys., 49, 3386 (1978).
27. J.M. Aitken and D.R. Young, Electron Trapping by Radiation-Induced Charge in MOS Devices, J. Appl. Phys., 47, 1196 (1976).
28. J.M. Aitken, Radiation-Induced Trapping Centers in Thin Silicon Dioxide Films, J. Non-Cryst. Solids, 40, 31 (1980).
29. A. Reisman, J.M. Aitken, A.K. Ray, M. Berkenblit, C.J. Merz, and R.P. Havreluk, On the Removal of Insulator Process-Induced Radiation Damage from Insulated Gate Field Effect Transistors at Elevated Pressure, J. Electrochem. Soc., 128, 1616 (1981).
30. A. Reisman, C.K. Williams, and J.R. Maldonado, Generation and Annealing of Defects in Silicon Dioxide, J. Appl. Phys., 62, 868 (1983).
31. M. Aslam, Common Origin for Electron and Hole Traps in MOS Devices, IEEE Trans. Electron Dev., ED-34, 2535 (1987).
32. M. Walters and A. Reisman, Radiation-Induced Neutral Electron Trap Generation in Electrically Biased Insulated Gate Field Effect Transistor Gate Insulators, J. Electrochem. Soc., 138, 2756 (1991).
33. M. DeLaus, Radiation Concerns in State-of-the-Art Processing Technologies, IEEE Nuclear and Space Radiation Effects Conference, Tutorial Short Course, Tucson, AZ (1994).
34. S.E. Thompson and T. Nishida, Tunneling and Thermal Emission of Electrons from a Distribution of Shallow Traps in SiO_2, Appl. Phys. Lett., 58, 1262 (1991).
35. C.C.-H. Hsu, T. Nishida, and C.-T. Sah, Observation of Threshold Oxide Electric Field for Trap Generation in Oxide Films on Silicon, J. Appl. Phys., 63, 5882 (1988).
36. D.M. Fleetwood, W.L. Warren, J.R. Schwank, P.S. Winokur, M.R. Shaneyfelt, and L.C. Riewe, Effects of Interface Traps and Border Traps on MOS Post-Irradiation Annealing Response, IEEE Tran. Nucl. Sci., NS-42,1698 (1995).
37. D.M. Fleetwood, Fast and Slow Border Traps in MOS Devices, IEEE Tran. Nucl. Sci., NS-43, 779 (1996).
38. S.P. Karna, A.C. Pineda, R.D. Pugh, W.M. Shedd, and T.R. Oldham, Electronic Structure Theory and Mechanisms of the Oxide Trapped Hole Annealing Process, IEEE Trans. Nucl. Sci., NS-47, 2316 (2000).
39. J. Nicklaw, Z.Y. Lu, D.M. Fleetwood, R.D.Schrimpf, and S.T. Pantelides, The Structure, Properties, and Dynamics of Oxygen Vacancies in Amorphous SiO_2, IEEE Trans. Nucl. Sci., NS-49, 2667 (2002).
40. B.E. Deal, The Current Understanding of Charges in Thermally Oxidized Silicon Structures, J. Electrochem. Soc., 121, 198C (1974).
41. W.L. Warren, D.M. Fleetwood, M.R. Shaneyfelt, J.R. Schwank, P.S. Winokur, R.A.B. Devine, and D. Mathiot, Links Between Oxide, Interface, and Border Traps in High-Temperature Annealed Si/SiO_2 Systems, Appl. Phys. Lett., 64, 3452 (1994).
42. D.J. DiMaria, Correlation of Trap Creation with Electron Heating in Silicon Dioxide, Appl. Phy. Lett., 51, 655 (1987).

43. T.H. DiStefano and M. Schatzkes, Dielectric Instability and Breakdown in SiO_2 Thin Films, J. Vac. Sci. Technol., 13, 50 (1976).
44. P. Solomon, Breakdown in Silicon Oxide, J. Vac. Sci. Technol., 14, 1122 (1977).
45. E. Harari, J. Appl. Phys., Dielectric Breakdown in Electrically Stressed Thin Films of Thermal SiO_2, 49, 2478 (1977).
46. E. Harari, Conduction and Trapping of Electrons in Highly Stressed Ultra-Thin Films of Thermal SiO_2, Appl. Phys. Lett., 30, 601 (1977).
47. F.J. Grunthaner, B.F. Lewis, N. Zamini, and J. Maserjian, XPS Studies of Structure-Induced Radiation Effects at the Si/SiO_2 Interface, IEEE Trans. Nucl. Sci., NS-27, 1640 (1980).
48. F.J. Grunthaner, P.J. Grunthaner, and J. Maserjian, Radiation-Induced Defects in SiO_2 as Determined with XPS, IEEE Trans. Nucl. Sci., NS-29, 1462 (1982).
49. F.J. Grunthaner and P.J. Grunthaner, Chemical and Electronic Structure of the SiO_2/Si Interface, Materials Science Reports, North Holland, Amsterdam, 1, 65 (1986).
50. S.K. Lai, Interface Trap Generation in Silicon Dioxide When Electrons are Captured by Trapped Holes, J. Appl. Phys., 54, 2540 (1983).
51. S.J. Wang, J.M. Sung, and S.A. Lyon, Relationship Between Hole Trapping and Interface State Generation in Metal-Oxide-Silicon Structures, Appl. Phy. Lett., 52, 1431 (1988).
52. T.R. Oldham, F.B. McLean, H.E. Boesch, Jr., and J. M. McGarrity, An Overview of Radiation-Induced Interface Traps in MOS Structures, Semicond. Sci. Technol., 4,986 (1989).
53. D.M. Fleetwood, "Border Traps" in MOS Devices, IEEE Trans. Nucl. Sci., NS-39, 269 (1992).
54. J.M. McGarrity, Considerations for Hardening MOS Devices and Circuits for Low Radiation Doses, IEEE Trans. Nucl.Sci., NS-27, 1739 (1980).
55. N.S. Saks, M.G. Ancona, and J.A. Modolo, Radiation Effects in MOS Capacitors with Very Thin Oxides at 80K, IEEE Trans. Nucl. Sci., NS-31, 1249 (1984).
56. J.M. Benedetto, H.E. Boesch, F.B. McLean, and J.P. Mize, Hole Removal in Thin Gate MOSFETs by Tunneling, IEEE Trans. Nucl.Sci., NS-32, 3916 (1985).
57. T.R. Oldham, A.J. Lelis, H.E. Boesch, Jr., J.M. Benedetto, F.B. McLean, and J.M. McGarrity, Post-Irradiation Effects in Field Oxide Isolation Structures, IEEE Trans. Nucl. Sci., NS-34, 1184 (1987).
58. J.M. Terrell, T.R. Oldham, A.J. Lelis, and J.M. Benedetto, Time-Dependent Annealing of Radiation-Induced Leakage Currents in CMOS Devices, IEEE Trans. Nucl. Sci., NS-36, 1808 (1989).
59. M.R. Shaneyfelt, P.E. Dodd, B.L. Draper, and R.S. Flores, Challenges in Hardening Technologies Using Shallow-Trench Isolation, IEEE Trans. Nucl. Sci., NS-45, 2584 (1998).
60. D.M. Fleetwood, S.L. Kosier, R.N. Nowlin, R.D. Schrimpf, R.A. Reber, Jr., M. DeLaus, P.S. Winokur, A. Wei, W.E. Combs, and R.L. Pease, Physical Mechanisms Contributing to Enhanced Bipolar Gain Degradation at Low Dose Rates, IEEE Trans. Nucl. Sci., NS-41, 1871 (1994).
61. D.M. Fleetwood, L.C. Riewe, J.R. Schwank, S.C. Witczak, and R.D. Schrimpf, Radiation Effects at Low Electric Fields in Thermal, SIMOX, and Bipolar-Base Oxides, IEEE Trans. Nucl. Sci., NS-43, 2537 (1996).
62. Scarpa, A. Paccagnella, F. Montera, G. Ghibaudo, G. Pananakakis, G. Ghidini, and P.G. Fuochi, Ionizing Radiation Induced Leakage Current on Ultra-Thin Gate Oxides, IEEE Trans. Nucl. Sci., NS-44, 1818 (1997).
63. M. Ceschia, A Paccagnella, A. Cester, A. Scarpa, and G. Ghidini, Radiation Induced Leakage Current and Stress Induced Leakage Current in Ultra-Thin Gate Oxides, IEEE Trans. Nucl. Sci., NS-45, 2375 (1998).

International Journal of High Speed Electronics and Systems
Vol. 14, No. 2 (2004) 605–623

ONLINE AND REALTIME DOSIMETRY USING OPTICALLY STIMULATED LUMINESCENCE

L.DUSSEAU

*CEM², University MontpellierII, CC083, Place E.Bataillon,
Montpellier CEDEX05, 34095 France
dusseau@cem2.univ-montp2.fr*

J.GASIOT

*CEM², University MontpellierII, CC083, Place E.Bataillon,
Montpellier CEDEX05, 34095 France
gasiot@cem2.univ-montp2.fr*

A new generation of Optically stimulated materials has been synthesized at the University Montpellier II. The very high sensitivity of these phosphors, the short time constant of the luminescence and the perfectly separated spectra enable many applications in real time and online dosimetry. Dosimetry can be considered as real-time when the dose change between two measurements is considered as low enough . For satellite applications, we have developed an integrated sensor to measure the dose received orbit by orbit. In radiotherapy, OSL has been used to control the dose deposited during intra-operative *in-vivo* irradiation. At CERN, we have proposed an online system to monitor the dose simultaneously in about one hundred locations inside the Compact Muon Solenoid (CMS) experiment.

Keywords: Optically stimulated luminescence, Real time Dosimetry.

1. Introduction

Electronic devices, like biological cells, are in a less extend sensitive to ionizing radiation. In Metal Oxide Semiconductor Field Effect Transistors (MOSFET), the irradiation results in a shift of the threshold voltage. In BJTs (Bipolar Junction Transistors) the current gain decreases drastically.[1]
In biological cells, radiation is known to damage DNA. Wherever radiation can be found, it is then useful to quantify its effect, hence to measure the dose deposited, in order to appreciate possible negative consequences. In the first attempts to measure the dose deposited by ionizing radiation, the use of calorimeters is very common. This absolute dosimetry consisted in measuring the temperature elevation of a given mass of an absorber material.[2] This system could be qualified as real time dosimetry but presented a few limitations, mainly due to the volume of the calorimeter itself. For low doses, the elevation of the temperature due to the irradiation was too low or too slow compared to thermal diffusion and hence, could not be measured. In the 60's a step forward was taken with Solid State Dosimetry. The synthesis of luminescent charge trapping materials enabled the developpment of small volume detectors. Two possibilities were offered to trigger the luminescence: (1) Thermo Luminescence (TL) consisted in heating the material, the energy necessary to release the carriers being brought by phonons; (2) The Optically Stimulated Luminescence process consisting in flashing the dosimeter with an appropriate wavelength while recording the luminescence emitted in a different wavelength.[3] In this case the energy was brought by photons. OSL, which didn't require the heating of the sample, was definitely faster and easier to use than TL. Unfortunately, the sensitivity of the OSL materials available at that time, in the Gy range, didn't allow applications in personnel dosimetry (μGy to mGy range). Therefore, TL was chosen to become the reference in dosimetry. At that stage, due to the use of heavy reading apparatus that could only provide a delayed measurement of the dose, it was no longer a question of Real Time dosimetry. In the 80's, a new generation of OSL materials were synthesized at the University of Montpellier.[4] Rare-earth-doped alkaline sulphides had been known for a long time for their Optically Stimulated Luminescent (OSL) properties.[5] Primarily devoted to infrared-visible conversion purposes, they were later studied intensively for applications in radiation dosimetry. The Sulphides synthesized at the University of Montpellier drastically increased the sensitivity, allowing photon dose measurements as low as 10μ Gy, with a 7 orders of magnitude dynamic range.[6] They have been printed on films, which made it possible to record not only an image, but a map of the dose deposited in a complex geometry.[7] Later, with the progress of photonic devices, these sulphides enabled the integration of an OSL coupled dosimeter-reader. The device, no larger than a sugar cube has the performance to compete with TLDs but can be read and reset online.[8] This article is a synthesis of the research led at the University of Montpellier on Online and Real-time Dosimetry using OSL. In section 2, the quantities used in dosimetry are reviewed. The concepts of "Real Time" and Online Dosimetry are also defined on the basis of practical examples and the main dosimeters are presented. Section 3 is devoted to Optically Stimulated Luminescence. After a brief review of the basic mechanism, the guidelines for designing an OSL dosimeter will be defined. Section 4 and 5 present three examples of achievement, the monitoring of the dose received by a satellite orbit by orbit, the in-vivo measurement of the dose deposited during a radiotherapy treatment and the online control of the dose delivered by the surrounding area of the Large Hadrons collider (LHC) accelerator at CERN.

2. Definitions

2.1. *Dose*

The key parameter of the radiation effects is the dose *D*, defined as the ratio *dE* over *dm*, where *dE* is the mean energy transferred by ionising radiation to a mass *dm* of a matter.[9]

$$D = \frac{dE}{dm} \qquad \text{(Gy)} \qquad (1)$$

The unit of dose ($J.kg^{-1}$), is the gray (Gy) ; the rad, is still being used in the radiation effect community and is equivalent to a cGy . Table 1 provides the range of doses corresponding to different cases of exposure and their subsequent effects.

Table 1. Range of doses corresponding to different exposure and their subsequent effects.

<table>
<tr><td>Dose</td><td>Typical exposure</td><td colspan="2">Number of Electron-hole pairs created and Typical Effect observed</td></tr>
<tr><td>1 mGy</td><td>A chest X-Ray or a year in the natural background</td><td colspan="2">$10^{12} / cm^3$
effect on insulators
(charge trapping)
Minor risk on biological Cells</td></tr>
<tr><td>4 Gy</td><td></td><td>$4.10^{15} / cm^3$
Transitory effects on semi-conductors</td><td>Single exposure lethal dose : leading after one month to death in 50% percent of the cases .</td></tr>
<tr><td>10Gy-20Gy</td><td colspan="3">Dose delivered to a tumour during radiotherapy</td></tr>
<tr><td>10-100Gy</td><td colspan="3">Annual dose received by a satellite</td></tr>
<tr><td rowspan="2">100 Gy</td><td rowspan="2"></td><td colspan="2">Voltage shift induced in the threshold of a power MOSFET</td></tr>
<tr><td colspan="2">The current gain of a BJT may be cut down by a factor of 10</td></tr>
<tr><td>1MGy</td><td>Dose expected in some Sub detectors at CERN</td><td colspan="2">Mechanical properties of the materials are altered.</td></tr>
</table>

The dose rate is defined as the derivative of the dose versus time. is expressed in $Gy.s^{-1}$ or in $rad.s^{-1}$.

$$\dot{D} = \frac{dD}{dt} \qquad (Gy.s^{-1}) \qquad (2)$$

The TID (Total Ionising Dose) as previously defined, does not depend on time. However, the effect of the same TID will be strongly dependant on the dose rate. For example, it has been shown that the degradation of the current gain in a BJTs will be several orders of magnitude higher if the dose is delivered at a very low dose rate.[10] Hence the concepts of "Real Time" and "On-line" dosimetry that will be covered in the following subsection.

2.2. *Real time and online dosimetry*

Wherever radiation can be found in order to quantify its effect, dose measurement can help to appreciate the possible negative consequences. The most commonly used passive dosimeters are TLDs (Thermo Luminescent Dosimeters) and radiological films.[11] They both provide an accurate value of the dose after an *a posteriori* processing. TLDs require a heavy and power consuming reader, the films chemical processing. This delayed reading is not a problem as long as the dose expected is low enough not to jeopardize the exposed people or the irradiated devices. The typical applications are: (1) personnel dosimetry that aims at verifying that the workers are not exposed to a dose that is not compatible with the principle of precaution; (2) quality control of irradiation facilities, in order to make sure that the dose actually delivered corresponds to the calibration and calculation. (3) long term monitoring of the dose in exposed or contaminated areas. Film and TID dosimetry is often associated with a device monitoring the flux of particles, acting as an advanced warning in case of a sudden increase in the dose rate.

Whenever the risk of over exposure is identified, delayed reading of the previous dosimeter may lead to a "too bad, too late" result which is not acceptable. Moreover, it is not always possible to reach the dosimeters after irradiation. Remote monitoring of the dose is called online dosimetry. The term "real-time" dosimetry implies that the dose is known at any instant of the on-going irradiation. However this terminology is often extended to the case of a recurrent measurement when the dose rate is low enough compared to the sampling period. These concerns are well illustrated by the following examples.

In hospital radiotherapy units, the physicians calculate the optimum dose to deliver to the tumour. Exceeding this dose is harmful to the patients health and leads to unnecessary exposure of the surrounding sound tissues. This is especially true with newly developed techniques such as intra-operative radiotherapy.[12] Real time dosimetry allows the physician to deliver step by step irradiation and stop exactly when the optimum dose has been reached.

The current trend in reducing the weight of satellites combined with new missions, more exposed to radiation than in the past, and the use of Commercial off the Shelf devices' may result in an increased vulnerability of systems to radiation effects in flight.[13] In space, knowledge of the dose received by onboard equipment is crucial to evaluate the degradation of the devices and optimise the life-time of the satellite. The space radiation environment, with its complex fluxes of particles, its low dose rate modulated by sudden increases due to solar events, is difficult to reproduce on the ground. Some techniques have been developed to predict the behaviour of electronic devices irradiated in flight, but they all rely on models of a fluctuating radiative environment.[14] Several recently published works put forward the way to idea that the dose received by the devices should be monitored in flight on commercial low orbit satellites.[15] [16] [17] In the case of a satellite, dosimetry has to be performed in real-time to give an advanced warning when the TID jeopardizes the onboard equipment. But in this case, dosimetry also has to be performed "online" since there is no reach a TLD or a film after the satellite is launched.

Another example of online real-time dosimetry can be found in highly radioactive areas such as those surrounding nuclear cores. Due to the high dose rate and high contamination, it may be no longer possible to reach the dosimeter after irradiation. Here again, the dose has to be monitored online and in quasi real time.[18]

2.3 *Online dosimeters*

The purpose of this section is not to provide an exhaustive catalogue of online dosimeters, but rather to give a few examples of well know detectors to compare with the OSL dosimeters.

2.3.1 *The Ion chamber*

The principle of the vacuum chamber was described by Greening in 1954. This theory is applicable to a cavity where pressure is so low that any gas ionisation is insignificant relative to the electron transfer between the electrodes.[19] Ion chambers are widely used to calibrate irradiation facilities and all other relative dosimeters. The field of application is limited by the cost, the dimensions, and the use of an electrometer to measure the charge collected which is not practical.

2.3.2 *The Radfets*

P-channel Metal Oxide Semiconductor (PMOS) transistors are the most commonly used devices to monitor the dose in flight. They rely on the trapping of charges in thick oxides. MOS dosimeters, stacked or not, constitute dose-integrating sensors.[20] On the basis of a minimum readable shift in the threshold voltage, (typically $1mV.cGy^{-1}$), they allow the measurement of the dose integrated over two or three orders magnitude range.[21] On the one hand, the permanent shift in the threshold voltage makes it possible to keep the information that can be read at any time during the flight. On the other hand, as the oxide is charged, it becomes difficult to make out small additional variations in the dose. Some attempts at performing a thermal reset of the oxide trapped charge have been performed. Recent results on MOS devices have shown that variations in operating conditions (dose rate, temperature) induce changes in the device signal to dose response.[22]

2.3.3 *UVPROMs*

More recently, a new sophisticated type of dosimeter quantifying the flip of UVPROM memory cells has been presented and flown on the MPTB satellite. The small size of the cells makes the UVPROM very suitable for micro dosimetry. A statistical approach has been developed which provide an accurate measurement of the macroscopic dose.[23] The overall sensitivity of the device is typically one Gy. In flight, the UVPROMS integrate doses in the range of 1Gy-1000Gy.

3. Generalities on the OSL

3.1. *Basic mechanism*

Optically Stimulated Luminescence results from a process starting with the irradiation of an insulator or a wide band gap semiconductor, as shown in (1) on Fig. 1. Ionising radiation creates in the material a large amount of trapped carriers. Some charges remain trapped on localized defects after irradiation for a time period depending on the temperature and the activation energy of the traps. Stimulating the material optically (2) will provide the energy necessary to release the trapped charge. A subsequent radiative recombination may be observed (3) as shown in Figure 1. Quantifying the amount of emitted light makes it possible to evaluate the energy absorbed by the dosimeter. It is interesting to note that, from an external point of view, OSL can be considered as an anti Stokes phenomenon since the wavelength stimulation is greater than the wavelength of the luminescence.

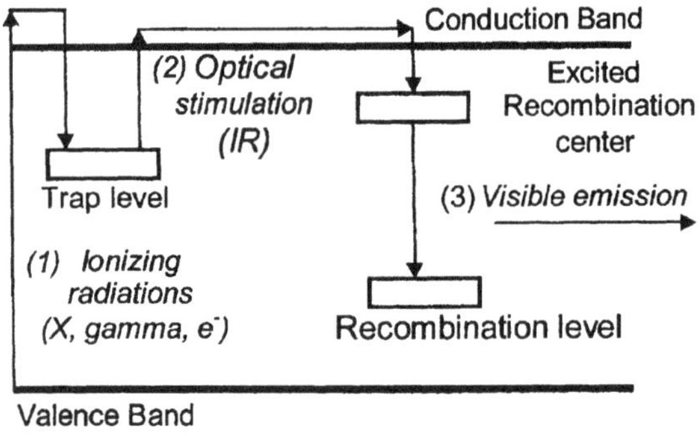

Fig. 1. Basic mechanism of the optically stimulated luminescence.

3.2 *The OSL materials*

3.2.1 *The main families of OSL materials.*

Several families of materials have been identified as emitting Optically Stimulated Luminescence. They differ by their chemical and optical properties and not all of them are appropriate for an application in dosimetry. The intrinsic ability to store charges for a large period of time and the efficiency of the luminescence is the major criterion for selecting OSL materials. This efficiency is strongly dependant on the doppants and their concentration. The major families, known for their high luminescence yield are listed in table 2 along with their excitation, stimulation and emission wavelengths.

Table 2. Summary of the major OSL materials and their characteristic wavelengths

Material	Stimulation (nm)	Emission (nm)
MgS:Ce,Sm	800 =>1500	500 =>700
SrS:Ce, Sm		
BaFX, X= Cl, Br, 1	470 => 630	350 => 450
NaCl:Cu	400 => 500	325 => 375
KCl:Eu	450 => 650	400 => 450
KBr:In	500 => 700	400 => 600
RbBr:Ti	600 => 800	300 => 450
RbI:In	600 => 800	350 => 450
α-Al_2O_3	300 => 450	350=> 500

Another important criterion can be deduced from table 2, the separation of the stimulation and the emission spectra. An overlap in the spectra makes it difficult or is impossible to discriminate the emission from the stimulation. It is the case for example for the α-Al_2O_3. Solutions have been found implying that the dosimeter is stimulated with a high intensity laser pulse on the edge of the spectrum. The low cross section is then compensated by the power of the laser. Another technique consists in stimulating the Al_2O_3 with a pulsed laser and recording the luminescence between the pulses. Both techniques lead to heavy reading systems, comparable in size to a TLD reader.[24] Nevertheless OSL reduces the reading time by a factor of 100 compared to thermo luminescence.

3.2.2 *Lenard's family*

The rare-earth-doped alkaline sulfides (MgS, CaS, BaS, SrS and BaS) have been identified by the German physicist P. Lenard as efficient phosphors[5]. They exhibit both TL and OSL. Primarily devoted to infrared-visible conversion, they were later studied intensively for applications in radiation dosimetry [25] A new generation of Sulphides has been synthesized at the University of Montpellier that make it possible to reach for the first time a sensitivity in the range of 10μGy.[4] The saturation has recently been investigated and found occur at 500Gy, ensuring a 7 orders of magnitude dynamic range. The stimulation wavelength, optimum at 1μm is easy to discriminate from the emission spectrum stretching from 450 to 650 nm.

3.3 *Guidelines to the design of an OSL dosimeter*

Any system based on OSL requires, besides the sensitive OSL phosphor, two additional devices, a stimulation source and a photo-detector, as shown on Fig. 2. The role of the stimulation source is to provide the energy necessary to de-trap the carriers.

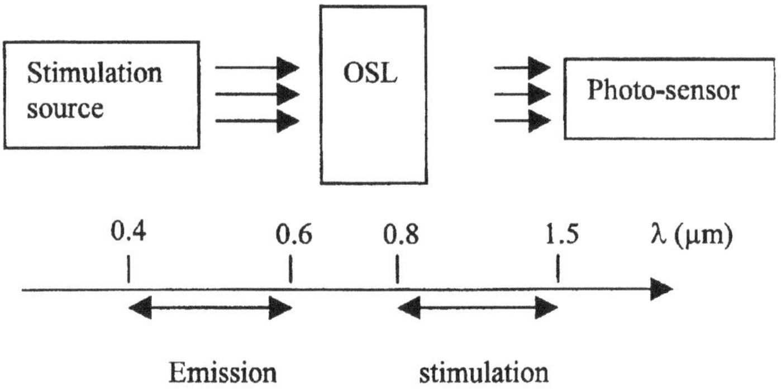

Fig. 2. generic structure of an OSL dosimeter using rare earth doped sulphides.

The sulphides doped with cerium and samarium used in this work allow a wide range of stimulation wavelength from 0.8 to 1.5 microns. Various systems can be used to collect the luminescence depending on the application targeted, providing they make it possible to discriminate the stimulation from the luminescence. From this point of view, silicon detectors are not suitable. In previously published works, stimulating the sample with a solid state laser emitting a 1.3 micron wavelength and collecting the OSL with a photo-multiplier tube (PMT) made it possible to discriminate emission from stimulation.[6] A typical emission spectrum of a calcium sulphide doped cerium samarium is given on figure 3. Such an emission is right in the PMT high sensitivity range and far away from the stimulating wavelength located at one micron and up.

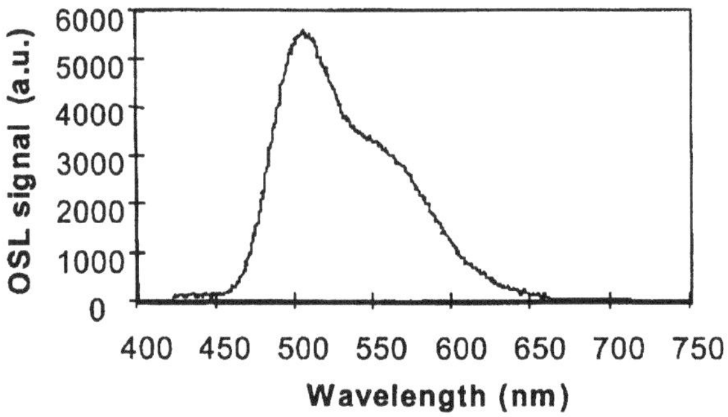

Fig. 3. Emission spectrum of the CaS:Ce,Sm.

3.4 *The OSL Signal*

Fig. 4 represents the typical signal shape of an OSL signal. The stimulation is switched on after 100ms which results in a quick luminescence. The decay of the signal can be easily modelled by considering a simple reaction rate equation providing an exponential solution.

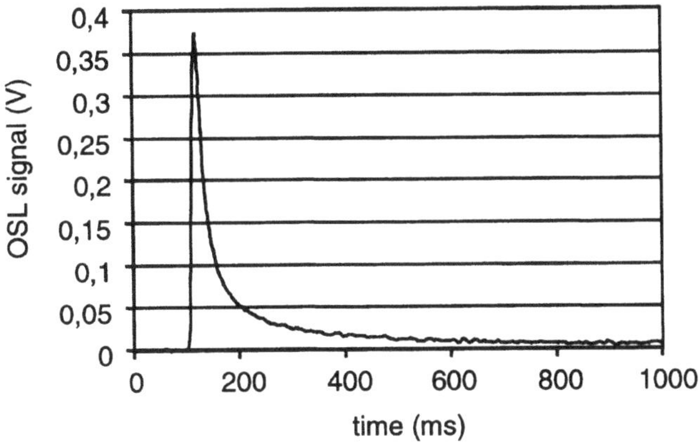

Fig. 4. typical shape of an OSL signal recorded with a GaAsP photodiode for a dose of 2mGy.

Let us consider irradiated OSL material, in which N charges have been trapped during the irradiation process. The number of charges depends on the dose D and thus is noted N(D). Under a constant stimulation flux F, an OSL signal (function of the time t) is observed. The OSL signal is noted $S_{osl}(t)$. The general shape of this signal has already been presented in Figure 4. Since the response of the OSL to a stimulation is obtained within a nanosecond the rise time of the luminescence can be considered as negligible.[26] In fact, it only depends on the time constant of the input amplifier stage. It is then reasonable to assume that the peak value of the signal is obtained for t=0. The decay of the OSL signal is then modeled with an exponential law,

$$S_{osl}(t) = K.N(D).F.\exp(-\sigma.F.t), \tag{3}$$

where K is a constant that depends on both the material and the detection set-up. σ is the cross section of the stimulated detrapping process. The time constant of the decay is then given by

$$\tau = \frac{1}{\sigma.F}. \tag{4}$$

At t = 0, the peak value of the signal is given by

$$S_{osl}(0) = S_{peak} = K.N(D).F \quad , \tag{5}$$

and is directly proportional to the dose D as long as the stimulation flux remains constant. Let us consider now the integrated OSL signal. It is given by integrating Eq.(2). It comes

$$\int_0^\infty S_{osl}(t).dt = \frac{K}{\sigma}.N(D). \tag{6}$$

The value given by Eq. (6) is no longer dependant on the stimulation flux but only proportional to the total amount of charge trapped within the material. In the following section, three examples of OSL online dosimeters are presented.

4. An OSL Sensor for Space Application

The University Montpellier II is developing a radiation sensor for the CNES (Centre National d'Etudes Spatiales).[27] The guideline for this project are to measure the dose deposited orbit by orbit, which implies a sensitivity in the mGy range (by cycling irradiation, reading and reset)?. Nevertheless, the eventuality of a sudden increase in the dose rate due to a solar event has to be taken into account, which requires, a wide dynamic range. Any equipment flying in space has to conform to requirements in terms of weight, consumption and hardness assurance. From this point of view, PMT were left aside in the design of the integrated sensor. Regular silicon photodiodes, were not considered either, since their extreme sensitivity to infrared light doesn't allow proper discrimination of the stimulation light from the OSL signal. Hence, wide gap composite material photodiodes such as GaAsP or GaP were preferred. A simple surface mount light emitting diode provides stimulation.

4.1 *Basic structure*

The basic cell of the OSL sensor is designed as the very simple structure presented in Figure 2, in which the OSL layer is directly sandwiched between the light emitting diode, LED, and the photodiode. Most of the infrared stimulation is absorbed by the OSL layer. The power consumption is limited to the current needed to bias the LED during stimulation, that is 50mA during 20 seconds under a 2V Bias voltage. The whole sensor can fit in a sugar cube volume. Figure 5. presents a OSL sensor, integrating also the read out circuitry and a feed back loop to control the current in the LED. The basic external circuitry needed to operate the sensor consists only of a dual operational amplifier that converts the current delivered by the photodiode into voltage. The signal is then amplified and digitised. A voltage source, triggered by a logic input is also needed to operate the stimulation source. Let us note that in the case of not hardened circuit, or for sensor test this additional circuitry can be shielded,. The degradation of the LED emission can be compensated by increasing the forward current through a feedback loop.[28] Adapted rad-hard devices have been developed.

Fig. 5 . compact OSL sensor integrating the reading circuitry for direct PCB mount.

4.2 *Calibration*

A first series of experiments have been performed using a ^{60}Co source in which irradiation and reading occur sequentially. The sensor is positioned on a tray in the irradiation field. The external reading circuitry remaining outside the exposed area. During irradiation, the system is switched off in order to avoid transient effects. From this point of view, the OSL device can be considered as a passive detector. During the reading phase, a dedicated circuit is switched on and the stimulation source operated. The OSL signal recorded is presented in Figure 4. Both the maximum and the total OSL emission are measured to monitor the dose. During the reading, the whole OSL layer is exposed to intense IR stimulation flux that completely resets the material after 10 seconds. The curve presented in Figure 6. was recorded by increasing the dose step by step, varying both the distance source to sample and the exposure time to explore a 3 orders of magnitude range of doses. The present prototype was developed to record the OSL in the 1 mGy to 1Gy Range. For higher doses the gain will have to be lowered to avoid the saturation of operational amplifier. The response is found to be linear on the investigated range of the dose. So far, the OSL Sensor has flown on the International Space Station as a part of , a radiation detection module designed by the CNES. An OSL Sensor has been integrated to the nano-satellite CUBESAT SACRED (Study on the Action of Cosmic Rays on Electronic Devices) experiment of ALCATEL Space. CNES is also planning to fly an improved version of the OSL on PROBA II.

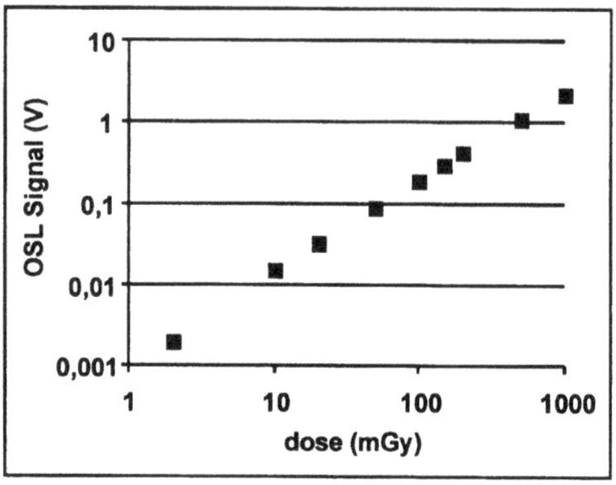

Fig. 6. Calibration curve of the OSL integrated sensor.

5. Remote Online OSL Dosimetry using optical fibers

In the previous example, the dosimeter benefits from a printed circuit board mount. Both the sensor and the electronics are exposed to radiation, the problem of the radiation induced damage is supposedly being addressed. It is not always possible or practical to use a sensor integrating both the OSL and the reading circuitry. The reason can be either a very harsh radiative environment, or the necessity to minimize the size of the sensor. in both cases, the use of optical fibres can provide an elegant solution. In this section, two examples of online dosimetry using OSL and optical fibres will be presented. The first application proposes a solution to the real-time monitoring of the dose delivered to a tumour during and Intra Operative Radio Therapy treatment. The second application addresses the measurement of very high doses in the hot environment of the Compact Muon Solenoid, one of the four experiments of the LHC at CERN.

5.1 In vivo dosimetry: principle and specific requirements

Intra-operative radiotherapy consists in irradiating the tumoural bed, in vivo during surgery. The aim of this technique is to eliminate the local residual abnormal cells while avoiding unnecessary exposure of the sound tissues. After removing the tumour, the surgeon and the radiotherapist install a cylindrical localizator in order to focus the electron beam straight on the tumoural volume as shown in Figure 7. After the irradiation is completed, the patient is carried back to the block and the surgery continues. The irradiation is carried out with electrons. A low energy, a few MeV, is preferred since the target area is directly in contact with the localizator not in depth. The dose delivered is typically a few tens of Gy. The dose rate is typically 200 or 300 monitor units per minute (1MU is equivalent to a cGy if the accelerator is correctly calibrated) . The optimum dose is computed by the physician, but a step by step monitoring of the dose actually delivered is necessary to make sure this dose is reached but not overstepped.

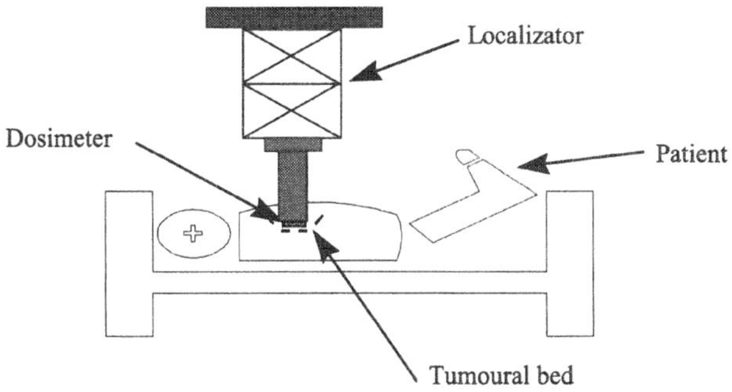

Fig. 7. configuration for an intra-operative radiotherapy treatment.

5.1.1. *Specifications*

Measuring the dose during the irradiation implies that the sensor is inserted below the localisator. The first mandatory concern is, of course, the detector size which has to be as small as possible. The dose is measured in the human tissue which effective atomic number Z is close to 7.2. As a consequence, the effective Z of the dosimeter will have to be as close as possible to this value in order not to affect the measurement. The energy deposited in the sensor has to come from secondary electrons emitted in the surrounding tissue and not in the sensor itself. Very small sensors, providing that their dimension is negligible compared to the mean free path of the secondary electrons, have an effective response close to the response of the surrounding material. (Bragg-Gray theorem). This assertion has two immediate consequences: (1) Metals or any other high Z material are prohibited for the design of the sensor (2) The sensor has to be small enough to verify the Bragg Gray Conditions.

5.1.2 *Experiments*

A prototype of online dosimeter was developed and tested at the Val d'Aurelle Hospital in Montpellier. The bloc diagram is described on figure 8-a. The OSL sensitive crystal, a cylinder of 1mg of magnesium sulphide, is enclosed in a polyethylene cap and positioned at the extremity of a multimode optical fibre. The role of the fibre is: (1) to lead the infrared stimulation provided by a fibered laser diode ; (2) convey the OSL signal toward a PMT. The different wavelengths are injected and sorted out in the fibre by means of optical couplers and filters. The OSL signal is then digitised and recorded. The sensitivity of the prototype was investigated using a ^{60}Co source at the University of Montpellier. The sensor is completely reset after measurement by a 3s intense stimulation. The minimum detectable dose was estimated to 100μGy. The dynamic range, limited in this arrangement only by the saturation of the output amplifier, covers 5 orders of magnitude. The use of 60Co was imposed by the impossibility to deliver low doses precisely with therapy electrons accelerators (5 to 20 M eV) at the hospital. Although the total dose delivered during the treatment usually rises up to tens of Gy, it is important to note that the OSL is meant to measure incremental doses in the cGy range. The curve presenting the response of the sensor as a function of the dose is presented on figure 8-b. As usually, we have to pay a special attention to the stability of the whole irradiated system with total dose.

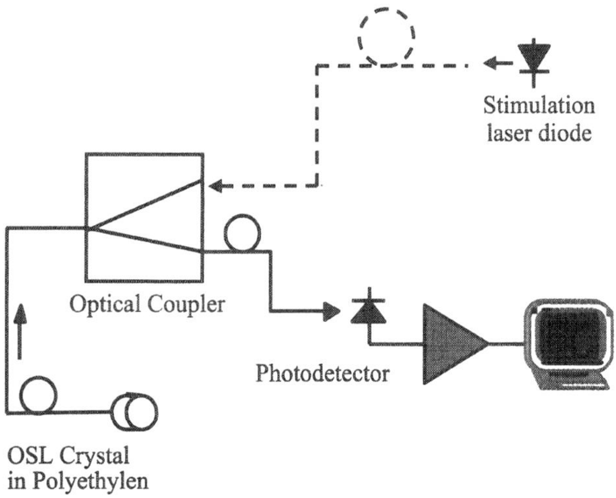

Fig. 8-a. block diagram of the OSL online dosimeter

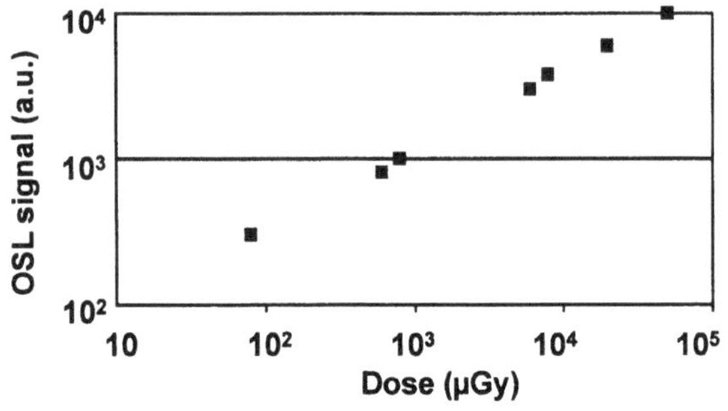

Fig. 8-b. Calibration curve of the OSL online dosimeter

5.2 *Monitoring of the dose deposited by the LPI 500 MeV electron beam with an online OSL sensor.*

5.2.1 *The CMS environment*

The radiation environment encountered in CERN experiments such as the Compact Muon Solenoid (CMS) differs completely from the previous medical and space, applications of the OSL dosimeters. In the Pixel sub detector of CMS, the total dose integrated over a 10 year period is estimated to 2 MGy with a particle fluence of 10^{14} per cm^2. In addition, the risk of accidental radiation bursts (that can damage the sub-detectors) due to beam-loss or bad tuning has also to be taken into account. Considering this potential risk and the uncertainties in the calculation, the idea of implementing a radiation monitoring system has to be considered. The present work was initiated by the CMS group who provided the beam time.

5.2.2 *Specific Modifications of the system*

In the medical application presented in section 5.1, the mandatory parameters where the sensitivity and the size of the sensor. Those requirement led to the choice of a PMT to collect the signal, and a single fibre carrying both the stimulation and the OSL signal. However, the optical coupler introduces a 10db attenuation on both wavelength. In the case of the dosimetry system proposed for CMS, very different specifications led to very different choices.[29]
First of all, the aim of this system is to measure the dose simultaneously in different locations of the detector. The number of sensor to be used is estimated in the order of one hundred, for CMS only, hence the necessity to simplify the system and reduce its cost. The range of dose to measure is also much higher and can reach 1MGy in the Pixel sub detector. To take into account those specificities, the system was modified as presented in figure 9. The sensor itself consists of an OSL crystal and an infrared LED positioned at the same extremity of a 50m long optical fibre. The LED provides the wavelength necessary for the optically stimulated reading. The resulting luminescence is collected by means of the optical fibre and guided toward a photo-multiplier tube. Due to the optical fibre, the photo-multiplier can be moved outside the irradiation area to prevent any radiation-induced damage. It is important to note that this design makes it possible to suppress the optical coupler, hence the 10db of attenuation. As a result, the signal intensity is now compatible with the use of a simple photodiode, similar to those used in the integrated sensor.

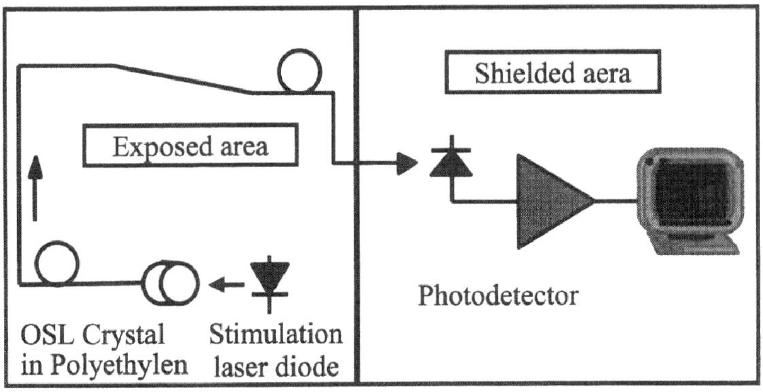

Fig. 9. Bloc diagram of the modified online system

5.2.3 *Experimental results*

During the irradiation period, 33 scans were performed at CERN with the LPI 500MeV electron beam. The typical fluence per scan was $1.8.10^{13}$ electrons per cm². The total dose delivered at each scan was estimated to 160Gy. OSL was read after each scan and the response recorded on a digital scope. The diagram on Fig.10 presents the evolution of both the OSL peak and the signal integrated over the reading period.

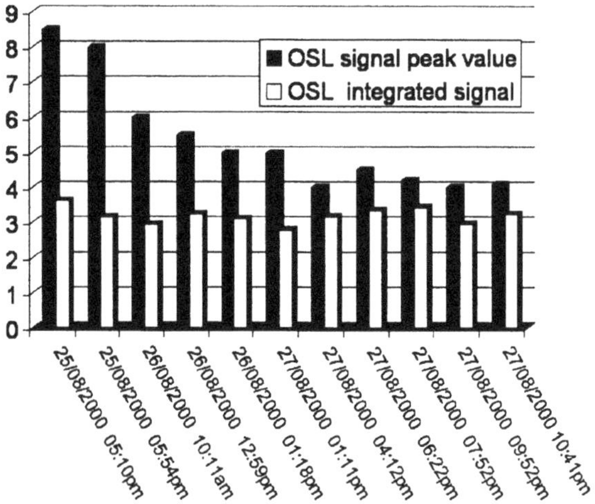

Fig. 10. Evolution of the peak integrated signal recorded during a series of irradiation runs at CERN with the 500MeV electrons of the LPI facility. The total dose for each run is 160Gy.

The online dosimeter developed for the remote monitoring of electron irradiation was designed in a way that exposes the stimulation LED as well as the OSL itself. The results presented in fig.10 indicate that the integrated signal remains constant within the uncertainties of the dose delivered, whereas the peak maximum is cut down by a factor of two after 5kGy. This result, demonstrates the decrease in the LED emission due to the displacement damage effect. It is in agreement with thee model presented in section 3.3 where the peak value of the OSL is directly bound to the stimulation intensity whereas the total amount of luminescence recorded only depends on the total dose.[29]

The OSL dosimetry has been recommended for the CMS dosimetry. The system actually proposed consists of about one hundred OSL crystals connected individually to the same amount of photodiodes by optical fibers. The reading is performed sequentially, by stimulating a given detector and multiplexing the signals toward a central data acquisition system.

6. Conclusion

A new generation of Optically stimulated materials has been synthesized at the University Montpellier II. The very high sensitivity of these phosphors, the quick time constant of the luminescence (less than 1ns) and the stimulation perfectly separated spectra enable many applications in real time and online dosimetry. We have developed an integrated sensor to measure the dose received orbit by orbit by a satellite. In radiotherapy, OSL has been used to control the dose deposited during intra-operative *in-vivo* irradiations. At CERN, we have proposed an online system to monitor the dose simultaneously in about 100 locations inside the Compact Muon Solenoid CMS experiment. Our OSL materials have also found many other applications shaped as films. They have made possible the mapping of a dose in complex geometries with a resolution of 50μm. OSL films have been used for the quality control of radio-surgery devices like the Gamma Knife. Beam profiles can be easily recorded with a variety particles ranging from protons, electrons, photons, and pions. Recently, the OSL materials have been adapted to neutron dosimetry. The future of the OSL relies on the fabrication of thin layer directly on active pixel sensors, or any other fast photo detector that could take advantage of the ns time constant, to push further the limits of what we call the real time dosimetry.

Acknowledgements

The authors would like to express their thanks to Robert Ecoffet from the CNES for its long term support on the OSL project. They also would like to acknowledge B.Camanzi, M.Glaser and P. Jarron, for their helpful contribution to the work done at CERN.

References

1. A. Holmes-Siedle and L. Adams "Handbook of Radiation Effects", Oxford University Press, Oxford, 1994, pp.95-199.
2. J. S. Laughlin and S. Genna , "Calorimetry", in Radiation Dosimetry Vol II: Instrumentation, eds F.H. Attix and W. C. Roesh, Academic press, New York,1966, pp. 389-448.
3. J. F. Fowler and F. H. Attix, " Solid State Integrating Dosimeters", in Radiation Dosimetry Vol II : Instrumentation, eds F.H. Attix and W. C. Roesh, Academic press, New York,1966, pp.269-283.
4. O. Missous, F. Loup, J. Fesquet, H. Prevost And J. Gasiot. "Optically Stimulated Luminescence of Rare-earth Doped Phosphors", *Eur. J. Solid State Inorg. Chem.* **28/s** (1991)163-166.
5. Ph. Lenard, F. Schmidt, R. Tomacshek, "Phosphoreszenz und Fluoreszenz", Handbuch der Experimental Physik, Leipzig, 1928.
6. O. Missous "Contribution à l'étude de la luminescence par stimulation optique dans le sulfure de magnésium dopé cérium, samarium en vue d'une utilisation en dosimetry", Ph.D. Thesis Université Montpellier II, 1992.
7. L. Dusseau, G. Polge, L. Albert, Y. Magnac, J.C Bessière, J. Fesquet, J. Gasiot, "Irradiated Integrated Circuits Dose-Attenuated Mapping Using Optically Stimulated Phosphors for Packaging Dosimetry", *IEEE Trans.on Nucl. Sci Vol.45*, **6** (1998) 2695-2699.
8. L. Dusseau, G. Ranchoux, G. Polge, D. Plattard, F. Saigné, J. C Bessière, J. Fesquet, J. L.Dusseau, D.Plattard, G.Polge, J-R Vaillé, G.Ranchoux, F.Saigné, J.Fesquet and J. Gasiot "An Integrated sensor Using Optically Stimulated Luminescence For In Flight Dosimétry" *IEEE Trans. Nucl. Sci. Vol.47*, **6** (2000) 2412-2415.
9. ICRU repport 10a "Radiation Quantities and Units" *National bureau of standards handbook 84*, US Government Printing office, Washington, D.C. 1962.
10. D. M. Fleetwood, S. L. Kosier, R. N. Nowlin, R.D. Schrimpf, R.A. Reber. Jr., M. DeLaus, P.S. Winokur, A. Wei, W.E. Combs and R.L. Pease, "Physical Mechanisms Contributing to Enhanced Bipolar Gain Degradation at Low Dose Rates", *IEEE Trans. Nucl. Sci.*, vol. 41, **6** (1994) 1871-1883.
11. R. A. Dudley, "Dosimetry with photographic emulsions", in Radiation Dosimetry Vol II: Instrumentation, eds F.H. Attix and W. C. Roesh, Academic press, New York,1966, pp. 326-382.
12. D. Beteille, R.Setzkorn, H. Prévost, L. Dusseau, J. Fesquet, J. Gasiot, "Laser Heating of thermoluminescentce plate: application to intraoperative radiotherapie" . *Medical physics* **23** (1996) 1-4.
13. F.W. Sexton, D.M. Fleetwood, C.C. Aldridge, G. Garrett, J.C. Pelletier and J.I. Gaona, "Qualifying Commercial Ics for space Total-Dose Environment", *IEEE Trans. Nucl. Sci. Vol.45*, **6** (1992) 1869-1875.
14. L. Dusseau, T.L. Randolph, R.D. Schrimpf, K.F. Galloway, F. Saigné, J. Fesquet, J. Gasiot and R. Ecoffet, "Prediction of low dose-rate effects in power Metal Oxide Semiconductor Field Effect Transistors based on isochronal annealing measurements" *Journal of Applied Physics*, **5** (1997) 2437-2441.
15. P.A. Aarnio and M. Huhtinen "hadron fluxes in inner parts of LHC detectors", *Nucl. Inst. and Meth*, **A336** (1993), 98-105.
16. J. W. Boag, "Ionization chambers", in *Radiation Dosimetry Vol II: Instrumentation*, eds F.H. Attix and W. C. Roesh, Academic press, New York,1966, pp. 1-67.
17. A.G. Holmes-Siedle and L.Adams "RADFETs: a review of the use of metal-oxide-silicon device as integrating dosimeters", *Rad.. Phys and Chem. 28*, **2** *(1986)* 235-244.
18. M.S. Gussenhoven, E.G Mullen, D.H. Brautigam, "Improved understanding of the Earth's radiation belts from the CRRES satellite", *IEEE. Trans. Nucl. Sci., Vol. 43* **2** (1996) 353-368, .
19. C.S. Dyer, C.L. Peerless, C.J. Watson, H.E. Evans, P. Knight, M. Cosby, C. Underwood, T. Cousins, R. Noulty, "Updated measurements from CREAM & CREDO implications for environment & shielding models", *IEEE Trans. Nucl. Sci., Vol. 45*, **3** (1998) 1584-1589..
20. E.J. Daly, J. Lemaire, D. Heynderickx, D. J. Rodgers, "Problems with models of the radiation belts", *IEEE Trans. Nucl. Sci.*, Vol. 43, **2**, (1996) 403-415.

21. A. Jaksik, G. Ristic, M. Pejovic, A. Mohammadzadeh, C. Sudre, and W. Lane "Gamma-Ray irradiation and post-irradiation responses of high dose range RADFETs", *Proc. IEEE-RADECS* (2001) 57-65.
22. B.O'Connell, C. Conneely, C.Mc.Carthy, J.Doyle, W.Lane and L.Adams "Electrical performances and radiation sensitivity of Stacked PMOS Dosimeters Under Bulkbias Control", *IEEE Trans. Nucl. Sci Vol.45.*, **6** (1998) 2689-2694.
23. L.Z.Sheick, P.J. Mc.Nulty and D.R.Roth "Dosimetry Based on the Erasure of the floating gates in the Natural Radiation environments in Space", *IEEE Trans. Nucl. Sci. Vol.45,* **6** (1998) 2681-2688.
24. G.Ranchoux "Dosimetrie par luminescence stimulée optiquement deportee par fibre optique – etude de capteurs à alumine alpha pour la dosimétrie du personnel temps réel" Ph.D. Thesis CEA-LETI Université Montpellier II, 2001.
25. E.N.Sandborn and E.L. BEARD, in Luminescence dosimerty, F.H. Attix, conf-650637, (1967), pp.57-61.
26. J. Gasiot, P. Braunlich and J.P. Fillard, "Nanosecond infrared laser stimulation of luminescence rare-earth doped AES" *Appl. Phys Lett.*, **5** (1982) 376-380.
27. L.Dusseau, D.Plattard, G.Polge, J-R Vaillé, G.Ranchoux, F.Saigné, J.Fesquet and J. Gasiot "An Integrated sensor Using Optically Stimulated Luminescence For In Flight Dosimétry" *IEEE Trans. Nucl. Sci. Vol. 47,* **6** (2000) 2412-2416.
28. J-R. Vaillé, S. Ducret, K. Idri, F. Saigné, , S. Matias, N.Iborra, R.Germanicus,, *R.* .Ecoffet and L. Dusseau, "Hardening of a radiation sensor based on Optically stimulated luminescence" IEEE-NSREC2003, july 20-24, Monterrey CA.
29. L. Dusseau, G. Polge, S. Mathias, J.R. Vaillé, R.Germanicus, R.Broadhead, B.Camanzi, M.Glaser, F. Saigné, J.Fesquet, J.Gasiot, "High Energy Particles Irradiation of Optically Stimulated Luminescent Films at CERN" *IEEE Trans. Nucl. Sci., Vol. 48*, **6** (2001) 2056-2060.

www.ingramcontent.com/pod-product-compliance
Ingram Content Group UK Ltd.
Pitfield, Milton Keynes, MK11 3LW, UK
UKHW020400150926
8624IPUK00004B/252

9 789812 389404